Viral Zoonoses: Interactions and Factors Driving Virus Transmission

Viral Zoonoses: Interactions and Factors Driving Virus Transmission

Editors

Myriam Ermonval
Serge Morand

Basel • Beijing • Wuhan • Barcelona • Belgrade • Novi Sad • Cluj • Manchester

Editors
Myriam Ermonval
Santé Globale
Institut Pasteur-Université
Paris Cité
Paris
France

Serge Morand
IRL HealthDEEP
CNRS - Kasetsart University –
Mahidol University
Bangkok
Thailand

Editorial Office
MDPI
St. Alban-Anlage 66
4052 Basel, Switzerland

This is a reprint of articles from the Special Issue published online in the open access journal *Viruses* (ISSN 1999-4915) (available at: www.mdpi.com/journal/viruses/special_issues/Viral_Zoonoses_Interactions_Factors_Driving_Virus_Transmission).

For citation purposes, cite each article independently as indicated on the article page online and as indicated below:

Lastname, A.A.; Lastname, B.B. Article Title. *Journal Name* **Year**, *Volume Number*, Page Range.

ISBN 978-3-0365-9900-7 (Hbk)
ISBN 978-3-0365-9899-4 (PDF)
doi.org/10.3390/books978-3-0365-9899-4

Contents

Editorial

Viral Zoonoses: Interactions and Factors Driving Virus Transmission

Myriam Ermonval [1,*] **and Serge Morand** [2,3,*]

[1] Unité Environnement et Risques Infectieux, Institut Pasteur, Université Paris-Cité, 75015 Paris, France
[2] IRL2021 HealthDEEP (Health, Disease Ecology, Environment and Policy), CNRS (Centre National de la Recherche Scientifique)–Kasetsart University–Mahidol University, Bangkok 10900, Thailand
[3] Faculty of Veterinary Technology, Kasetsart University, Bangkok 10900, Thailand
* Correspondence: myriam.ermonval@pasteur.fr (M.E.); serge.morand@cnrs.fr (S.M.)

Citation: Ermonval, M.; Morand, S. Viral Zoonoses: Interactions and Factors Driving Virus Transmission. *Viruses* **2024**, *16*, 9. https://doi.org/10.3390/v16010009

Received: 12 December 2023
Revised: 15 December 2023
Accepted: 17 December 2023
Published: 20 December 2023

1. Introduction

The beginning of the 21st century was marked by an increase in the number of emerging/reemerging infectious diseases detected worldwide and by the challenging COVID-19 pandemic. Most of these emerging diseases are caused by viruses that are primarily RNA viruses of animal origin, with a long history of adaptation to their natural hosts, becoming pathogenic when crossing species barriers [1]. In humans, they can cause serious illness that can be sometimes fatal, with some virus species having a high mortality rate in the groups of neuropathies (Nipah virus, Rabies virus, etc.), of acute pulmonary syndromes (SARS-CoV-2, Influenza viruses), or of viral hemorrhagic fevers, as exemplified by some species of *Bunyavirales* (hantavirus, arenavirus), *Flaviviridae* (Dengue virus), and *Filoviridae* (Ebola virus) [2]. The conditions for viral persistence in animal reservoirs, particularly among the very diverse species of rodents and bats [3], and pathogenicity in humans are not always elucidated [4]. Meanwhile, outbreaks are influenced by human activities that disrupt ecosystems and increase contact between infected animals and humans. Therefore, population dynamics are of great importance, with domestic animals playing key roles as intermediate hosts in the transmission of viruses [5,6]. Determining interactions at different levels (molecular, cellular, systemic) of viruses with their different human and animal hosts [7–9] and their persistence in the environment [10] is of great importance for assessing transmission risks. Above all, the contribution of different disciplinary approaches integrated into the "One Health" concept is now recognized as being fundamental for the implementation of measures to prevent new viral emergences [11]. In this context, the spillover to other species, even from humans to domestic or wild animals in the event of a major epidemic episode, as recently demonstrated by the spillback of SARS-CoV-2 to farmed minks [12], white-tailed deer [13], and rodents [14,15], constitutes a major risk to be considered.

2. Contents

The contributions to this Special Issue on Viral Zoonoses cover general aspects of the pathophysiology of zoonotic viruses in different hosts and assess the role of certain factors favoring their transmission, therefore, as reviewed in Agusi et al. 2022, addressing the importance of One Health approaches, including a broad community of scientists working in different fields.

The published studies highlight the requirement for:

(i). Good surveillance (epidemiology) of the viruses present in wild and domestic animals: Hamel et al. 2023 discovered a new flavivirus infecting mosquitoes in rural areas of Thailand. The circulation of the highly pathogenic H5N1 avian flu virus responsible for sporadic outbreaks is presented in a seroprevalence and meta-analysis study carried out by Ntakiyisumba et al. 2023, supporting the necessity of monitoring the avian influenza virus to protect farm birds from contamination.

(ii). Pertinent techniques to identify and diagnose viruses and their possible transmission to humans: To obtain insights into inter-host adaptation, Embregts et al. 2022 used NGS to analyze virus populations within specific hosts and tissues by isolating the Rabies virus from a broad range of CNS and non-CNS samples of mouse and human origin. The comparative study of the different detection assays used to detect Nipah virus infection, as reviewed by Garbuglia et al. 2023, illustrates the importance of high sensitivity tests for the management of epidemics.

(iii). Ecological and anthropological approaches to assess the risk and consequences of epidemics: The influence of climate change on outbreaks due to henipaviruses circulating in bats was evaluated in an article by Latinne and Morand 2022. A study conducted by Rojas Sereno et al. 2022 exploring the factors associated with the spatial expansion of bats carrying the Rabies virus in Colombia shows the importance of these data for reducing epidemics by improving vaccination.

Furthermore, understanding the transmission factors between different animal species, for example, between wild and domestic animals, through spillover between different animal species or from animals to humans, as well as from humans to farm animals, is essential to predict and limit the risk of new pandemics. Such questions are addressed in some of the publications in the Special Issue: the implication of contact between wild and domestic animals and the increasing risk of emergence is shown by the work of Morcatty et al. 2022 carried out to identify viruses in different wild animals close to domestic ones on sale in Indonesian wildlife markets. Virus transmission between animal species and possible spillback from humans to animals are examined in the studies of Souza et al. 2023 and Kimble et al. 2022 that describe the circulation of influenza viruses in farm pigs, and in the study carried out by Vandegrift et al. 2022 on SARS-CoV-2 found in white-tailed deer.

3. Conclusions

Faced with the increase in emerging infectious diseases, their occurrence on a global scale, and the damage caused to ecosystems, mainly by human activities, thereby increasing the contact between wild animals, domestic animals, and humans, the global "One Health" approach is essential. The articles published in the Special Issue "Viral Zoonoses: Interactions and Factors Driving Virus Transmission" contribute to this reflection.

Acknowledgments: We are grateful to all of the contributors to this Special Issue published in the journal *Viruses*.

Conflicts of Interest: The authors declare no conflict of interest.

List of Contributions:

1. Agusi, E.R.; Allendorf, V.; Eze, E.A.; Asala, O.; Shittu, I.; Dietze, K.; Busch, F.; Globig, A.; Meseko, C.A. SARS-CoV-2 at the Human–Animal Interface: Implication for Global Public Health from an African Perspective. *Viruses* **2022**, *14*, 2473. https://doi.org/10.3390/v14112473.
2. Embregts, C.W.E.; Farag, E.A.B.A.; Bansal, D.; Boter, M.; van der Linden, A.; Vaes, V.P.; Berg, I.v.M.-V.D.; Ijpelaar, J.; Ziglam, H.; Coyle, P.V.; et al. Rabies Virus Populations in Humans and Mice Show Minor Inter-Host Variability within Various Central Nervous System Regions and Peripheral Tissues. *Viruses* **2022**, *14*, 2661. https://doi.org/10.3390/v14122661.
3. Garbuglia, A.R.; Lapa, D.; Pauciullo, S.; Raoul, H.; Pannetier, D. Nipah Virus: An Overview of the Current Status of Diagnostics and Their Role in Preparedness in Endemic Countries. *Viruses* **2023**, *15*, 2062. https://doi.org/10.3390/v15102062.
4. Hamel, R.; Vargas, R.E.M.; Rajonhson, D.M.; Yamanaka, A.; Jaroenpool, J.; Wichit, S.; Missé, D.; Kritiyakan, A.; Chaisiri, K.; Morand, S.; et al. Identification of the Tembusu Virus in Mosquitoes in Northern Thailand. *Viruses* **2023**, *15*, 1447. https://doi.org/10.3390/v15071447.
5. Kimble, J.B.; Souza, C.K.; Anderson, T.K.; Arendsee, Z.W.; Hufnagel, D.E.; Young, K.M.; Lewis, N.S.; Davis, C.T.; Thor, S.; Baker, A.L.V. Interspecies Transmission from Pigs to Ferrets of Antigenically Distinct Swine H1 Influenza A Viruses with Reduced Reactivity to Candi-

date Vaccine Virus Antisera as Measures of Relative Zoonotic Risk. *Viruses* **2022**, *14*, 2398. https://doi.org/10.3390/v14112398.

6.   Latinne, A.; Morand, S. Climate Anomalies and Spillover of Bat-Borne Viral Diseases in the Asia–Pacific Region and the Arabian Peninsula. *Viruses* **2022**, *14*, 1100. https://doi.org/10.3390/v14051100.

7.   Morcatty, T.Q.; Pereyra, P.E.R.; Ardiansyah, A.; Imron, M.A.; Hedger, K.; Campera, M.; Nekaris, K.A.-I.; Nijman, V. Risk of Viral Infectious Diseases from Live Bats, Primates, Rodents and Carnivores for Sale in Indonesian Wildlife Markets. *Viruses* **2022**, *14*, 2756. https://doi.org/10.3390/v14122756.

8.   Ntakiyisumba, E.; Lee, S.; Park, B.-Y.; Tae, H.-J.; Won, G. Prevalence, Seroprevalence and Risk Factors of Avian Influenza in Wild Bird Populations in Korea: A Systematic Review and Meta-Analysis. *Viruses* **2023**, *15*, 472. https://doi.org/10.3390/v15020472.

9.   Rojas-Sereno, Z.E.; Streicker, D.G.; Medina-Rodríguez, A.T.; Benavides, J.A. Drivers of Spatial Expansions of Vampire Bat Rabies in Colombia. *Viruses* **2022**, *14*, 2318. https://doi.org/10.3390/v14112318.

10.   Souza, C.K.; Kimble, J.B.; Anderson, T.K.; Arendsee, Z.W.; Hufnagel, D.E.; Young, K.M.; Gauger, P.C.; Lewis, N.S.; Davis, C.T.; Thor, S.; et al. Swine-to-Ferret Transmission of Antigenically Drifted Contemporary Swine H3N2 Influenza A Virus Is an Indicator of Zoonotic Risk to Humans. *Viruses* **2023**, *15*, 331. https://doi.org/10.3390/v15020331.

11.   Vandegrift, K.J.; Yon, M.; Nair, M.S.; Gontu, A.; Ramasamy, S.; Amirthalingam, S.; Neerukonda, S.; Nissly, R.H.; Chothe, S.K.; Jakka, P.; et al. SARS-CoV-2 Omicron (B.1.1.529) Infection of Wild White-Tailed Deer in New York City. *Viruses* **2022**, *14*, 2770. https://doi.org/10.3390/v14122770.

References

1.   Marston, H.D.; Folkers, G.K.; Morens, D.M.; Fauci, A.S. Emerging viral diseases: Confronting threats with new technologies. *Sci. Transl. Med.* **2014**, *6*, 253ps10. [CrossRef] [PubMed]

2.   Belhadi, D.; El Baied, M.; Mulier, G.; Malvy, D.; Mentre, F.; Laouenan, C. The number of cases, mortality and treatments of viral hemorrhagic fevers: A systematic review. *PLoS Negl. Trop. Dis.* **2022**, *16*, e0010889. [CrossRef] [PubMed]

3.   Plowright, R.K.; Peel, A.J.; Streicker, D.G.; Gilbert, A.T.; McCallum, H.; Wood, J.; Baker, M.L.; Restif, O. Transmission or Within-Host Dynamics Driving Pulses of Zoonotic Viruses in Reservoir-Host Populations. *PLoS Negl. Trop. Dis.* **2016**, *10*, e0004796. [CrossRef] [PubMed]

4.   Ermonval, M.; Baychelier, F.; Tordo, N. What Do We Know about How Hantaviruses Interact with Their Different Hosts? *Viruses* **2016**, *8*, 223. [CrossRef] [PubMed]

5.   Vourc'h, G.; Moutou, F.; Morand, S.; Jourdain, E. *Zoonoses the Ties That Bind Humans to Animals*, 1st ed.; Editions Quae: Versailles, France, 2022.

6.   Wells, K.; Morand, S.; Wardeh, M.; Baylis, M. Distinct spread of DNA and RNA viruses among mammals amid prominent role of domestic species. *Glob. Ecol. Biogeogr.* **2020**, *29*, 470–481. [CrossRef] [PubMed]

7.   Gallo, G.; Kotlik, P.; Roingeard, P.; Monot, M.; Chevreux, G.; Ulrich, R.G.; Tordo, N.; Ermonval, M. Diverse susceptibilities and responses of human and rodent cells to orthohantavirus infection reveal different levels of cellular restriction. *PLoS Negl. Trop. Dis.* **2022**, *16*, e0010844. [CrossRef] [PubMed]

8.   Kell, A.M. Innate Immunity to Orthohantaviruses: Could Divergent Immune Interactions Explain Host-specific Disease Outcomes? *J. Mol. Biol.* **2022**, *434*, 167230. [CrossRef] [PubMed]

9.   Subudhi, S.; Rapin, N.; Misra, V. Immune System Modulation and Viral Persistence in Bats: Understanding Viral Spillover. *Viruses* **2019**, *11*, 192. [CrossRef] [PubMed]

10.   Labadie, T.; Batejat, C.; Leclercq, I.; Manuguerra, J.C. Historical Discoveries on Viruses in the Environment and Their Impact on Public Health. *Intervirology* **2020**, *63*, 17–32. [CrossRef] [PubMed]

11.   Leifels, M.; Khalilur Rahman, O.; Sam, I.C.; Cheng, D.; Chua, F.J.D.; Nainani, D.; Kim, S.Y.; Ng, W.J.; Kwok, W.C.; Sirikanchana, K.; et al. The one health perspective to improve environmental surveillance of zoonotic viruses: Lessons from COVID-19 and outlook beyond. *ISME Commun.* **2022**, *2*, 107. [CrossRef] [PubMed]

12.   Wolters, W.J.; de Rooij, M.M.T.; Molenaar, R.J.; de Rond, J.; Vernooij, J.C.M.; Meijer, P.A.; Oude Munnink, B.B.; Sikkema, R.S.; van der Spek, A.N.; Spierenburg, M.A.H.; et al. Manifestation of SARS-CoV-2 Infections in Mink Related to Host-, Virus- and Farm-Associated Factors, The Netherlands 2020. *Viruses* **2022**, *14*, 1754. [CrossRef] [PubMed]

13.   Roundy, C.M.; Nunez, C.M.; Thomas, L.F.; Auckland, L.D.; Tang, W.; Richison, J.J., 3rd; Green, B.R.; Hilton, C.D.; Cherry, M.J.; Pauvolid-Correa, A.; et al. High Seroprevalence of SARS-CoV-2 in White-Tailed Deer (Odocoileus virginianus) at One of Three Captive Cervid Facilities in Texas. *Microbiol. Spectr.* **2022**, *10*, e0057622. [CrossRef] [PubMed]

14. Lewis, J.; Zhan, S.; Vilander, A.C.; Fagre, A.C.; Kiaris, H.; Schountz, T. SARS-CoV-2 infects multiple species of North American deer mice and causes clinical disease in the California mouse. *bioRxiv* **2022**. [CrossRef]
15. Ueha, R.; Ito, T.; Ueha, S.; Furukawa, R.; Kitabatake, M.; Ouji-Sageshima, N.; Uranaka, T.; Tanaka, H.; Nishijima, H.; Kondo, K.; et al. Evidence for the spread of SARS-CoV-2 and olfactory cell lineage impairment in close-contact infection Syrian hamster models. *Front. Cell. Infect. Microbiol.* **2022**, *12*, 1019723. [CrossRef] [PubMed]

Review

SARS-CoV-2 at the Human–Animal Interface: Implication for Global Public Health from an African Perspective

Ebere Roseann Agusi [1,2,3,†], Valerie Allendorf [2,*,†], Emmanuel Aniebonam Eze [3], Olayinka Asala [1], Ismaila Shittu [1], Klaas Dietze [2], Frank Busch [2], Anja Globig [2,*,†] and Clement Adebajo Meseko [1,4,†]

[1]   National Veterinary Research Institute, Vom 930001, Nigeria
[2]   Institute of International Animal Health/One Health, Friedrich-Loeffler-Institut, 17493 Greifswald-Insel Riems, Germany
[3]   Department of Microbiology, University of Nigeria Nsukka, Enugu 410001, Nigeria
[4]   College of Veterinary Medicine, University of Minnesota, Minneapolis, MN 55455, USA
*   Correspondence: anja.globig@fli.de (A.G.); valerie.allendorf@fli.de (V.A.)
†   These authors contributed equally to this work.

Abstract: The coronavirus disease 2019 (COVID-19) pandemic has become the most far-reaching public health crisis of modern times. Several efforts are underway to unravel its root cause as well as to proffer adequate preventive or inhibitive measures. Zoonotic spillover of the causative virus from an animal reservoir to the human population is being studied as the most likely event leading to the pandemic. Consequently, it is important to consider viral evolution and the process of spread within zoonotic anthropogenic transmission cycles as a global public health impact. The diverse routes of interspecies transmission of SARS-CoV-2 offer great potential for a future reservoir of pandemic viruses evolving from the current SARS-CoV-2 pandemic circulation. To mitigate possible future infectious disease outbreaks in Africa and elsewhere, there is an urgent need for adequate global surveillance, prevention, and control measures that must include a focus on known and novel emerging zoonotic pathogens through a one health approach. Human immunization efforts should be approached equally through the transfer of cutting-edge technology for vaccine manufacturing throughout the world to ensure global public health and one health.

Keywords: SARS-CoV-2; COVID-19; zoonosis; one health; emerging infectious disease; Africa

Citation: Agusi, E.R.; Allendorf, V.; Eze, E.A.; Asala, O.; Shittu, I.; Dietze, K.; Busch, F.; Globig, A.; Meseko, C.A. SARS-CoV-2 at the Human–Animal Interface: Implication for Global Public Health from an African Perspective. *Viruses* 2022, *14*, 2473. https://doi.org/10.3390/v14112473

Academic Editors: Myriam Ermonval and Serge Morand

Received: 28 September 2022
Accepted: 7 November 2022
Published: 9 November 2022

1. Introduction

The magnitude of morbidity and mortality of the severe acute respiratory syndrome coronavirus 2 (SARS-CoV-2) pandemic has led to serious and far-reaching impacts on healthcare systems, societies, economies, and politics worldwide [1]. To prepare for and potentially prevent the occurrence of such an event, research was initially centered on the probable origin of the zoonotic virus and its mechanism of spillover to humans [2]. Nonetheless, the increasing reports on zoonotic transmissions from humans to animals (anthroponosis) [3] and evidence of bi-directional transmissions at the human–animal interface [4] point to potential risks in the modification, adaptation, and perpetuation of the pathogen in nature. Consequently, human–animal interfaces with the potential for spillover infections from animals to humans and vice versa pose a significant and continuous global public health threat.

As in the example of SARS-CoV-2, it is estimated that over 75% of emerging human infectious diseases have animals as the primary source [5]. Age-long close contact between humans and domesticated species enabled the early transmission and co-evolution (e.g., measles, smallpox) of the most adaptable pathogens to humans [6]. In comparison with domestic animals, opportunities for close contact between humans and wildlife are relatively rare. Yet, the recent emergence of many diseases, such as severe acute respiratory syndrome (SARS) [7], Ebola [8], monkeypox [9], or Nipah virus encephalitis [10], demonstrate the increased risk of a spillover of wildlife-sourced pathogens into the human population [11]. The continuous growth of the global population leads to increasing demand for food and natural resources [12,13]. This, in turn, has led to changes in land use, continuous human encroachment, and changes in ecosystems, resulting in overlapping habitats of wildlife, domestic animals, and humans, which are regarded as the main drivers of zoonotic pathogen emergence [14].

Key characteristics evaluated for associations with high-risk disease emergence include host plasticity [15] spatio-temporal distribution [16], and human-to-human transmissibility [17]. In addition, human practices that facilitate zoonotic and inter-human transmission act synergistically to promote viral emergence [18]. Predominantly respiratory viruses with the capacity for rapid human-to-human transmission may then succeed in broad geographic spread [18].

The key objective of this review is to assess the extent of potential anthropogenic transmission of SARS-CoV-2 to domestic and wild animals from an African perspective and its implication on viral evolution and global public health. In the derivation of information and data used for this review, a qualitative approach was adopted. Research keywords such as *zoonoses, SARS-CoV-2, emerging diseases in Africa*, and *one health* and journal articles published between the years 2001 to 2022 were mostly considered. Over 100 peer-reviewed and pre-print articles as well as statistical data from governmental websites, published until March 2022, were the sources adopted for this work.

2. The Probable Origin of SARS-CoV-2 Points to Animals

Phylogenetic analysis of the genome of SARS-CoV-2 shows its affiliation to the species *Severe acute respiratory syndrome-related coronavirus* (SARSr-CoV) of the subgenus *Sarbecovirus* within the *Betacoronavirus* genus [19,20]. Members of this genus were found in African and Asian bats [21,22] as well as Asian pangolins [23]. The closest relative of SARS-CoV-2 was isolated from a horseshoe bat (*Rhinolophus macrotis*) in a Laotian bat cave in 2021 [24]. Although it is possible that spillover occurred through direct bat-to-human contact, the first reported COVID-19 cases in December 2019 were associated with Wuhan wet market activities [25], where live-trapped carnivores, such as raccoon dogs and badgers, were offered for sale, but no bat species [26]. Different animals farmed for food or fur, including civet cats [27], foxes [28], minks [29], and raccoon dogs [30], all proved highly susceptible to sarbecoviruses. Taken together, these circumstances suggest a live intermediate host as the primary source of the SARS-CoV-2 progenitor that humans were repeatedly exposed to, as was the case with the origin of SARS-CoV [31]. Nonetheless, clinical and laboratory confirmations of field infections with SARS-CoV-2-related viruses in such animals before the outbreak of COVID-19 are yet to be made.

In places where environmental and anthropogenic activities predispose to spillover transmissions, the risk of the emergence of a novel zoonotic pathogen is likely to be high [32]. Unfortunately, wild animal hosts and high-risk interfaces facilitating spillover are vastly understudied—and most likely under-reported, in particular in sub-Saharan Africa [33]. Consequently, enhanced monitoring and control efforts beginning from local towards a global strategic one-health surveillance as well as a significant increase in funding for basic and applied research in high-risk countries are being recommended [34–36].

2.1. Mutations of SARS-CoV-2 Are a Common Feature Facilitating the Crossing of Interspecies Barriers

As with many RNA viruses, the genetic diversity of coronaviruses is due to the high frequency of homologous recombination and accumulation of mutations [37], which supports the breaking of interspecies barriers, varying tissue tropism, and adapting to biological variations [38]. For example, the recombination of the genomes of SARSr-CoV in the spike glycoprotein (S) region is responsible for the mediation of initial cross-species transmission from bats to other mammals [39].

Studies by Lytras et al. (2021) [19] and Temmam et al. (2021) [24] suggested that homologous recombination of a sarbecovirus of unknown origin and a bat coronavirus within the S-region resulted in the progenitor of SARS-CoV-2 [40]. The authors found that the entire genome of SARS-CoV-2 is very similar to SARSr-Ra-BatCoV RaTG13, isolated in 2013 from a horseshoe bat in China, except for the sequence coding for the receptor-binding domain. Viruses genetically related to SARS-CoV-1 were also detected in Ghanaian and Nigerian leaf-nosed bats in 2009 [41] and 2010 [42], respectively, indicating deep-rooted genetic and ecological factors in sub-Saharan Africa yet to be fully explored. Generally, these results point to the complexities of the origin of SARS-CoV-2, in which recombination enabled further evolutionary selection of strains from distinct host species before its spillover to humans. Repetitive close contact between animals and humans, enabling spillover and reverse transmission, may further promote the genetic diversity of coronaviruses [43].

2.2. Frequent Transmissions of SARS-CoV-2 to Animals may Create Future Animal Reservoir Hosts

The evolutionary selection of viruses with a greater ability to rapidly adapt to new hosts co-selects for viruses capable of effective intra-species transmission in the new host [17], underpinning disease emergence theory [44–46]. Transmission of SARS-CoV-2 from humans to susceptible animal hosts and onwards is likely to amplify mutations that could, in turn, re-infect humans with altered virus variants [47,48], potentially leading to renewed pandemic spread. The susceptibility of animal species depends on several factors, including the compatibility between the viral spike protein and the host receptor ACE2 [49] or alternative receptors such as Neuropilin-1 [50,51] or CD147 [52], the species, and the capacity of the virus to escape the immune system and restriction factors of the new host [47]. For an animal species to act further on as a successful reservoir host, the virus has to become established in the animal population by efficient intra-species transmission, leading to transiently infected animals with prolonged or repetitive phases of shedding infectious particles [53].

Experimental studies suggest that animal species such as cats, dogs [47], ferrets [53], raccoon dogs [54], crab-eating macaques [30], rhesus macaques [55], white-tailed deer [56], rabbits [57], and Syrian hamsters [58] are susceptible to SARS-CoV-2 infection. Moreover, it was found that onward cat-to-cat and ferret-to-ferret [59] as well as hamster-to-hamster transmission [60] can occur through physical or airborne contact.

The worldwide geographical distribution of SARS-CoV-2 in animals according to reports submitted to the OIE is summarized in Figure 1. As of July 2022, 36 countries have reported a total of 679 occurrences of SARS-CoV-2 infections in 24 different animal species of carnivores, primates, ungulates, and rodents [3]. These anecdotally reported cases of natural transmission were mainly directly from infected humans in close contact with animals, particularly to (i) companion animals such as cats and dogs [61] as well as (ii) captive wild animals such as lions, tigers, pumas, snow leopards, and gorillas in zoos [62] and (iii) farmed fur animals such as minks and ferrets. Further on, natural infection was shown in wild white-tailed deer in the U.S. with comparatively high seroprevalences of 37% in the investigated population, implying rapid and efficient spread among this abundant wildlife species [63]. Conversely, animal-to-human transmissions have been demonstrated only in a few reports. Nonetheless, mink-to-human transmission in a mink farm in Denmark [4], the hamster-to-human cluster reported from Hong Kong [64], and the most recently reported case of human infection with a highly divergent SARS-CoV-2 deriving from a wild-tailed deer in Canada [65] demonstrate the potential for further zoonotic transmission cycles accompanied by alterations in nucleotide and amino acid patterns, potentially resulting in new pandemic variants.

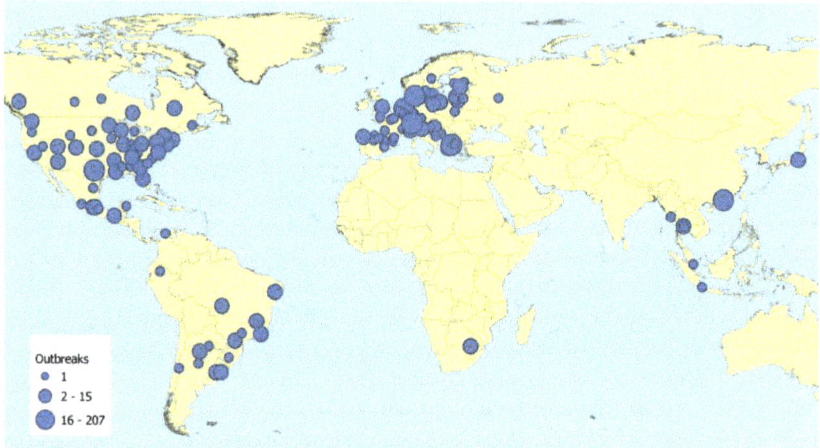

Figure 1. Global distribution of animal infections with SARS-CoV-2. (reprinted with permission from WOAH [3], source: https://www.woah.org/app/uploads/2022/02/sars-cov-2-situation-report-9.pdf, CC BY-NC-ND 4.0 (accessed on 19 November 2021).

However, further investigation is required to determine whether SARS-CoV-2 or other related betacoronaviruses can, in turn, transform new animal hosts into virus reservoirs [66].

3. Effective Spillover at the Human–Animal Interface Marks the Entry Point of Novel Infectious Diseases

Spillover events from animal reservoirs into humans are not uncommon in locations with a high frequency of human–animal contact [67]. Serological studies demonstrated evidence of SARSr-CoV-specific antibodies in human residents in rural locations, and even higher rates were recorded in humans living near bat caves in China [68]. Spillover risks increase with human encroachment into rural areas and activities resulting from economic networks around and between rural and urban areas. When a densely packed and immuno-logically naïve human population in an urban area is then exposed to a novel pathogen, spillover events have a much higher likelihood of resulting in extensive spread [69]. Urbanization characterized by rapid intensification of agriculture, socioeconomic change, and ecological fragmentation can have profound impacts on the epidemiology of infectious diseases [70].

Wildlife populations are heterogeneously distributed, and certain species group in spatial aggregations with livestock and humans, creating interfaces that might be important for the transmission of zoonotic agents [71]. Consequently, anthropogenic pressures can create diverse wildlife–livestock–human interfaces, representing a critical point for cross-species transmission and the emergence of pathogens into new host populations.

Many zoonotic viruses have recently emerged from bats. It has been argued that bats may have an immune system that allows them to coexist with viruses from different virus families [72], hence making them viral reservoirs for filoviruses (e.g., Ebola virus [73], Marburg virus [74]), paramyxoviruses (Hendra henipavirus [75], Nipah henipavirus [76]), rhabdoviruses [77], arenaviruses (*Tacaribe mammarenavirus* has a bat host) [78], and Sarbecoviruses. Spillover to humans happened directly via close contact or bushmeat consumption [79] or indirectly via an intermediate animal species [80]. The complexity of urban systems as networks of physical interfaces across which pathogens can be transmitted between humans and animals exists within the context of societal, cultural, and policy interfaces. Therefore, the investigation of the conditions and the reasons for human–animal contact potentially leading to the emergence of zoonotic diseases requires a multisectoral approach [81,82].

Spillover Transmission Is Driven by Human Activity

Interspecies transmission of zoonotic pathogens (Figure 2) is mainly driven by human behaviors having a direct or indirect influence on ecosystems and human–animal interactions [83]. Vector-independent transmitted pathogens such as SARS-CoV-2 are distributed through close contact with an infected individual (e.g., droplets, aerosols, fecal-oral) or by contact with contaminated fomites [83].

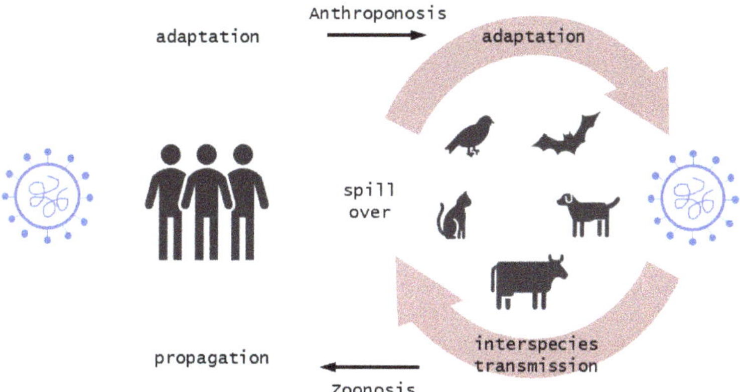

Figure 2. Transmission pathways between humans and animals (image created in Microsoft Power-Point 2019).

Direct transmission of zoonotic pathogens applies bi-directionally almost solely regarding the transmission from and to domestic animals in a variety of settings as well as regarding farmed animals or those kept in zoos, where direct contact is inevitable and a substantial part of the human–animal relationship. In the wildlife domain, the direct transmission pathway must be considered one-directional to humans from wildlife hunted and traded for meat consumption or traditional medicine and bi-directional between humans and synanthropic species such as pest rodents [84] or frugivorous bats [85,86]. Other direct contacts between wildlife and humans potentially leading to bi-directional direct transmission may occur during conservation interventions or field research [87]. Remarkably, with the increasing demand for exotic pet species and their trade around the world, the range of pathogens potentially being transmitted within this domain is broadened [88,89].

Fomite transmission from humans to animals pertains whenever waste or sewage can be accessed by domestic or wild animals. For example, SARS-CoV-2 survives on surfaces of personal protective equipment and other household materials for up to three days under normal conditions [90]. Inappropriate discarding or disposal of contaminated personal protective equipment (PPE) such as face masks and gloves as well as tissue wipes, etc., are regarded as a source of COVID-19 infection, especially among stray and wild animals [91]. Even the most elaborate waste management or sewage systems may still be accessed by rodents or birds and therefore cannot prevent onward transmission and spread of pathogens. In the animal-to-human direction, the main transmission pathways to consider are contaminated food of animal origin [92] as well as unwashed manured crops [93]. Furthermore, the use or consumption of contaminated water can be the source of infection with zoonotic pathogens in both directions [94].

4. Compounding a Bad Situation: The Impact of Zoonotic Diseases in Africa with a Focus on Nigeria

As experienced with the SARS-CoV-2 pandemic, zoonotic diseases have the potential to threaten human and animal health globally, potentially destabilizing our local economy and impacting food security. In addition, countries of the African continent are still seriously challenged by neglected zoonotic diseases (NZDs), causing huge economic losses and mortality (World Bank, 2018) [95]. The World Health Organization (WHO) has identified eight NZDs: anthrax, bovine tuberculosis, brucellosis, cysticercosis, echinococcosis, leishmaniasis, rabies, and human African trypanosomiasis. These diseases are termed "neglected", as they mainly affect poor populations who live near domestic or wild animals, often in areas where there are little or no adequate or healthy sanitary conditions [96]. Furthermore, they are neglected due to underestimation of the disease burden, which is usually concentrated in developing countries with ineffective diagnostics and deprived healthcare delivery systems [97]. Nonetheless, NZDs are known to have devastating impacts on human health and welfare, on domestic animal and wildlife health, as well as on risks associated with disease emergence [98]. Additionally, it can be assumed that their presence and persistence in African populations are linked to well-established transmission cycles between humans and animals that may further be used by other or novel zoonotic pathogens with similar transmission modes.

Relatedly, according to a WHO report, while there has been a 63% jump in the number of zoonotic outbreaks in the African region in the decade from 2012–2022 compared to 2001–2011, about 30% of 63% of the substantiated public health events recorded in this region were zoonotic disease outbreaks [99]. The burden of infectious diseases in Africa ranges from newly evolved strains of pathogens (e.g., multi-drug-resistant tuberculosis and chloroquine-resistant malaria) to pathogens that have recently entered human populations for the first time (e.g., HIV-1, Ebola virus, SARS-CoV-2) as well as pathogens that have been historically present in humans but have recently increased in incidence (e.g., Lyme disease, Lassa fever) [100]. With the presence and emergence of many zoonotic diseases, a noticeable trend can be observed. Increasing ecological changes and economical challenges further enhance the development and intensification of factors of (re-)emergence, such as urbanization or socio-cultural behavior such as farming, hunting, and tourism [101].

4.1. SARS-CoV-2 Infection Detection Rate in the African Population Is Lower Compared to the Rest of the World

In 2020, the WHO warned in their projection that Africa could likely be the next epicenter of SARS-CoV-2, with an estimated 44 million infections and 190,000 deaths in the first year of the pandemic [102]. Up to July 2022, with 8.7 million cumulated cases and 173,248 deaths reported from the African region, the SARS-CoV-2-attributed mortality in Africa was projected to fall by nearly 94% in 2022 [103]. Based on this record, the enduring mystery of COVID-19 is why the pandemic has not hit low-income African nations as hard as wealthy countries in North America and Europe [104]. The adoption of heightened disease surveillance systems, such as mandatory screening at ports of entry, setting up isolation and quarantine centers, activation of disease surveillance mechanisms from previous influenza and Ebola surveillance systems, as well as contact tracing to swiftly detect and respond to the outbreak, are discussed as potential explanations [105]. However, most African governments are still struggling with strict implementation of containment measures such as border and travel restrictions, bans on large gatherings, social distancing, as well as poor uptake on free vaccinations [106]. Furthermore, the general shortage of diagnostic tests for SARS-CoV-2 on the continent threatens to possibly contribute to a massive underestimation of the true burden of the disease [107].

Concurrently, factors intrinsic to the African population are being discussed to be causative for a lower morbidity and mortality on the continent, such as the general younger demographic of the population, lack of diagnosis in cases of deaths, genetic factors [108], the increased circulation of other coronaviruses having a protective effect against critical COVID-19 [108,109], or the higher burden of other infectious diseases such as malaria leading to an increased alertness of the innate immune system and consequently a lower morbidity and mortality [110].

4.2. Socioeconomic and Ecological Factors in Africa Enhance the Probability of Anthropozoonotic Transmission of SARS-CoV-2

Despite the relatively low numbers of SARS-CoV-2 in Africa, particularly in Nigeria, the impending shock on the public health sector could have a devastating impact on the country's previously strained and fragile health system and could quickly turn into a socioeconomic emergency. The limited availability of diagnostic tests makes the detection of asymptomatic patients nearly impossible and deepens the uncertainty of the potential impact of SARS-CoV-2 infections. Beyond these health and social risks, there are several factors that impact negatively on Nigeria's economy, including lower trade and foreign investment in the immediate term, falling demands linked with the lockdowns or travel bans, and lastly, a continental supply shock affecting domestic and intra-African trade [111].

Affecting Nigeria's economic growth, the crisis—together with the effects of climate change—has a significant impact on the overall well-being of people and the number of people living in poverty. According to United Nations (UN) estimates, approximately 30 million more people worldwide could fall into poverty, with a significant rise in the number of acutely food-insecure people [112]. Inadvertently, poverty and food insecurity also have a great impact on human migration, encroachment, as well as changes in human–wildlife interfaces [113,114]. Therefore, policymakers are saddled with the responsibility of enacting measures that will be consistent in tackling realities posed by the SARS-CoV-2 pandemic.

4.3. Limited Access to Vaccines Increases the Vulnerability of the Population

The WHO has set a global target of 70% of the population of all countries to be vaccinated by mid-2022 [112]. As of mid-2022, more than 5.33 billion people have received at least one dose of a COVID-19 vaccine globally [115]. However, the vaccination coverage differs strongly in different regions of the globe. In the African region, about 27% of the population was vaccinated with at least one dose [115]. In comparison, in upper-middle- to high-income countries, 81% of the inhabitants have already received at least one dose of vaccine [116]. This inequitable vaccine distribution is not only leaving millions of people vulnerable, but it is also allowing novel, perhaps even more virulent variants of the virus to emerge and subsequently spread across the globe.

As the African region grapples to meet the rising demand for essential vaccination commodities, less than 10% of its nations are projected to hit the year-end target of fully vaccinating 40% of their people by 2022. [117]. The main factors aside from skepticism and hesitancy of some members of the population to be vaccinated [117] are the limited access to vaccines, the disruption of the cold chain, and the shortage in associated medical care [118]. One of the most glaring revelations of the COVID-19 pandemic is the realization that human vaccine manufacturing is almost nonexistent in the entire African continent [118].

4.4. Human-to-Animal Transmission May Happen Undetected due to Limited Implementation of Biosecurity and Active Surveillance in Animal Holdings

At global and national levels, veterinary professionals, their representative associations, and animal health regulatory bodies have played various roles in protecting the public's health during the SARS-CoV-2 pandemic [119]. Controlling highly contagious animal diseases on a large scale has been a recurring challenge for veterinarians for decades [120]. In addition, they understand the epidemiology of the zoonotic disease as well as the risk of potential spillover at the various human–animal interfaces [120].

The Food and Agriculture Organisation of the United Nations (FAO) recommends the systematic and close surveillance of farmed livestock as well as domestic animals, as the transmission to these animals may lead to the further uncontrolled spread of the virus and the potential formation of new reservoirs [121]. In already existing systems in countries in Europe and North America, where the ownership especially of livestock is meticulously registered and monitored by the authorities, the recommended surveillance strategies were rapidly implemented [121]. In countries of Africa without such systems, the implementation of monitoring and control measures is vastly restricted. While SARS-COV-2 transmission from humans to animals was reported in most parts of the world, evidence is scant in Africa, and this may be related to limited surveillance [121].

5. Conclusions: Mitigating Future Respiratory Virus Pandemics

In an increasingly globalized world, a spillover of a zoonotic pathogen poses major risks to the global society and economy. This has been well-demonstrated by the ongoing coronavirus disease SARS-CoV-2 pandemic, which has resulted in an unprecedented global public health, social, and economic crisis. Despite our experiences with emerging zoonotic diseases such as SARS, Ebola, Lassa fever, or influenza and subsequently improved national and global surveillance systems, humanity is probably not able to totally prevent the emergence of zoonotic pathogens. Nonetheless, the SARS-CoV-2 pandemic has emphasized the importance of sophisticated global preparedness measures, including monitoring and early detection, prevention of spread, and strategic response mechanisms. Especially in locations where there is a high overlap of wildlife, increasing livestock or agricultural production, and human encroachment, the implementation of early warning systems is needed. Consequently, spotting and mitigating the risks of future spillovers involves working closely with the communities in hotspots for disease emergence and appropriate risk communication. High-risk animal-to-human interfaces must be given special attention. This includes companion animals whose owners have been infected, farmed animals that are often reared in large numbers and are predisposed to viral spread, as well as wildlife in close proximity to human and domestic animal populations. Other high-risk interfaces with human-animal-environmental activities such as land-use change, hunting, and agricultural practices that have been identified to have facilitated viral spillover events should be a focus for education and interventions directed at disease prevention.

Monitoring and control measures limiting zoonotic and onward transmissions should be established and publicly enforced. Particularly in developing countries, farming communities remain vulnerable. Often farms are operated by women and children with limited access to medical, veterinary, and animal production services. The lack of food safety control systems most likely prevents adequate responses to emerging and resurgent zoonotic diseases. Non-pharmaceutical measures, such as physical distancing, strict hand hygiene, respiratory etiquette, appropriate use of facemasks, and home isolation of suspected or confirmed cases as the first line of defense in mitigating potential outbreaks may go along with heavy losses in people's livelihoods.

Once available, vaccination can provide direct protection in reducing susceptibility among the uninfected and indirect protection in reducing viral spread among the population [122]. Accessibility to an effective vaccine remains a significant challenge in the intervention against SARS-CoV-2. If the effective vaccination rate is satisfactorily high, herd immunity generated by a transmission-blocking vaccine will help to control or even eliminate the spread. Achieving this goal may become more difficult when external constraints affect the deployment of vaccines or when vaccinal and natural immunity is inadequate due to further-evolving immune escape variants.

To successfully tackle zoonotic spillover and its related issues, the resolutions displayed in Table 1 can be considered not just exclusively for the African setting:

Table 1. Resolutions suggested to tackle zoonotic spillover infections and related issues.

Measure	Explanation
Enactments	As a precaution, the public is urged to keep safe distances from wildlife, particularly species that are known to be susceptible to SARS-CoV-2 infections. This can be achieved through enactments such as emergency orders and temporary bans on hunting, trading, and non-essential contact with wildlife. Relatedly, prohibitive measures may be taken to discourage bushmeat hunting and the consumption of raw, unprocessed meat. In all of these, alternative sources of low-cost nutrition should be provided in resource limited communities in order to spare wildlife hunting.
Education	To highlight issues of public health concern, awareness must be raised among the population regarding the possibility of disease transmission from animals to humans and vice versa. While some cultural or religious traditions may predispose to direct contact with wildlife, there is a need to effectively communicate the risks of disease transmission. It may be helpful to recall traditional folk tales that highlight the inherent danger in handling (deceased) wildlife as a learning tool in local public teaching. Many such folklores abound among the native Yorubas of Nigeria and may be applicable in other tribal settings in Africa [123]. Furthermore, early education of pupils on the dangers of emerging or neglected zoonotic diseases and the precautionary measures to be taken can be a significant way to bridge the knowledge gap.
National One Health Strategic Plan	A national one health strategic plan should be actively adopted across the different organizational frontiers (in Africa) to safeguard a holistic impact. From joint (national) research projects to research projects by independent donors, combined efforts must achieve the objective of containing virus spread. The capabilities of self-determined and independent, top-level research should be continuously increased via access to funding, training, and education. This enables local and independent monitoring and rapid identification of novel emerging pathogens.
Vaccine equity	The autonomous development and production of vaccines in Africa must be actively supported. This will create the enabling conditions to achieve a rapid response to emerging or pandemic pathogens and increase the global vaccination coverage while continually developing vaccines for mutant strains of the virus [117].
Heightened Monitoring	Gathering data on human practices as well as contact with animals in settings with diverse host assemblages will ensure effective analysis regarding potential spillover risks. Environmental, veterinary and medical scientists may further investigate the pathogens present and exchanged at the identified human–animal interfaces of risk. This will give an informed direction towards critical points for disease control and behavior change interventions aimed at prevention.

Author Contributions: Conceptualization, E.R.A., V.A., E.A.E., A.G. and C.A.M.; methodology, E.R.A.; writing—original draft preparation, E.R.A. and V.A.; writing—review and editing, A.G., C.A.M., E.A.E., O.A., I.S., K.D. and F.B.; visualization, V.A.; supervision, A.G., C.A.M. and E.A.E.; funding acquisition, A.G. and K.D. All authors have read and agreed to the published version of the manuscript.

Funding: This research was conducted within the project "Nigeria Addressing COVID-19 Through a One Health Approach", funded by the German Federal Ministry of Health on the basis of a decision by the German Bundestag within the "Global Health Protection Programme" (grant number ZMI1-2521GHP904).

Institutional Review Board Statement: Not applicable.

Informed Consent Statement: Not applicable.

Data Availability Statement: Not applicable.

Acknowledgments: We acknowledge the National Veterinary Research Institute, Vom, and the Friedrich-Loeffler-Institute, Germany. We wish to thank the two anonymous reviewers for spending their time and for their critical but fair comments helping us to improve the manuscript.

Conflicts of Interest: The authors declare no conflict of interest.

References

1. Shang, Y.; Li, H.; Zhang, R. Effects of Pandemic Outbreak on Economies: Evidence from Business History Context. *Front. Public Health* **2021**, *9*, 632043. [CrossRef] [PubMed]
2. World Health Organization. WHO-Convened Global Study of Origins of SARS-CoV-2: China Part Joint WHO-China Study. Available online: https://www.who.int/publications/i/item/who-convened-global-study-of-origins-of-sars-cov-2-china-part (accessed on 17 July 2022).
3. World Organization for Animal Health (WOAH). SARS-CoV-2 in Animals—Situation. 2022. Available online: https://www.woah.org/app/uploads/2022/02/sars-cov-2-situation-report-9.pdf (accessed on 3 August 2022).
4. Fenollar, F.; Mediannikov, O.; Maurin, M.; Devaux, C.; Colson, P.; Levasseur, A.; Fournier, P.E.; Raoult, D. Mink, SARS-CoV-2, and the Human-Animal Interface. *Front. Microbiol.* **2021**, *12*, 663815. [CrossRef] [PubMed]
5. Taylor, L.H.; Latham, S.M.; Woolhouse, M.E. Risk factors for human disease emergence. *Phil. Trans. R. Soc. Lond. B* **2001**, *356*, 983–989. [CrossRef] [PubMed]
6. Wolfe, N.D.; Dunavan, C.P.; Diamond, J. Origins of major human infectious diseases. *Nature* **2007**, *447*, 279–283. [CrossRef] [PubMed]
7. Demmler, G.J.; Ligon, B.L. Severe acute respiratory syndrome (SARS): A review of the history, epidemiology, prevention, and concerns for the future. *Semin. Pediatr. Infect. Dis.* **2003**, *14*, 240–244. [CrossRef]
8. Stawicki, S.P.; Arquilla, B.; Galwankar, S.C.; Hoey, B.A.; Jahre, J.A.; Kalra, S.; Kelkar, D.; Papadimos, T.J.; Sabol, D.; Sharpe, R.P. The emergence of ebola as a global health security threat: From 'lessons learned' to coordinated multilateral containment efforts. *J. Glob. Infect. Dis.* **2014**, *6*, 164–177. [CrossRef]
9. Ligon, B.L. Monkeypox: A review of the history and emergence in the Western hemisphere. *Semin. Pediatr. Infect. Dis.* **2004**, *15*, 280–287. [CrossRef]
10. Epstein, J.H.; Field, H.E.; Luby, S.; Pulliam, J.R.; Daszak, P. Nipah virus: Impact, origins, and causes of emergence. *Curr. Infect. Dis. Rep.* **2006**, *8*, 59–65. [CrossRef]
11. Morens, D.M.; Fauci, A.S. Emerging infectious diseases: Threats to human health and global stability. *PLoS Pathog.* **2013**, *9*, e1003467. [CrossRef]
12. Gu, D.; Andreev, K.; Dupre, M.E. Major Trends in Population Growth Around the World. *China CDC Wkly.* **2021**, *3*, 604–613. [CrossRef]
13. Byrnes, B.; Bumb, B. Population Growth, Food Production and Nutrient Requirements. *J. Crop. Prod.* **2008**, *1*, 1–27. [CrossRef]
14. Daszak, P.; Cunningham, A.A.; Hyatt, A.D. Emerging infectious diseases of wildlife—Threats to biodiversity and human health. *Science* **2000**, *287*, 443–449. [CrossRef]
15. Johnson, C.K.; Hitchens, P.L.; Evans, T.S.; Goldstein, T.; Thomas, K.; Clements, A.; Joly, D.O.; Wolfe, N.D.; Daszak, P.; Karesh, W.B.; et al. Spillover and pandemic properties of zoonotic viruses with high host plasticity. *Sci. Rep.* **2015**, *5*, 14830. [CrossRef]
16. Cuadros, F.D.; Xiao, Y.; Mukandavire, Z.; Correa-Agudelo, E.; Hernández, A.; Kim, H.; MacKinnon, N.J. Spatiotemporal transmission dynamics of the COVID-19 pandemic and its impact on critical healthcare capacity. *Health Place* **2020**, *64*, 102404. [CrossRef]
17. Li, C.; Ji, F.; Wang, L.; Wang, L.; Hao, J.; Dai, M.; Liu, Y.; Pan, X.; Fu, J.; Li, L.; et al. Asymptomatic and Human-to-Human Transmission of SARS-CoV-2 in a 2-Family Cluster, Xuzhou, China. *Emerg. Infect. Dis.* **2020**, *26*, 1626–1628. [CrossRef]
18. Burrell, C.J.; Howard, C.R.; Murphy, F.A. Epidemiology of Viral Infections. *Fenner White's Med. Virol.* **2017**, *5*, 185–203. [CrossRef]
19. Lytras, S.; Xia, W.; Hughes, J.; Jiang, X.; Robertson, D.L. The animal origin of SARS-CoV-2. *Science* **2021**, *373*, 968–970. [CrossRef]
20. Coronaviridae. Chapter Version: ICTV Ninth Report; 2009 Taxonomy Release. Available online: https://ictv.global/report_9th/RNApos/Nidovirales/Coronaviridae (accessed on 1 August 2022).
21. Leopardi, S.; Oluwayelu, D.; Meseko, C.; Marciano, S.; Tassoni, L.; Bakarey, S.; Monne, I.; Cattoli, G.; De Benedictis, P. The close genetic relationship of lineage D Betacoronavirus from Nigerian and Kenyan straw-colored fruit bats (*Eidolon helvum*) is consistent with the existence of a single epidemiological unit across sub-Saharan Africa. *Virus Genes* **2016**, *52*, 573–577. [CrossRef]
22. Ge, X.-Y.; Wang, N.; Zhang, W.; Hu, B.; Li, B.; Zhang, Y.-Z.; Zhou, J.-H.; Luo, C.-M.; Yang, X.-L.; Wu, L.-J.; et al. Coexistence of multiple coronaviruses in several bat colonies in an abandoned mineshaft. *Virol. Sin.* **2016**, *31*, 31–40. [CrossRef]
23. Peng, M.S. The high diversity of SARS-CoV-2-related coronaviruses in pangolins alerts potential ecological risks. *Zool. Res.* **2021**, *42*, 834–844. [CrossRef]
24. Temmam, S.; Vongphayloth, K.; Baquero, E.; Munier, S.; Bonomi, M.; Regnault, B.; Douangboubpha, B.; Karami, Y.; Chrétien, D.; Sanamxay, D.; et al. Bat coronaviruses related to SARS-CoV-2 and infectious for human cells. *Nature* **2022**, *604*, 330–336. [CrossRef] [PubMed]
25. Allam, Z. The first 50 days of COVID-19: A detailed chronological timeline and extensive review of literature documenting the pandemic. In *Surveying the Covid-19 Pandemic and its Implications. Urban Health, Data Technology and Political Economy*; Elsevier: Amsterdam, The Netherlands, 2020; pp. 1–7. [CrossRef]
26. McCarthy, S. Coronavirus: Thousands of Wild Animals Were Sold at Wuhan Markets in Months before COVID-19 Outbreak, New Report Finds. *South China Morning Post*, 2021. Available online: https://www.scmp.com/news/china/science/article/3136674/thousands-wild-animals-were-sold-wuhan-markets-months-covid-19 (accessed on 23 November 2021)

27.   Center for Genomic Regulation. Ferrets, cats and civets most susceptible to coronavirus infection after humans: Ducks, rats, mice, pigs and chickens had lower or no susceptibility to infection. *Science Daily*, 2021. Available online: www.sciencedaily.com/releases/2020/12/201210112147.htm(accessed on 11 September 2021).
28.   Porter, S.M.; Hartwig, A.E.; Bielefeldt-Ohmann, H.; Bosco-Lauth, A.M.; Root, J.J. Susceptibility of wild canids to severe acute respiratory syndrome coronavirus 2 (SARS-CoV-2). *BioRxiv* **2022**. *preprint*. [CrossRef]
29.   Oreshkova, N.; Molenaar, R.J.; Vreman, S.; Harders, F.; Munnink, B.B.O.; Der Honing, R.W.H.-V.; Gerhards, N.; Tolsma, P.; Bouwstra, R.; Sikkema, R.S.; et al. SARS-CoV-2 infection in farmed minks, the Netherlands, April and May 2020. *Euro Surveill.* **2020**, *25*, 2001005. [CrossRef] [PubMed]
30.   Freuling, C.M.; Breithaupt, A.; Müller, T.; Sehl, J.; Balkema-Buschmann, A.; Rissmann, M.; Klein, A.; Wylezich, C.; Höper, D.; Wernike, K.; et al. Susceptibility of Raccoon Dogs for Experimental SARS-CoV-2 Infection. *Emerg. Infect. Dis.* **2020**, *26*, 2982–2985. [CrossRef] [PubMed]
31.   Gao, G.; Liu, W.; Liu, P.; Lei, W.; Jia, Z.; He, X.; Liu, L.; Shi, W.; Tan, Y.; Zou, S.; et al. Surveillance of SARS-CoV-2 in the environment and animal samples of the Huanan Seafood Market. *Res. Sq.* **2022**. Available online: https://assets.researchsquare.com/files/rs-1370392/v1_covered.pdf?c=1645813311 (accessed on 17 March 2021).
32.   Allen, T.; Murray, K.A.; Zambrana-Torrelio, C.; Morse, S.S.; Rondinini, C.; Di Marco, M.; Breit, N.; Olival, K.J.; Daszak, P. Global hotspots and correlates of emerging zoonotic diseases. *Nat. Commun.* **2017**, *8*, 1124. [CrossRef]
33.   Houe, H.; Nielsen, S.S.; Nielsen, L.R.; Ethelberg, S.; Mølbak, K. Opportunities for improved disease surveillance and control by use of integrated data on animal and human health. *Front. Vet. Sci.* **2019**, *6*, 301. [CrossRef]
34.   Grange, S.L.; Goldstein, T.; Johnson, C.K.; Anthony, S.; Gilardi, K.; Daszak, P.; Olival, K.J.; O'Rourke, T.; Murray, S.; Olson, S.H.; et al. Ranking the risk of animal-to-human spillover for newly discovered viruses. *Proc. Natl. Acad. Sci. USA* **2021**, *118*, e2002324118. [CrossRef]
35.   Jones, K.E.; Patel, N.G.; Levy, M.A.; Storeygard, A.; Balk, D.; Gittleman, J.L.; Daszak, P. Global trends in emerging infectious diseases. *Nature* **2008**, *451*, 990–993. [CrossRef]
36.   Li, X.; Luk, H.K.; Lau, S.P.; Woo, P.C. Human Coronaviruses: General Features. In *Reference Module in Biomedical Sciences*; Elsevier: Amsterdam, The Netherlands, 2019. [CrossRef]
37.   Amer, H.M. Bovine-like coronaviruses in domestic and wild ruminants. *Anim. Health Res. Rev.* **2018**, *19*, 113–124. [CrossRef]
38.   Mousavizadeh, L.; Ghasemi, S. Genotype and phenotype of COVID-19: Their roles in pathogenesis. *J. Microbiol. Immunol. Infect.* **2021**, *54*, 159–163. [CrossRef]
39.   Ji, W.; Wang, W.; Zhao, X.; Zai, J.; Li, X. Cross-species transmission of the newly identified coronavirus 2019-nCoV. *J. Med. Virol.* **2020**, *92*, 433–440. [CrossRef]
40.   Lau, S.K.P.; Luk, H.K.H.; Wong, A.C.P.; Li, K.S.M.; Zhu, L.; He, Z.; Fung, J.; Chan, T.T.Y.; Fung, K.S.C.; Woo, P.C.Y. Possible Bat Origin of Severe Acute Respiratory Syndrome Coronavirus 2. *Emerg. Infect. Dis.* **2020**, *26*, 1542–1547. [CrossRef]
41.   Pfefferle, S.; Oppong, S.; Drexler, J.F.; Gloza-Rausch, F.; Ipsen, A.; Seebens, A.; Müller, M.A.; Annan, A.; Vallo, P.; Adu-Sarkodie, Y.; et al. Distant relatives of severe acute respiratory syndrome coronavirus and close relatives of human coronavirus 229E in bats, Ghana. *Emerg. Infect. Dis.* **2009**, *15*, 1377–1384. [CrossRef]
42.   Quan, P.-L.; Firth, C.; Street, C.; Henriquez, J.A.; Petrosov, A.; Tashmukhamedova, A.; Hutchison, S.K.; Egholm, M.; Osinubi, M.O.V.; Niezgoda, M.; et al. Identification of a severe acute respiratory syndrome coronavirus-like virus in a leaf-nosed bat in Nigeria. *mBio* **2010**, *1*, e00208-10. [CrossRef]
43.   Parrish, C.R.; Holmes, E.C.; Morens, D.M.; Park, E.-C.; Burke, D.S.; Calisher, C.H.; Laughlin, C.A.; Saif, L.J.; Daszak, P. Cross-species virus transmission and the emergence of new epidemic diseases. *Microbiol. Mol. Biol. Rev.* **2008**, *72*, 457–470. [CrossRef]
44.   Cleaveland, S.; Laurenson, M.K.; Taylor, L.H. Diseases of humans and their domestic mammals: Pathogen characteristics, host range and the risk of emergence. *Philos. Trans. R. Soc. B* **2001**, *356*, 991–999. [CrossRef]
45.   Woolhouse, M.E. Population biology of emerging and re-emerging pathogens. *Trends Microbiol.* **2002**, *10*, S3–S7. [CrossRef]
46.   Rendon-Marin, S.; Martinez-Gutierrez, M.; Whittaker, G.R.; James, J.A.; Ruiz-Saenz, J. SARS CoV 2 spike protein in silico interaction with ACE2 receptors from wild and domestic species. *Front. Genet.* **2021**, *12*, 571707. [CrossRef]
47.   Van Aart, A.E.; Velkers, F.C.; Fischer, E.A.; Broens, E.M.; Egberink, H.; Zhao, S.; Engelsma, M.; Hakze-van der Honing, R.W.; Harders, F.; de Rooij, M.M.; et al. SARS-CoV-2 infection in cats and dogs in infected mink farms. *Transbound Emerg. Dis.* **2021**, *69*, 3001–3007. [CrossRef]
48.   Gusev, E.; Sarapultsev, A.; Solomatina, L.; Chereshnev, V. SARS-CoV-2-Specific Immune Response and the Pathogenesis of COVID-19. *Int. J. Mol. Sci.* **2022**, *23*, 1716. [CrossRef]
49.   Nelson, D.D.; Duprau, J.L.; Wolff, P.L.; Evermann, J.F. Persistent Bovine Viral Diarrhea Virus Infection in Domestic and Wild Small Ruminants and Camelids Including the Mountain Goat (*Oreamnos americanus*). *Front. Microbiol.* **2016**, *6*, 1415. [CrossRef]
50.   Daly, J.L.; Simonetti, B.; Klein, K.; Chen, K.-E.; Kavanagh Williamson, M.; Antón-Plágaro, C.; Shoemark, D.K.; Simón-Gracia, L.; Bauer, M.; Hollandi, R.; et al. Neuropilin-1 is a host factor for SARS-CoV-2 infection. *Science* **2020**, *370*, 861–865. [CrossRef]
51.   Cantuti-Castelvetri, L.; Ojha, R.; Pedro, L.D.; Djannatian, M.; Franz, J.; Kuivanen, S.; Van Der Meer, F.; Kallio, K.; Kaya, T.; Anastasina, M.; et al. Neuropilin-1 facilitates SARS-CoV-2 cell entry and infectivity. *Science* **2020**, *370*, 856–860. [CrossRef]
52.   Wang, K.; Chen, W.; Zhang, Z.; Deng, Y.; Lian, J.-Q.; Du, P.; Wei, D.; Zhang, Y.; Sun, X.-X.; Gong, L.; et al. CD147-spike protein is a novel route for SARS-CoV-2 infection to host cells. *Signal Transduct. Target. Ther.* **2020**, *5*, 283. [CrossRef] [PubMed]

53. Strumillo, S.T.; Kartavykh, D.; de Carvalho, F.F., Jr.; Cruz, N.C.; de Souza Teodoro, A.C.; Sobhie, D.R.; Curcio, M.F. Host-virus interaction and viral evasion. *Cell Biol. Int.* **2021**, *45*, 1124–1147. [CrossRef] [PubMed]
54. Račnik, J.; Kočevar, A.; Slavec, B.; Korva, M.; Rus, K.R.; Zakotnik, S.; Zorec, T.M.; Poljak, M.; Matko, M.; Rojs, O.Z.; et al. Transmission of SARS-CoV-2 from Human to Domestic Ferret. *Emerg. Infect. Dis.* **2021**, *27*, 2450–2453. [CrossRef]
55. Salguero, F.J.; White, A.D.; Slack, G.S.; Fotheringham, S.A.; Bewley, K.R.; Gooch, K.E.; Longet, S.; Humphries, H.E.; Watson, R.J.; Hunter, L.; et al. Comparison of rhesus and cynomolgus macaques as an infection model for COVID-19. *Nat. Commun.* **2021**, *12*, 1260. [CrossRef] [PubMed]
56. Hale, V.L.; Dennis, P.M.; McBride, D.S.; Nolting, J.M.; Madden, C.; Huey, D.; Ehrlich, M.; Grieser, J.; Winston, J.; Lombardi, D.; et al. SARS-CoV-2 infection in free-ranging white-tailed deer. *Nature* **2022**, *602*, 481–486. [CrossRef]
57. Mykytyn, A.Z.; Lamers, M.M.; Okba, N.M.A.; Breugem, T.I.; Schipper, D.; Doel, P.B.V.D.; van Run, P.; van Amerongen, G.; de Waal, L.; Koopmans, M.P.G.; et al. Susceptibility of rabbits to SARS-CoV-2. *Emerg. Microbes Infect.* **2021**, *10*, 1–7. [CrossRef]
58. Francis, M.E.; Goncin, U.; Kroeker, A.; Swan, C.; Ralph, R.; Lu, Y.; Etzioni, A.L.; Falzarano, D.; Gerdts, V.; Machtaler, S.; et al. SARS-CoV-2 infection in the Syrian hamster model causes inflammation as well as type I interferon dysregulation in both respiratory and non-respiratory tissues including the heart and kidney. *PLoS Pathog.* **2021**, *17*, e1009705. [CrossRef]
59. Shi, J.; Wen, Z.; Zhong, G.; Yang, H.; Wang, C.; Huang, B.; Liu, R.; He, X.; Shuai, L.; Sun, Z.; et al. Susceptibility of ferrets, cats, dogs, and other domesticated animals to SARS-coronavirus 2. *Science* **2020**, *368*, 1016–1020. [CrossRef]
60. Sia, S.F.; Yan, L.-M.; Chin, A.W.H.; Fung, K.; Choy, K.-T.; Wong, A.Y.L.; Kaewpreedee, P.; Perera, R.A.P.M.; Poon, L.L.M.; Nicholls, J.M.; et al. Pathogenesis and transmission of SARS-CoV-2 in golden hamsters. *Nature* **2020**, *583*, 834–838. [CrossRef]
61. De Morais, H.A.; dos Santos, A.P.; do Nascimento, N.C.; Kmetiuk, L.B.; Barbosa, D.S.; Brandão, P.E.; Guimarães, A.M.S.; Pettan-Brewer, C.; Biondo, A.W. Natural Infection by SARS-CoV-2 in Companion Animals: A Review of Case Reports and Current Evidence of Their Role in the Epidemiology of COVID-19. *Front. Vet. Sci.* **2020**, *7*, 1–10. [CrossRef]
62. Koeppel, K.N.; Mendes, A.; Strydom, A.; Rotherham, L.; Mulumba, M.; Venter, M. SARS-CoV-2 Reverse Zoonoses to Pumas and Lions, South Africa. *Viruses* **2022**, *14*, 120. [CrossRef]
63. Palermo, P.M. SARS-CoV-2 neutralizing antibodies in white-tailed deer from Texas. *Vector Borne Zoonotic Dis.* **2022**, *22*, 62–64. [CrossRef]
64. Yen, H.L.; Sit, T.H.; Brackman, C.J.; Chuk, S.S.; Gu, H.; Tam, K.W.; Law, P.Y.; Leung, G.M.; Peiris, M.; Poon, L.L.; et al. Transmission of SARS-CoV-2 delta variant (AY.127) from pet hamsters to humans, leading to onward human-to-human transmission: A case study. *Lancet* **2022**, *399*, 1070–1078. [CrossRef]
65. Pickering, B.; Lung, O.; Maguire, F.; Kruczkiewicz, P.; Kotwa, J.D.; Buchanan, T.; Gagnier, M.; Guthrie, J.L.; Jardine, C.M.; Marchand-Austin, A.; et al. Highly divergent white-tailed deer SARS-CoV-2 with potential deer-to-human transmission. *BioRxiv* **2022**. *preprint*. [CrossRef]
66. Tan, C.C.S.; Lam, S.D.; Richard, D.; Owen, C.J.; Berchtold, D.; Orengo, C.; Nair, M.S.; Kuchipudi, S.V.; Kapur, V.; van Dorp, L.; et al. Transmission of SARS-CoV-2 from humans to animals and potential host adaptation. *Nat. Commun.* **2022**, *13*, 2988. [CrossRef]
67. Ellwanger, J.H.; Chies, J.A.B. Zoonotic spillover: Understanding basic aspects for better prevention. *Genet. Mol. Biol.* **2021**, *44* (Suppl. S1). [CrossRef]
68. Wang, N.; Li, S.-Y.; Yang, X.-L.; Huang, H.-M.; Zhang, Y.-J.; Guo, H.; Luo, C.-M.; Miller, M.; Zhu, G.; Chmura, A.A.; et al. Serological evidence of bat SARS-related coronavirus infection in humans, China. *Virol. Sin.* **2018**, *33*, 104–107. [CrossRef] [PubMed]
69. Li, H.; Mendelsohn, E.; Zong, C.; Zhang, W.; Hagan, E.; Wang, N.; Li, S.; Yan, H.; Huang, H.; Zhu, G.; et al. Human-animal interactions and bat coronavirus spill-over potential among rural residents in Southern China. *Biosaf. Health* **2019**, *1*, 84–90. [CrossRef] [PubMed]
70. Hassell, J.M. Urbanization and Disease Emergence: Dynamics at the Wildlife–Livestock–Human Interface. *Trends Ecol. Evol.* **2017**, *32*, 55–67. [CrossRef] [PubMed]
71. Wood, J.L.; Leach, M.; Waldman, L.; MacGregor, H.; Fooks, A.R.; Jones, K.E. A framework for the study of zoonotic disease emergence and its drivers: Spill-over of bat pathogens as a case study. *Phil. Trans. R. Soc. B* **2012**, *367*, 2881–2892. [CrossRef] [PubMed]
72. Banerjee, A.; Baker, M.L.; Kulcsar, K.; Misra, V.; Plowright, R.; Mossman, K. Novel Insights into Immune Systems of Bats. *Front. Immunol.* **2020**, *11*, 26. [CrossRef] [PubMed]
73. Health Organization (WHO). Ebola Virus Disease. 2021. Available online: https://www.who.int/news-room/fact-sheets/detail/ebola-virus-disease (accessed on 4 December 2021).
74. World Health Organization (WHO). Marburg Virus Disease. 2021. Available online: https://www.who.int/health-topics/marburg-virus-disease (accessed on 4 December 2021).
75. World Health Organization (WHO). Hendra Virus Disease. 2021. Available online: https://www.who.int/health-topics/hendra-virus-disease (accessed on 4 December 2021).
76. World Health Organization (WHO). Nipah Virus. 2021. Available online: https://www.who.int/news-room/fact-sheets/detail/nipah-virus (accessed on 4 December 2021).
77. Shope, R.E. Rabies-related viruses. *Yale J. Biol. Med.* **1982**, *55*, 271–275.
78. Centers for Disease Control and Prevention (CDC). Arenaviruses. 2021. Available online: https://www.cdc.gov/vhf/virus-families/arenaviridae.html (accessed on 17 March 2022).

79. Kurpiers, L.A.; Schulte-Herbrüggen, B.; Ejotre, I.; Reeder, D.M. Bushmeat and Emerging Infectious Diseases: Lessons from Africa. *Probl. Wildl.* **2015**, *21*, 507–551. [CrossRef]

80. Saéz, A.M.; Weiss, S.; Nowak, K.; Lapeyre, V.; Zimmermann, F.; Düx, A.; Kühl, H.S.; Kaba, M.; Regnaut, S.; Merkel, K.; et al. Investigating the zoonotic origin of the West African Ebola epidemic. *EMBO Mol. Med.* **2015**, *7*, 17–23. [CrossRef]

81. Guan, Y.J.; Zheng, B.J.; He, Y.Q.; Liu, X.L.; Zhuang, Z.X.; Cheung, C.L.; Luo, S.W.; Li, P.H.; Zhang, L.J.; Butt, K.M.; et al. Isolation and characterization of viruses related to the SARS coronavirus from animals in southern China. *Science* 2003, *302*, 276–278. [CrossRef]

82. Sánchez, C.A.; Li, H.; Phelps, K.L.; Zambrana-Torrelio, C.; Wang, L.-F.; Zhou, P.; Shi, Z.-L.; Olival, K.J.; Daszak, P. A strategy to assess spillover risk of bat SARS-related coronaviruses in Southeast Asia. *Nat. Commun.* **2022**, *13*, 4380. [CrossRef]

83. World Health Organization. Transmission of SARS-CoV-2: Implications for Infection Prevention Precautions. *Sci. Brief.* **2020**. Available online: https://www.who.int/news-room/commentaries/detail/transmission-of-sars-cov-2-implications-for-infection-prevention-precautions (accessed on 11 July 2022).

84. Bonwitt, J.; Saez, A.M.; Lamin, J.; Ansumana, R.; Dawson, M.; Buanie, J.; Lamin, J.; Sondufu, D.; Borchert, M.; Sahr, F.; et al. At Home with Mastomys and Rattus: Human-Rodent Interactions and Potential for Primary Transmission of Lassa Virus in Domestic Spaces. *Am. J. Trop. Med. Hyg.* **2017**, *96*, 935–943. [CrossRef]

85. Gurley, E.S.; Hegde, S.T.; Hossain, K.; Sazzad, H.M.S.; Hossain, M.J.; Rahman, M.; Sharker, M.A.Y.; Salje, H.; Islam, M.S.; Epstein, J.H.; et al. Convergence of Humans, Bats, Trees, and Culture in Nipah Virus Transmission, Bangladesh. *Emerg. Infect. Dis.* **2017**, *23*, 1446–1453. [CrossRef]

86. Altizer, S.; Becker, D.J.; Epstein, J.H.; Forbes, K.M.; Gillespie, T.R.; Hall, R.J.; Hawley, D.M.; Hernandez, S.M.; Martin, L.B.; Plowright, R.K.; et al. Food for contagion: Synthesis and future directions for studying host-parasite responses to resource shifts in anthropogenic environments. *Philos. Trans. R. Soc. Lond. B. Biol. Sci.* **2018**, *373*, 20170102. [CrossRef]

87. Shervani, Z.; Khan, I.; Siddiqui, N.Y.; Khan, T.; Qazi, U.Y. Risk of SARS-CoV-2 transmission from humans to pets and vice versa. *Eur. J. Med. Health Sci.* **2021**, *1*, 34–38. [CrossRef]

88. Cadar, D.; Allendorf, V.; Schulze, V.; Ulrich, R.G.; Schlottau, K.; Ebinger, A.; Hoffmann, B.; Hoffmann, D.; Rubbenstroth, D.; Ismer, G.; et al. Introduction and spread of variegated squirrel bornavirus 1 (VSBV-1) between exotic squirrels and spill-over infections to humans in Germany. *Emerg. Microbes Infect.* **2021**, *10*, 602–611. [CrossRef]

89. Reynolds, M.G.; Davidson, W.B.; Curns, A.T.; Conover, C.S.; Huhn, G.; Davis, J.P.; Wegner, M.; Croft, D.R.; Newman, A.; Obiesie, N.N.; et al. Spectrum of infection and risk factors for human monkeypox, United States, 2003. *Emerg. Infect. Dis.* **2007**, *13*, 1332–1339. [CrossRef]

90. Kampf, G.; Lemmen, S.; Suchomel, M. Ct values and infectivity of SARS-CoV-2 on surfaces. *Lancet Infect. Dis.* **2021**, *21*, e141. [CrossRef]

91. Fadare, O.O.; Okoffo, E.D. COVID-19 face masks: A potential source of microplastic fibers in the environment. *Sci. Total Environ.* **2020**, *737*, 140279. [CrossRef]

92. Magwedere, K.; Bishi, A.; Tjipura-Zaire, G.; Eberle, G.; Hemberger, Y.; Hoffman, L.C.; Dziva, F. Brucellae through the food chain: The role of sheep, goats and springbok (*Antidorcus marsupialis*) as sources of human infections in Namibia. *J. S. Afr. Vet. Assoc.* **2011**, *82*, 205–212. [CrossRef]

93. Buchholz, U.; Bernard, H.; Werber, D.; Böhmer, M.M.; Remschmidt, C.; Wilking, H.; Deleré, Y.; der Heiden, M.A.; Adlhoch, C.; Dreesman, J.; et al. German outbreak of Escherichia coli O104:H4 associated with sprouts. *N. Engl. J. Med.* **2011**, *365*, 1763–1770. [CrossRef] [PubMed]

94. Sophia, G.; de Vries, B.J.; Visser, I.M.; Nagel, M.G.A.; Goris, R.A.; Hartskeerl, M.P.G. Leptospirosis in Sub-Saharan Africa: A systematic review. *Inter. J. Infect. Dis.* **2014**, *28*, 47–64. [CrossRef]

95. Berthe, F.C.J.; Bouley, T.; Karesh, W.B.; Le Gall, F.G.; Machalaba, C.C.; Plante, C.A.; Seifman, R.M. *Operational Framework for Strengthening Human, Animal and Environmental Public Health Systems at Their Interface (English)*; World Bank Group: Washington, DC, USA, 2018; Available online: http://documents.worldbank.org/curated/en/703711517234402168/Operational-framework-for-strengthening-human-animal-and-environmental-public-health-systems-at-their-interface (accessed on 7 July 2022).

96. World Health Organization (WHO). In Africa, 63% Jump in Diseases Spread from Animals to People Seen in Last Decade. 2022. Available online: https://www.afro.who.int/news/africa-63-jump-diseases-spread-animals-people-seen-last-decade (accessed on 12 August 2022).

97. Mableson, H.E.; Okello, A.; Picozzi, K.; Welburn, S.C. Neglected zoonotic diseases-the long and winding road to advocacy. *PLoS Negl. Trop. Dis.* **2014**, *8*, e2800. [CrossRef] [PubMed]

98. Grace, D.; Songe, M.; Knight-Jones, T. Impact of Neglected Diseases on Animal Productivity and Public Health in Africa. Africa—OIE Regional Commission. 2015. Available online: https://www.oie.int/app/uploads/2021/03/2015-afr1-grace-a.pdf (accessed on 17 August 2022).

99. Elelu, N.; Aiyedun, J.O.; Mohammed, I.G.; Oludairo, O.; Odetokun, I.A.; Mohammed, K.M.; Bale, J.O.; Nuru, S. Neglected zoonotic diseases in Nigeria: Role of the public health veterinarian. *Pan Afr. Med. J.* **2019**, *32*, 36. [CrossRef] [PubMed]

100. Rist, C.L.; Arriola, C.S.; Rubin, C. Prioritizing zoonoses: A proposed One Health tool for collaborative decision-making. *PLoS ONE* **2014**, *9*, e109986. [CrossRef]

101. Grace, D.; Mutua, F.; Ochungo, P.; Kruska, R.; Jones, K.; Brierley, L.; Lapar, L.; Said, M.; Herrero, M.; Phuc, P.M.; et al. *Mapping of Poverty and Likely Zoonoses Hotspots*; Zoonoses Project 4, Report to the UK Department for International Development; ILRI: Nairobi, Kenya, 2012.
102. World Health Organization. New WHO Estimates: Up to 190,000 People Could Die of COVID-19 in Africa If Not Controlled. 2020. Available online: https://www.afro.who.int/news/new-who-estimates-190-000-people-could-die-covid-19-africa-if-not-controlled (accessed on 12 November 2021).
103. Cabore, J.W.; Karamagi, H.C.; Kipruto, H.K.; Mungatu, J.K.; Asamani, J.A.; Droti, B.; Titi-Ofei, R.; Seydi, A.B.W.; Kidane, S.N.; Balde, T.; et al. COVID-19 in the 47 countries of the WHO African region: A modelling analysis of past trends and future patterns. *Lancet Glob. Health* **2022**, *10*, e1099–e1114. [CrossRef]
104. Jones, B.A.; Grace, D.; Kock, R.; Alonso, S.; Rushton, J.; Said, M.Y.; McKeever, D.; Mutua, F.; Young, J.; McDermott, J.; et al. Zoonosis emergence linked to agricultural intensification and environmental change. *Proc. Natl. Acad. Sci. USA* **2013**, *110*, 8399–8404. [CrossRef]
105. Kapata, N.; Ihekweazu, C.; Ntoumi, F.; Raji, T.; Chanda-Kapata, P.; Mwaba, P.; Mukonka, V.; Bates, M.; Tembo, J.; Corman, V.; et al. Is Africa prepared for tackling the COVID-19 (SARS-CoV-2) epidemic. Lessons from past outbreaks, ongoing pan-African public health efforts, and implications for the future. *Int. J. Infect. Dis.* **2020**, *93*, 233–236. [CrossRef]
106. Agusi, E.R.; Ijoma, S.I.; Nnochin, C.S.; Njoku-Achu, N.O.; Nwosuh, C.I.; Meseko, C.A. The COVID-19 pandemic and social distancing in Nigeria: Ignorance or defiance. *Pan Afr. Med.* **2020**, *35* (Suppl. S2), 52. [CrossRef]
107. Torti, C.; Mazzitelli, M.; Trecarichi, E.M.; Darius, O. Potential implications of SARS-CoV-2 epidemic in Africa: Where are we going from now? *BMC Infect. Dis.* **2022**, *20*, 412. [CrossRef]
108. Nordling, L. Africa's pandemic puzzle: Why so few cases and deaths? *Science* **2020**, *369*, 756–757. [CrossRef]
109. Soko, N.D.; Dlamini, S.; Ntsekhe, M.; Dandara, C. The COVID-19 Pandemic and Explaining Outcomes in Africa: Could Genomic Variation Add to the Debate? *OMICS J. Integr. Biol.* **2022**. [CrossRef]
110. Altable, M.; de la Serna, J.M. Protection against COVID-19 in African population: Immunology, genetics, and malaria clues for therapeutic targets. *Virus Res.* **2021**, *299*, 198347. [CrossRef]
111. The World Bank. Nigeria's Economy Faces Worst Recession in Four Decades, Says New World Bank Report. 2020. Available online: https://www.worldbank.org/en/news/press-release/2020/06/25/nigerias-economy-faces-worst-recession-in-four-decades-says-new-world-bank-report (accessed on 23 May 2022).
112. World Health Organization. Achieving 70% COVID-19 Immunization Coverage by Mid-2022. 2021. Available online: https://www.who.int/news/item/23-12-2021-achieving-70-covid-19-immunization-coverage-by-mid-2022 (accessed on 19 November 2021).
113. Food and Agriculture Organization of the United Nations. *The State of Food Security and Nutrition in the World*; FAO: Rome, Italy, 2021.
114. Orjuela-Grimm, M.; Deschak, C.; Gama, C.A.A.; Carreño, S.B.; Hoyos, L.; Mundo, V.; Bojorquez, I.; Carpio, K.; Quero, Y.; Xicotencatl, A.; et al. Migrants on the Move and Food (In)security: A Call for Research. *J. Immigr Minor. Health* **2022**, *24*, 1318–1327. [CrossRef]
115. Holder, J. Tracking Coronavirus Vaccinations Around the World. *New York Times Newspaper*, 2022. Available online: https://www.nytimes.com/interactive/2021/world/covid-vaccinations-tracker.html(accessed on 5 September 2022).
116. World Health Organization. Less than 10% of African Countries to Hit Key COVID-19 Vaccination Goal. 2021. Available online: https://www.afro.who.int/news/less-10-african-countries-hit-key-covid-19-vaccination-goal (accessed on 19 November 2021).
117. Mutombo, P.N.; Fallah, M.P.; Munodawafa, D.; Kabel, A.; Houeto, D.; Goronga, T.; Mweemba, O.; Balance, G.; Onya, H.; Kamba, R.S.; et al. COVID-19 vaccine hesitancy in Africa: A call to action. *Lancet Glob. Health* **2022**, *10*, e320–e321. [CrossRef]
118. Nature Editorial. Africa is bringing vaccine manufacturing home. *Nature* **2022**, *602*, 184. [CrossRef]
119. Chan, O.S.K.; Bradley, K.C.F.; Grioni, A.; Lau, S.K.P.; Li, W.-T.; Magouras, I.; Naing, T.; Padula, A.; To, E.M.W.; Tun, H.M.; et al. Veterinary Experiences can Inform One Health Strategies for Animal Coronaviruses. *Ecohealth* **2021**, *18*, 301–314. [CrossRef]
120. Ducrot, C.; Bed'Hom, B.; Béringue, V.; Coulon, J.-B.; Fourichon, C.; Guérin, J.-L.; Krebs, S.; Rainard, P.; Schwartz-Cornil, I.; Torny, D.; et al. Issues and special features of animal health research. *Vet. Res.* **2011**, *42*, 96. [CrossRef]
121. The Food and Agriculture Organization of the United Nations. *The Impact of Disasters and Crises on Agriculture and Food Security*; FAO: Rome, Italy, 2021. [CrossRef]
122. He, S.; Han, J.; Lichtfouse, E. Backward transmission of COVID-19 from humans to animals may propagate reinfections and induce vaccine failure. *Environ. Chem. Lett.* **2021**, *19*, 763–768. [CrossRef]
123. Ajayi, E.A.; Olatumile, A. Indigenous folklores as a tool of transformative learning for environmental sustainability in Nigeria. *Andragoške Studije* **2018**, *11*, 29–44. [CrossRef]

Article

Rabies Virus Populations in Humans and Mice Show Minor Inter-Host Variability within Various Central Nervous System Regions and Peripheral Tissues

Carmen W. E. Embregts [1,*], Elmoubashar A. B. A. Farag [2], Devendra Bansal [2], Marjan Boter [1], Anne van der Linden [1], Vincent P. Vaes [1], Ingeborg van Middelkoop-van den Berg [1], Jeroen. IJpelaar [1], Hisham Ziglam [3], Peter V. Coyle [3], Imad Ibrahim [3,4], Khaled A. Mohran [5,6], Muneera Mohammed Saleh Alrajhi [5], Md. Mazharul Islam [5], Randa Abdeen [5], Abdul Aziz Al-Zeyara [5], Nidal Mahmoud Younis [5], Hamad Eid Al-Romaihi [2], Mohammad Hamad J. AlThani [2], Reina S. Sikkema [1], Marion P. G. Koopmans [1], Bas B. Oude Munnink [1,†] and Corine H. GeurtsvanKessel [1,†]

[1] Department of Viroscience, Erasmus Medical Centre, 3015 GD Rotterdam, The Netherlands
[2] Ministry of Public Health, Doha P.O. Box 42, Qatar
[3] Hamad Medical Corporation, Doha P.O. Box 3050, Qatar
[4] Biomedical Research Centre, Qatar University, Doha P.O. Box 2713, Qatar
[5] Department of Animal Resources, Ministry of Municipality, Doha P.O. Box 35081, Qatar
[6] Biotechnology Departments ERC, Animal Health Research Institute, Dokki 12611, Egypt
* Correspondence: c.embregts@erasmusmc.nl
† These authors contributed equally to this work.

Citation: Embregts, C.W.E.; Farag, E.A.B.A.; Bansal, D.; Boter, M.; van der Linden, A.; Vaes, V.P.; van Middelkoop-van den Berg, I.; IJpelaar, J.; Ziglam, H.; Coyle, P.V.; et al. Rabies Virus Populations in Humans and Mice Show Minor Inter-Host Variability within Various Central Nervous System Regions and Peripheral Tissues. *Viruses* **2022**, *14*, 2661. https://doi.org/10.3390/v14122661

Academic Editors: Myriam Ermonval and Serge Morand

Received: 18 October 2022
Accepted: 25 November 2022
Published: 28 November 2022

Publisher's Note: MDPI stays neutral with regard to jurisdictional claims in published maps and institutional affiliations.

Abstract: Rabies virus (RABV) has a broad host range and infects multiple cell types throughout the infection cycle. Next-generation sequencing (NGS) and minor variant analysis are powerful tools for studying virus populations within specific hosts and tissues, leading to novel insights into the mechanisms of host-switching and key factors for infecting specific cell types. In this study we investigated RABV populations and minor variants in both original (non-passaged) samples and in vitro-passaged isolates of various CNS regions (hippocampus, medulla oblongata and spinal cord) of a fatal human rabies case, and of multiple CNS and non-CNS tissues of experimentally infected mice. No differences in virus populations were detected between the human CNS regions, and only one non-synonymous single nucleotide polymorphism (SNP) was detected in the fifth in vitro passage of virus isolated from the spinal cord. However, the appearance of this SNP shows the importance of sequencing newly passaged virus stocks before further use. Similarly, we did not detect apparent differences in virus populations isolated from different CNS and non-CNS tissues of experimentally infected mice. Sequencing of viruses obtained from pharyngeal swab and salivary gland proved difficult, and we propose methods for improving sampling.

Keywords: rabies virus; NGS; minor variants; CNS

1. Introduction

Rabies is a viral encephalitis that is caused by viruses of the genus *Lyssavirus*, and an estimated 99% of human rabies cases are caused by rabies virus (RABV) transmitted by dogs [1]. Lyssaviruses are transmitted through the saliva of lyssavirus-infected animals, and rapid post-exposure prophylaxis is essential to prevent development of the disease. There are, however, no treatment options after the onset of neurological symptoms, which makes rabies the deadliest zoonosis worldwide with a near 100% case-fatality rate. Every year at least 59,000 human rabies fatalities are reported worldwide, but the true burden is expected to be much higher than that due to diagnostic difficulties, especially in rabies-endemic regions [2,3]. Furthermore, loss of livestock (cattle, sheep, goats) due to RABV infections is a major problem in South America [4].

Lyssaviruses are neurotropic and can infect peripheral nerve endings, after which they travel to the spinal cord via the ventral or dorsal root ganglions. After reaching the brain and infecting local neurons, the virus further spreads to the salivary gland and other organs and is finally secreted through the saliva. Despite their neurotrophic nature, lyssaviruses are capable of infecting muscle cells at the initial infection site, various cell types during the centrifugal spread, and acinar cells in the salivary gland, of which the latter step is essential for virus secretion and infection of the next host [5,6].

RABV, along with other lyssaviruses, are negative single-stranded RNA viruses (-ssRNA). These RNA viruses are generally known to display high rates of evolutionary change by both natural selection and recombination, enabling them to rapidly adapt to new hosts and cell types [7–9]. Interestingly, RABV shows a rather low mutation rate when compared to other -ssRNA viruses [10]. While most lyssaviruses are thought to be restricted to specific hosts, RABV is transmitted by multiple species of the orders Carnivora and Chiroptera (bats) [11]. In addition, rare spill-over infections have been documented in a variety of other mammals, including camels and kudus [12,13]. Studying adaptability of RABV within different environments is of immense importance, given its ability to thrive in wide range of hosts by infecting different types of cells throughout its infection cycle [14–16]. Comparing virus populations in different organs, both inside and outside of the CNS during different phases of infection, is a first step in understanding the genetic plasticity of RABV.

While most viral sequencing is performed on brain tissue specimens, given the presence of high viral loads, there is less information about viral populations in other organs. In this study we investigated RABV populations in three different central nervous system (CNS) regions (hippocampus, medulla oblongata and spinal cord) of a fatal human rabies case [12], and the adaptation of the isolated virus after in vitro passages in mouse neuroblastoma cells. In parallel, we investigated RABV populations in various CNS (brain, trigeminal ganglion, dorsal and ventral spinal cord) and non-CNS (salivary gland, tongue, nuchal skin and pharyngeal swab) biopsies of C57BL/6 mice that were experimentally infected with a silver-haired bat rabies virus (SHBRV) reference strain [17,18]. By using next-generation sequencing (NGS) and minor variant analysis, we were able to describe the virus population within various CNS biopsies of a fatal human RABV case, as well as the virus population within various CNS and non-CNS tissue biopsies of mice infected with a bat-related RABV strain, in great detail. While minor variants were found to play an insignificant role in RABV host switch events [19], investigating minor variants may provide crucial insights into where mutations take place, and which variants might become dominant.

Pharyngeal swabs contain secreted virus and are therefore an important specimen for studying virus evolution. However, the low virus loads obtained from pharyngeal swabs severely limit the possibility of investigating virus populations. The parotid salivary gland is easy to sample and contains saliva that is ready to be secreted. Therefore, we also assessed the suitability of using salivary gland biopsies as a read-out for secreted virus.

2. Materials and Methods

2.1. Human Materials and Cell Culture

Human CNS biopsies (medulla oblongata, hippocampus, spinal cord) were obtained from a fatal human rabies case in Qatar described previously [12]. Samples were stored in virus transport medium and stored at $-80\,^{\circ}C$ until processing. Samples were homogenized in DMEM (Lonza), centrifuged for five minutes at $5000\times g$, after which 200 µL of the supernatant was filtered using a 0.45 µm filter and was incubated on mouse neuroblastoma astrocytes (MNA) for 1 h. After removal of the inoculum the cells were incubated with supplemented DMEM medium at $37\,^{\circ}C$ in the presence of 5% CO_2. Virus was passaged when cytopathic effect was observed; the cell culture supernatant was centrifuged and filtered, and 200 µL was inoculated onto freshly seeded MNA cells. The remainder of the filtered virus was stored at $-80\,^{\circ}C$.

2.2. Animal Materials

Six-to-eight-week-old mice (C57BL/6) were intramuscularly inoculated with 10^5 or 10^6 $TCID_{50}$ of silver-haired bat rabies virus (SHBRV) in the left hind leg. This experiment was designed and performed to validate an in vivo infection model, and the samples used in this study were taken after the initial experiment had finished. Mice were euthanized upon showing neurological symptoms, which appeared between day six and eight. Biopsies of target organs and tissues of interest (brain, trigeminal ganglion, spinal cord dorsal and ventral horn, parotid salivary gland and tongue), as well as a pharyngeal swab were taken, homogenized in 1 mL of virus-transport medium, and stored at $-80\,^{\circ}$C until further processing. In parallel, organs were fixed in 10% neutral-buffered formalin to verify virus protein expression by immunohistochemistry (IHC). Samples were taken from 10 mice in total, from which the samples of four animals were used to investigate differences in viral populations within the salivary gland and pharyngeal swab isolates, and the samples of the remaining six animals were used to compare virus populations throughout different organs (brain, trigeminal ganglion, spinal cord dorsal and ventral horn, tongue, nuchal skin, and parotid salivary gland). All animal experiments were performed in compliance with Dutch legislation for the protection of animals used for scientific purposes (implementing EU Directive 2010/63) and other relevant regulations. The research was conducted under a project license (AVD1010020187204) approved by the competent Dutch authority. The specific study protocol (18−7204−01) was approved by the institutional Animal Welfare Body.

2.3. Sample Preparation and RNA Isolation

Isolates from original materials from the human RABV patient, the third and fifth passage of these materials on MNA cells, and the samples taken from the infected mice, were included for RNA isolation and sequencing. All samples were centrifuged and filtered through a 0.45 µm filter, after which 100 µL of the filtrate was mixed with lysis buffer (Qiagen). The animal samples were pre-processed with Omnicleave (Lucigen) to cleave present host RNA, and all samples were further processed using the High Pure RNA isolation kit (Roche) following the manufacturers guidelines, including the on-column DNA digestion step. RT-PCR was performed as an initial investigation on the presence of viral RNA, using a RABV genotype 1 RT-PCR [20] on all isolates obtained from mice tissues, and a pan-lyssavirus RT-PCR [21] on the isolates and passages obtained from the human materials.

2.4. cDNA Library Preparation and Sequencing

cDNA was made using random primers (Thermo Fisher) and SuperScript IV (Thermo Fisher) and dsDNA was made using Klenow (NEBNext). The KAPA Hyper Plus kit (Roche) was used to prepare the library for sequencing with minor adjustments: fragmentation time was reduced to 3 min and the adapters were diluted 1:10. Targeted enrichment was performed using VirCapSeq [22] and all 50 samples were pooled equimolarly and sequenced on an Illumina MiSeq v3 flow cell (2 × 300 bp).

2.5. Minor Variant Analysis

Consensus sequences were generated using a reference-based alignment against the previously sequenced SHBRV-18 strain (AY705373.1) or the RABV_Nepal_2018 strain (MN534894.1). The reads were re-aligned to the newly generated consensus sequences and minor variants with a 20% frequency cutoff were determined using Geneious version 9.1.8 using default settings.

2.6. Immunohistochemistry and Immunofluorescence

Presence of viral proteins in formalin-fixed paraffin-embedded tissue biopsies was verified by immunohistochemistry (IHC), and presence of virus in the cell culture passages was verified by immunofluorescence (IF). Briefly, 3 µM slides were cut from paraffin-embedded tissues of both human and murine samples, and the RABV nucleoprotein (RABV-N) was detected using the 5DF12 antibody (kindly provided by P. Koraka). Images were acquired using an Olympus BX51 microscope. MNA cultures were fixed with ice-cold 80% acetone and were stained with the FITC-conjugated anti-RABV-N antibody (Fujirebio). A nuclear counterstain with Evans blue was included and images were acquired using a Zeiss AX10 Colibri 7 fluorescence microscope.

3. Results

NGS of RABV isolated from three regions of the CNS of a fatal human case (medulla oblongata, hippocampus and spinal cord), as well as the following in vitro passages in mouse neuroblastoma (MNA) cells (illustrated in Figure 1), was successful. No single nucleotide polymorphisms (SNP) were detected in the consensus sequence or a minor variant of original materials of the three different tissues, nor in the first in vitro passage (Table 1). One non-synonymous SNP within the matrix (M2) gene was first observed in the third in vitro passage of the spinal cord which became dominant in the fifth passage. Different dominant SNPs were found in the third and fifth passage of the virus isolated from medulla oblongata and the fifth passage of the virus isolated from the hippocampus, but these did not lead to codon changes. One minor variant was detected in the third in vitro passage of virus isolated from the spinal cord, but it decreased below the 20% threshold in the fifth in vitro passage. No minor variants were detected throughout the passages of the hippocampus, and only in passage three, three minor variants were detected in virus isolated from the medulla oblongata.

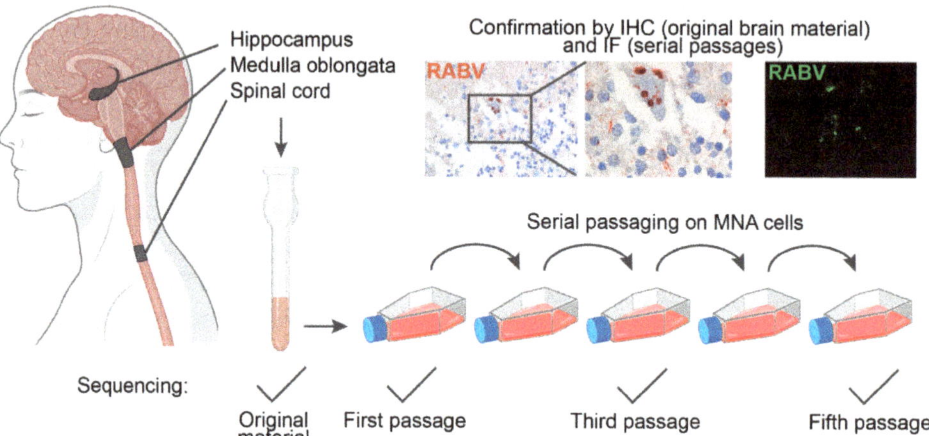

Figure 1. Overview of the human central nervous system (CNS) biopsy isolates used for in vitro passaging and sequencing. The grey areas in the CNS depicted on the left indicate the three areas that were included in our study. The expression of viral proteins within these samples was verified with immunohistochemistry (IHC, red staining). Serial passaging of the isolates was performed on mouse neuroblastoma cells (MNA), and the first, third and fifth passage was used for sequencing. Presence of virus within the passages was confirmed by PCR and immunofluorescence (IF, green staining top right picture). The figure, except the IHC and IF pictures, was generated using BioRender.com.

To investigate the possible occurrence of organ-specific lyssavirus adaptations, we sequenced virus isolated from different CNS tissues (brain, trigeminal ganglion, spinal cord) and non-CNS tissues (nuchal skin, tongue epithelium, salivary gland and pharyngeal swab) of C57BL/6 mice that were experimentally infected with the silver-haired bat rabies virus strain (SHBRV, Figure 2A), a strain for which the disease progression has been well described before in our lab [17,18]. In contrast to the abundant presence of viral proteins in CNS tissues, only very few RABV-positive cells were found in non-CNS tissues (Figure 2B), which is reflected in the higher Ct values of non-CNS tissue biopsies (Figure 2C and Supplementary Table S1).

Table 1. Overview of the SNPs and minor variants found in RABV isolated from different human CNS regions (original material, OM) and in vitro passages (P1, P3, P5) of these materials. The presence and corresponding percentages of the specific minor variants are indicated in blue; orange indicates that the specific minor variant is presented as a SNP in another tissue. Grey indicates an ongoing nucleotide shift (49.6%C, 47.4%T).

Passage	Tissue	SNPs (>50%)	Minor Variants (>20%)				
			A2397G	A2413C	T2699C	T2977Y	T3445G
OM	Medulla oblongata						
	Spinal cord						
	Hippocampus						
P1	Medulla oblongata						
	Spinal cord						
	Hippocampus						
P3	Medulla oblongata						
	Spinal cord	T2977Y (M2: S185P)		21.6			
	Hippocampus	A2399G (non-coding)					
P5	Medulla oblongata	G2397A (non-coding)	43.4		20.8		20.1
	Spinal cord	T2977C (M2: S185P)				>50	
	Hippocampus	A2399G (non-coding)					

RABV is predominantly secreted via saliva, making saliva the preferred sample for investigating secreted RABV populations. However, obtaining enough saliva sample to successfully culture or sequence secreted virus is often a challenge, especially when working with small experimental animals. We therefore investigated if the salivary gland tissue is a reliable source for characterizing secreted virus populations by performing pairwise collection and sequencing of viruses isolated from pharyngeal swabs and salivary gland biopsies obtained from mice 1–4. Despite the presence of RABV as detected by PCR, only one of these pairwise comparisons (mouse 3) resulted in successful viral sequences of both samples. In this pairwise sample comparison, we found one SNP and three minor variants in the salivary gland isolate and two different SNPs in the pharyngeal swab isolate, of which one resulted in an amino acid change in the polymerase gene (Table 2). The second sequence obtained from a pharyngeal swab isolate contained one non-synonymous SNP in the matrix protein, a mutation that was not observed in any other sample. No other samples were sequenced from mice 1–4, and therefore we cannot conclude if the observed mutations are organ-specific. Moreover, no specific SNPs were detected in the salivary gland isolates of mice eight and nine.

To compare virus populations in CNS tissues with non CNS tissues, isolates from both CNS (brain, trigeminal ganglion, spinal cord ventral and dorsal) and non-CNS (tongue epithelium, nuchal skin, salivary gland) tissues from mice 5–10 were collected and used for sequencing. Successful sequencing required a Ct value below ~28–29 (Figure 2C and Supplementary Table S2). Due to low viral loads in the non-CNS tissues, only the isolates obtained from tissues of mice eight and nine resulted in successful sequencing of all seven different tissues. No major differences were detected between samples of the individual mice. The only SNP detected in virus isolated from the trigeminal ganglion and tongue epithelial was present as a minor variant in all tissue isolates except the salivary gland of mouse eight. Additionally, all tissue isolates from mouse nine, but none of the other samples included in the experiment, contained the same SNPs. The pattern that SNPs in our dataset group together on the individual host (mouse) level, and not on organ level, was strengthened by the fact that only one SNP was observed in more than one mice. A non-synonymous SNP in the RABV glycoprotein protein gene was observed in the nuchal skin sample from mouse nine.

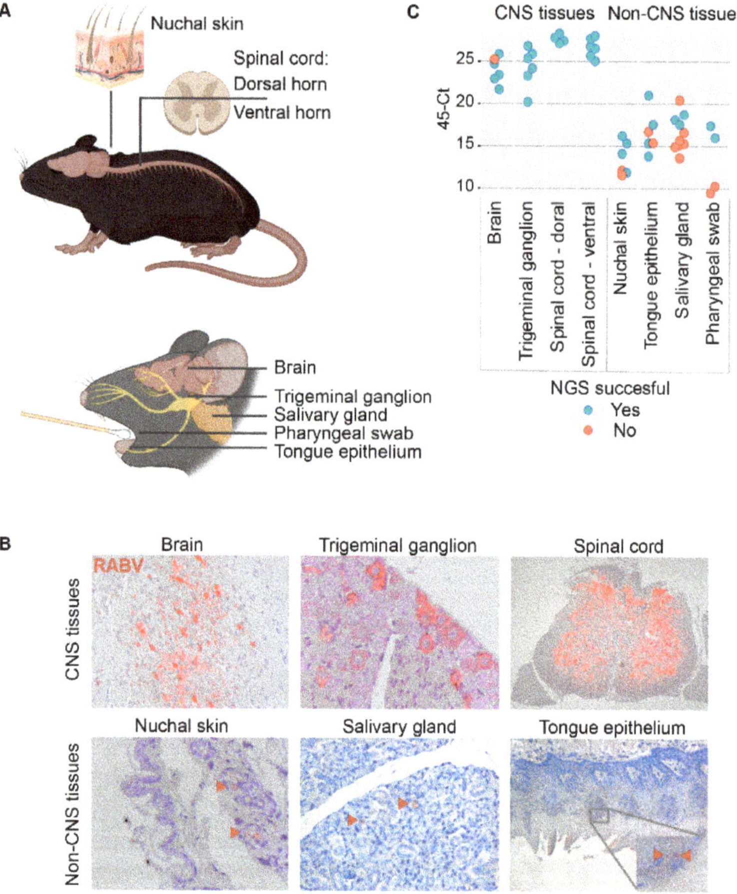

Figure 2. Overview of the mice tissue isolates used for sequencing, and verification of the infection by IHC and PCR. All samples processed for PCR and subsequent sequencing are shown in (**A**). Presence of viral proteins was verified by IHC (red staining) (**B**). Red arrows in the lower panel show the sparse distribution of RABV-positive cells in non-CNS tissues. The table in (**C**) presents the 45-Ct values indicating the viral loads of each specified organ. Green cells resulted in a successful sequencing result, grey cells yielded incomplete sequencing runs and were therefore excluded for further analysis. Panel (**A**) was generated using BioRender.com.

A limited number of minor variants were detected in this dataset, similar to the SNPs. However, the majority of them were found in single animals and did not appear to be tissue-specific, as they were observed as SNP in other isolates of the same mouse.

Table 2. Overview of SNPs and minor variants in the sequenced mice tissues. Samples are ordered per mice, the SNPs (>50% abundance) are included in the third column, with the indicated amino acid change (if present) in bold. Minor variants, with a cutoff of 20%, are shown from column 3 onwards. The presences of the specific minor variants are indicated in blue, orange indicates that the specific minor variants are presented as a SNP in another tissue. An overview of the abundance (%) of the minor variants can be found in Supplementary Table S2.

Mouse ID	Organ	SNPs (>50%)	Minor Variants (>20%)															
			A247C	C2406A	T4600C (G:R456W)	A5315G (L:G32D)	T5898G (L:D226E)	A6771G	A7704G	G7704A	A8313G	T10668C	C10668T	G10785A	C11031T	T11031C	A11775G	G11775A
2	Pharyngeal swab	A1617C (M1:L62T), C10668T											●					
3	Salivary gland	C1783T																
3	Pharyngeal swab	C1702T, T5898G (L:D226E)					●											
5	Trigeminal ganglion	G3131T																
5	Spinal cord—dorsal	G3131T																
5	Spinal cord—ventral	G3131T																
6	Brain	C247A																
6	Trigeminal ganglion	C247A																
6	Spinal cord—dorsal	C247A																
6	Spinal cord—ventral	C247A																
6	Nuchal skin	C247A, C2552T																
7	Brain	A2406C, C5246T																
7	Trigeminal ganglion	A2406C, C5246T																
7	Spinal cord—dorsal	A2406C, C5246T																
7	Spinal cord—ventral	A2406C, C5246T																
7	Nuchal skin	A2406C																
8	Brain																	
8	Trigeminal ganglion	C10668T											●					
8	Spinal cord—dorsal																	
8	Spinal cord—ventral																	
8	Nuchal skin																	
8	Tongue epithelium	C10668T											●					
8	Salivary gland																	
9	Brain																	
9	Trigeminal ganglion																	
9	Spinal cord—dorsal																	
9	Spinal cord—ventral																	
9	Nuchal skin	T4600C (G:R456W)			●													
9	Tongue epithelium																	
9	Salivary gland																	
10	Brain																	
10	Trigeminal ganglion																	
10	Spinal cord—dorsal	A7704G, C11031T, G11775A							●						●			●

4. Discussion

In the presented study we investigated virus populations isolated from human CNS tissues, and from several CNS- and non-CNS tissues from experimentally infected mice. Although we did not detect RABV evolution within the analyzed human CNS specimen, multiple SNPs and minor variants were detected within few in vitro passages. Rapid virus adaptation to in vitro culture has been described for RABV previously [14,15,23], and mutations were observed both during homologous and heterologous culture conditions [14]. The observed changes might be indicative of either cell-culture or host adaptations, given that the in vitro passages were performed on mouse neuroblastoma cells and not on cells of human origin. Irrespective of the nature of the mutations, their presence indicates the need for using low-passage stocks for infection experiments, as well as the necessity of sequencing every in vitro passage before using them in experiments.

While the Ct levels in the salivary gland extracts were lower than in pharyngeal swabs of mice, only one of the pairwise comparisons between salivary gland and pharyngeal swab isolates resulted in successful viral sequences of both samples. Given the low number of successful sequences obtained from salivary gland isolates, we cannot definitively conclude if the salivary gland is a reliable source for studying secreted lyssaviruses. Most likely the high level of host RNA in salivary glands resulted in this failure to retrieve complete viral sequences. Furthermore, high levels of RNAses present in the parotid salivary gland are known to drastically reduce the yield of high-quality RNA isolated from its tissue [24]. Given the simplicity of collecting a pharyngeal swab, we propose an optimization of the sample collection protocol to allow for successful sequencing in future experiments. This includes lowering the volume of virus-transport medium in order to reduce dilution of the sample, storing the swab directly in lysis buffer, and using a swab of different materials, since foam swabs were found to be the optimal choice for virus collection [25,26]. The presence of these minor variants in the salivary gland, but not in the pharyngeal swab, might be explained by the presence of virus originating from (para-)sympathetic nerves that innervate the salivary gland [27].

In line with a previous study [14], no differences were found in virus populations isolated from CNS and salivary gland biopsies. Besides the salivary gland, no differences were detected in virus population isolated from other non-CNS tissues, indicating that RABV does not require adaptations in order to spread between the different tissues and cell types of its host. While the virus can infect various extra-neuronal tissues and cells of non-neuronal origin, the secreted virus needs to be capable of infecting (peripheral) nerves again. Therefore, large organ-specific adaptations are not to be expected, since this would not be beneficial to the spread and infectivity of the virus. Throughout the dataset we detected one SNP in the RABV glycoprotein, an essential protein for binding to host receptors. Changes in the glycoprotein gene sequence are commonly associated with changes in virulence and/or adaptations to a novel host [28–31]. However, this SNP was detected in virus isolated from nuchal skin, a tissue that is normally not actively involved in virus dissemination.

In conclusion, this is the first study that performed NGS and minor variant analysis on virus isolated from a broad range of CNS and non-CNS samples of human and mouse origin. The mice experiment was performed independently from the collection and in vitro passaging of the human CNS isolates. While this resulted in insights of inter-host adaptations of both a dog-related RABV strain (human case) and a bat-related RABV strain (mice experiment), more in-depth insights into the dog-related strain could be obtained by performing a specifically designed animal experiment that uses a dog-related RABV isolate. Altogether, better understanding of the evolution and adaptation of RABV in the host will be valuable in increasing the understanding of the spread of RABV and related lyssaviruses within different hosts.

Supplementary Materials: The following information can be downloaded at: https://www.mdpi.com/article/10.3390/v14122661/s1 Table S1: Ct values of the independent samples selected for viral sequencing. Table S2: Abundance of the observed minor variants (in percentages) detected in the various mice tissues.

Author Contributions: Conceptualization and methodology: C.W.E.E., C.H.G., B.B.O.M. and M.P.G.K.; Investigation: C.W.E.E., V.P.V., I.v.M.-v.d.B., J.I., M.B., A.v.d.L. and B.B.O.M.; Resources: E.A.B.A.F., D.B., H.Z., P.V.C., I.I., C.H.G., K.A.M., M.M.S.A., M.M.I., R.A., A.A.A.-Z., N.M.Y., H.E.A.-R., M.H.J.A., R.S.S. and M.P.G.K.; Formal analysis: B.B.O.M. Visualization: C.W.E.E.; Writing—original draft preparation: C.W.E.E., B.B.O.M., C.H.G. and M.P.G.K.; Writing—review and editing: C.W.E.E., E.A.B.A.F., D.B., M.B., A.v.d.L., V.P.V., I.v.M.-v.d.B., J.I., H.Z., P.V.C., I.I., K.A.M., M.M.S.A., M.M.I., R.A., A.A.A.-Z., N.M.Y., H.E.A.-R., M.H.J.A., R.S.S., M.P.G.K., B.B.O.M. and C.H.G. All authors have read and agreed to the published version of the manuscript.

Funding: This work was funded through an Erasmus MC Fellowship 2019, project number 110581. CE is funded by a VENI Grant from The Netherlands Organization for Scientific Research (NWO-VENI 09150162010181).

Institutional Review Board Statement: Ethical approval for the collection and analysis of the human CNS samples was obtained by the Health Research Governance Department at the Ministry of Public Health in Qatar under the protocol ID MRC- 04-19-387. All animal experiments were performed in compliance with the Dutch legislation for the protection of animals used for scientific purposes (implementing EU Directive 2010/63) and other relevant regulations. The research was conducted under a project license (AVD1010020187204) approved by the Dutch competent authority. The specific study protocol (18-7204-01) was approved by the institutional Animal Welfare Body.

Informed Consent Statement: Informed consent for post-mortem CNS sample collection was obtained from relatives.

Data Availability Statement: The data presented in this study are available on request from the corresponding author.

Acknowledgments: We would like to thank the EDC team, and especially Vincent Duiverman and Rianne Stam, for their excellent support during the animal experiments. Georgina Aron is acknowledged for her help with culturing the patient materials.

Conflicts of Interest: The authors declare no conflict of interest. The funders had no role in the design of the study; in the collection, analyses, or interpretation of data; in the writing of the manuscript; or in the decision to publish the results.

References

1. WHO. *Expert Consultation on Rabies-Third Report*; WHO: Geneva, Switzerland, 2018.
2. Hampson, K.; Coudeville, L.; Lembo, T.; Sambo, M.; Kieffer, A.; Attlan, M.; Barrat, J.; Blanton, J.D.; Briggs, D.J.; Cleaveland, S.; et al. Estimating the Global Burden of Endemic Canine Rabies. *PLoS Negl. Trop. Dis.* **2015**, *9*, e0003709. [CrossRef]
3. Taylor, L.H.; Hampson, K.; Fahrion, A.; Abela-Ridder, B.; Nel, L.H. Difficulties in Estimating the Human Burden of Canine Rabies. *Acta Trop.* **2017**, *165*, 133–140. [CrossRef] [PubMed]
4. Benavides, J.A.; Rojas Paniagua, E.; Hampson, K.; Valderrama, W.; Streicker, D.G. Quantifying the Burden of Vampire Bat Rabies in Peruvian Livestock. *PLoS Negl. Trop. Dis.* **2017**, *11*, e0006105. [CrossRef]
5. Charlton, K.M.; Casey, G.A. Experimental Rabies in Skunks: Immunofluorescence Light and Electron Microscopic Studies. *Lab. Investig.* **1979**, *41*, 36–44. [PubMed]
6. Allendorf, S.D.; Cortez, A.; Heinemann, M.B.; Harary, C.M.A.; Antunes, J.M.A.; Peres, M.G.; Vicente, A.F.; Sodré, M.M.; da Rosa, A.R.; Megid, J. Rabies Virus Distribution in Tissues and Molecular Characterization of Strains from Naturally Infected Non-Hematophagous Bats. *Virus Res.* **2012**, *165*, 119–125. [CrossRef]
7. Delmas, O.; Holmes, E.C.; Talbi, C.; Larrous, F.; Dacheux, L.; Bouchier, C.; Bourhy, H. Genomic Diversity and Evolution of the Lyssaviruses. *PLoS ONE* **2008**, *3*, e2057. [CrossRef] [PubMed]
8. Troupin, C.; Dacheux, L.; Tanguy, M.; Sabeta, C.; Blanc, H.; Bouchier, C.; Vignuzzi, M.; Duchene, S.; Holmes, E.C.; Bourhy, H. Large-Scale Phylogenomic Analysis Reveals the Complex Evolutionary History of Rabies Virus in Multiple Carnivore Hosts. *PLoS Pathog.* **2016**, *12*, e1006041. [CrossRef] [PubMed]
9. Sooksawasdi Na Ayudhya, S.; Meijer, A.; Bauer, L.; Oude Munnink, B.; Embregts, C.; Leijten, L.; Siegers, J.Y.; Laksono, B.M.; van Kuppeveld, F.; Kuiken, T.; et al. Enhanced Enterovirus D68 Replication in Neuroblastoma Cells Is Associated with a Cell Culture-Adaptive Amino Acid Substitution in VP1. *Msphere* **2020**, *5*, e00941-20. [CrossRef] [PubMed]
10. Nadin-Davis, S.A.; Fehlner-Gardiner, C. Chapter 5 Lyssaviruses-Current Trends. *Adv. Virus Res.* **2008**, *71*, 207–250. [CrossRef]
11. Marston, D.A.; Banyard, A.C.; McElhinney, L.M.; Freuling, C.M.; Finke, S.; de Lamballerie, X.; Müller, T.; Fooks, A.R. The Lyssavirus Host-Specificity Conundrum—Rabies Virus—the Exception Not the Rule. *Curr. Opin. Virol.* **2018**, *28*, 68–73. [CrossRef]

12. Munnink, B.B.O.; Abd Farag, E.A.B.; GeurtsvanKessel, C.; Schapendonk, C.; van der Linden, A.; Kohl, R.; Arron, G.; Ziglam, H.; Goravey, W.G.M.; Coyle, P.V.; et al. First Molecular Analysis of Rabies Virus in Qatar and Clinical Cases Imported into Qatar, a Case Report. *Int. J. Infect. Dis.* **2020**, *96*, 323–326. [CrossRef] [PubMed]
13. Scott, T.P.; Fischer, M.; Khaiseb, S.; Freuling, C.; Höper, D.; Hoffmann, B.; Markotter, W.; Müller, T.; Nel, L.H. Complete Genome and Molecular Epidemiological Data Infer the Maintenance of Rabies among Kudu (Tragelaphus strepsiceros) in Namibia. *PLoS ONE* **2013**, *8*, e58739. [CrossRef] [PubMed]
14. Bonnaud, E.M.; Troupin, C.; Dacheux, L.; Holmes, E.C.; Monchatre-Leroy, E.; Tanguy, M.; Bouchier, C.; Cliquet, F.; Barrat, J.; Bourhy, H. Comparison of Intra- and Inter-Host Genetic Diversity in Rabies Virus during Experimental Cross-Species Transmission. *PLOS Pathog.* **2019**, *15*, e1007799. [CrossRef] [PubMed]
15. Kissi, B.; Badrane, H.; Audry, L.; Lavenu, A.; Tordo, N.; Brahimi, M.; Bourhy, H. Dynamics of Rabies Virus Quasispecies during Serial Passages in Heterologous Hosts. *J. Gen. Virol.* **1999**, *80*, 2041–2050. [CrossRef] [PubMed]
16. Holmes, E.C.; Woelk, C.H.; Kassis, R.; Bourhy, H. Genetic Constraints and the Adaptive Evolution of Rabies Virus in Nature. *Virology* **2002**, *292*, 247–257. [CrossRef] [PubMed]
17. Koraka, P.; Martina, B.E.; Smreczak, M.; Orlowska, A.; Marzec, A.; Trebas, P.; Roose, J.M.; Begeman, L.; Gerhauser, I.; Wohlsein, P.; et al. Inhibition of Caspase-1 Prolongs Survival of Mice Infected with Rabies Virus. *Vaccine* **2019**, *37*, 4681–4685. [CrossRef]
18. Marosi, A.; Dufkova, L.; Forró, B.; Felde, O.; Erdélyi, K.; Širmarová, J.; Palus, M.; Hönig, V.; Salát, J.; Tikos, R.; et al. Combination Therapy of Rabies-Infected Mice with Inhibitors of pro-Inflammatory Host Response, Antiviral Compounds and Human Rabies Immunoglobulin. *Vaccine* **2019**, *37*, 4724–4735. [CrossRef]
19. Marston, D.A.; Horton, D.L.; Nunez, J.; Ellis, R.J.; Orton, R.J.; Johnson, N.; Banyard, A.C.; McElhinney, L.M.; Freuling, C.M.; Fırat, M.; et al. Genetic Analysis of a Rabies Virus Host Shift Event Reveals Within-Host Viral Dynamics in a New Host. *Virus Evol.* **2017**, *3*, vex038. [CrossRef]
20. Wakeley, P.R.; Johnson, N.; McElhinney, L.M.; Marston, D.; Sawyer, J.; Fooks, A.R. Development of a Real-Time, Differential RT-PCR TaqMan® Assay for Lyssavirus Genotypes 1, 5 and 6. *Dev. Biol.* **2006**, *126*, 227–236. [CrossRef]
21. Wadhwa, A.; Wilkins, K.; Gao, J.; Condori Condori, R.E.; Gigante, C.M.; Zhao, H.; Ma, X.; Ellison, J.A.; Greenberg, L.; Velasco-Villa, A.; et al. A Pan-Lyssavirus Taqman Real-Time RT-PCR Assay for the Detection of Highly Variable Rabies Virus and Other Lyssaviruses. *PLoS Negl. Trop. Dis.* **2017**, *11*, e0005258. [CrossRef]
22. Briese, T.; Kapoor, A.; Mishra, N.; Jain, K.; Kumar, A.; Jabado, O.J.; Ian Lipkina, W. Virome Capture Sequencing Enables Sensitive Viral Diagnosis and Comprehensive Virome Analysis. *MBio* **2015**, *6*, e01491-15. [CrossRef]
23. Morimoto, K.; Hooper, D.C.; Carbaugh, H.; Fu, Z.F.; Koprowski, H.; Dietzschold, B. Rabies Virus Quasispecies: Implications for Pathogenesis. *Proc. Natl. Acad. Sci. USA* **1998**, *95*, 3152–3156. [CrossRef] [PubMed]
24. Watermann, C.; Valerius, K.P.; Wagner, S.; Wittekindt, C.; Klussmann, J.P.; Baumgart-Vogt, E.; Karnati, S. Step-by-Step Protocol to Perfuse and Dissect the Mouse Parotid Gland and Isolation of High-Quality RNA from Murine and Human Parotid Tissue. *Biotechniques* **2016**, *60*, 200–203. [CrossRef]
25. Jansson, L.; Akel, Y.; Eriksson, R.; Lavander, M.; Hedman, J. Impact of Swab Material on Microbial Surface Sampling. *J. Microbiol. Methods* **2020**, *176*, 106006. [CrossRef] [PubMed]
26. Roque, M.; Proudfoot, K.; Mathys, V.; Yu, S.; Krieger, N.; Gernon, T.; Gokli, K.; Hamilton, S.; Cook, C.; Fong, Y. A Review of Nasopharyngeal Swab and Saliva Tests for SARS-CoV-2 Infection: Disease Timelines, Relative Sensitivities, and Test Optimization. *J. Surg. Oncol.* **2021**, *124*, 465–475. [CrossRef]
27. Segal, K.; Lisnyansky, I.; Nageris, B.; Feinmesser, R. Parasympathetic Innervation of the Salivary Glands. *Oper. Tech. Otolaryngol.-Head Neck Surg.* **1996**, *7*, 333–338. [CrossRef]
28. Borucki, M.K.; Chen-Harris, H.; Lao, V.; Vanier, G.; Wadford, D.A.; Messenger, S.; Allen, J.E. Ultra-Deep Sequencing of Intra-Host Rabies Virus Populations during Cross-Species Transmission. *PLoS Negl. Trop. Dis.* **2013**, *7*, e2555. [CrossRef] [PubMed]
29. de Novaes Oliveira, R.; de Souza, S.P.; Lobo, R.S.; Castilho, J.G.; Macedo, C.I.; Carnieli, P., Jr.; Fahl, W.O.; Achkar, S.M.; Scheffer, W.; Kotait, I.; et al. Rabies Virus in Insectivorous Bats: Implications of the Diversity of the Nucleoprotein and Glycoprotein Genes for Molecular Epidemiology. *Virology* **2010**, *405*, 352–360. [CrossRef]
30. Yang, Y.J.; Zhao, P.S.; Wu, H.X.; Wang, H.L.; Zhao, L.L.; Xue, X.H.; Gai, W.W.; Gao, Y.W.; Yang, S.T.; Xia, X.Z. Production and Characterization of a Fusion Peptide Derived from the Rabies Virus Glycoprotein (RVG29). *Protein Expr. Purif.* **2014**, *104*, 7–13. [CrossRef]
31. Nitschel, S.; Zaeck, L.M.; Potratz, M.; Nolden, T.; Kamp, V.; Franzke, K.; Höper, D.; Pfaff, F.; Finke, S. Point Mutations in the Glycoprotein Ectodomain of Field Rabies Viruses Mediate Cell Culture Adaptation through Improved Virus Release in a Host Cell Dependent and Independent Manner. *Viruses* **2021**, *13*, 1989. [CrossRef]

Review

Nipah Virus: An Overview of the Current Status of Diagnostics and Their Role in Preparedness in Endemic Countries

Anna Rosa Garbuglia [1,*], **Daniele Lapa** [1], **Silvia Pauciullo** [1], **Hervé Raoul** [2] and **Delphine Pannetier** [3]

1   Laboratory of Virology, National Institute for Infectious Diseases "Lazzaro Spallanzani" (IRCCS), 00149 Rome, Italy; daniele.lapa@inmi.it (D.L.); silvia.pauciullo@inmi.it (S.P.)
2   French National Agency for Research on AIDS—Emerging Infectious Diseases (ANRS MIE), Maladies Infectieuses Émergentes, 75015 Paris, France; herve.raoul@inserm.fr
3   Institut National de la Santé et de la Recherche Médicale, Jean Mérieux BSL4 Laboratory, 69002 Lyon, France; delphine.pannetier@inserm.fr
*   Correspondence: argarbuglia@iol.it or annarosa.garbuglia@inmi.it; Tel.: +39-0655-170-692

Abstract: Nipah virus (NiV) is a paramyxovirus responsible for a high mortality rate zoonosis. As a result, it has been included in the list of Blueprint priority pathogens. Bats are the main reservoirs of the virus, and different clinical courses have been described in humans. The Bangladesh strain (NiV-B) is often associated with severe respiratory disease, whereas the Malaysian strain (NiV-M) is often associated with severe encephalitis. An early diagnosis of NiV infection is crucial to limit the outbreak and to provide appropriate care to the patient. Due to high specificity and sensitivity, qRT-PCR is currently considered to be the optimum method in acute NiV infection assessment. Nasal swabs, cerebrospinal fluid, urine, and blood are used for RT-PCR testing. N gene represents the main target used in molecular assays. Different sensitivities have been observed depending on the platform used: real-time PCR showed a sensitivity of about 103 equivalent copies/reaction, SYBRGREEN technology's sensitivity was about 20 equivalent copies/reaction, and in multiple pathogen card arrays, the lowest limit of detection (LOD) was estimated to be 54 equivalent copies/reaction. An international standard for NiV is yet to be established, making it difficult to compare the sensitivity of the different methods. Serological assays are for the most part used in seroprevalence studies owing to their lower sensitivity in acute infection. Due to the high epidemic and pandemic potential of this virus, the diagnosis of NiV should be included in a more global One Health approach to improve surveillance and preparedness for the benefit of public health. Some steps need to be conducted in the diagnostic field in order to become more efficient in epidemic management, such as development of point-of-care (PoC) assays for the rapid diagnosis of NiV.

Keywords: Nipah virus; zoonosis; One Health; molecular diagnosis; infection

Citation: Garbuglia, A.R.; Lapa, D.; Pauciullo, S.; Raoul, H.; Pannetier, D. Nipah Virus: An Overview of the Current Status of Diagnostics and Their Role in Preparedness in Endemic Countries. *Viruses* **2023**, *15*, 2062. https://doi.org/10.3390/v15102062

Academic Editors: Myriam Ermonval and Serge Morand

Received: 20 September 2023
Revised: 2 October 2023
Accepted: 5 October 2023
Published: 7 October 2023

1. General Aspects

Nipah virus (NiV) infection is a viral disease that has emerged in Southeast Asia and is caused by a negative single-stranded RNA virus of 18,000 nucleotides in length, which belongs to the Paramyxoviridae family and to the *Henipavirus* genus. This genus also includes other species that can infect humans such as Ghanaian bat virus, Mojiang virus, and Hendra virus (HeV) [1]. First detected in the 1998–1999 outbreaks in Malaysia and Singapore, NiV is responsible for a zoonosis with mainly severe respiratory and neurological clinical manifestations in humans and regular outbreaks in Bangladesh, India, or the Philippines, and the morbidity and the mortality rate are related to the viral strain [2]. The high fatality rates in humans associated with NiV of up to 70% [3] has led to their classification as risk-group 4 pathogens, restricting work on these viruses to Biosafety Level 4 (BSL-4) facilities. The BSL categories take into account the lethality of the disease and the availability of preventive and therapeutic treatments, which do not currently exist in the case of NiV or HeV. NiV and HeV genomes share about 80% nucleotide identity [4], so diagnostics tests can be cross-reactive between these two viruses depending on the RNA sequence targeted. Similar to all paramyxoviruses, the genome codes for the following proteins: nucleocapsid protein (N), phosphoprotein (P), matrix protein (M), glycoprotein F (F), glycoprotein G (G), and RNA polymerase, which represents the large protein (L). The non-structural proteins C, V, and W, which play a key role in the pathogenicity of NiV, are encoded by the P gene [5]. The RNA genome is associated with the viral proteins of the replicative complex, which include nucleoprotein, phosphoprotein, and the polymerase L enclosed by a lipid bilayer envelope containing the attachment protein G and the fusion F protein [6]. The receptor of NiV is ephrin-B2, present in the endothelial cells and neurons [7–9]. The NiV-G and -F proteins are necessary for the binding and fusion to the host cells and for budding [10,11] (Figure 1).

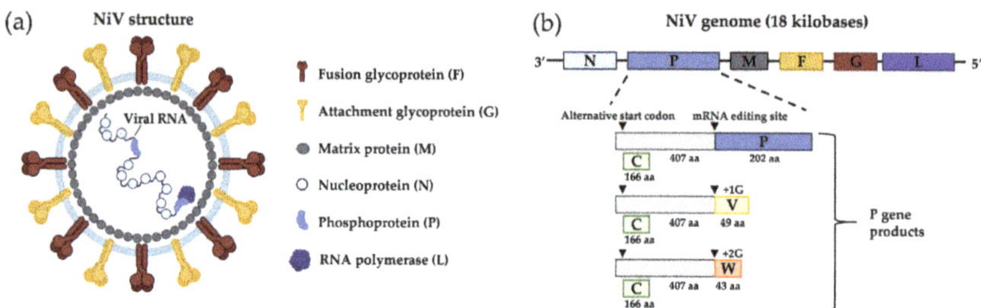

Figure 1. Nipah virus (NiV) structure and genome organization. (**a**) Enveloped NiV virion comprises different structural proteins: nucleoprotein (N), phosphoprotein (P), matrix protein (M), fusion glycoprotein (F), and attachment glycoprotein (G). Viral RNA polymerase (L) and N proteins are associated with the viral genome (negative-sense single-strand RNA). They are created by BioRender. (**b**) Schematic representation of NiV genome organization. Genes encoding N, P, M, F, G, and L proteins are shown. P gene encodes accessory proteins using an alternative start codon (C protein) or using mRNA editing (V and W proteins).

2. Epidemiology

The first human cases of NiV were observed in Malaysia in 1998–1999, near Sungai Nipah (Nipah River village). In this outbreak, close contacts with pigs, pig excreta, or fruit bats were shown to be the main risk factors for disease transmission [12,13].

2.1. Fruit Bats

Pteropus bats were identified to be a reservoir of NiV infection in Malaysia [14]. NiV neutralizing antibodies were detected in blood, and NiV RNA was detected in urine, saliva, serum, and different organs of *Pteropus* bats in many regions of Asia, and also in countries where no human infections had been described [15–17]. The *Pteropus* bats are widely distributed in Southeast Asia, the Indian Ocean, Oceania, and Sub-Saharan Africa [18,19]. In India, in May 2018, 18 NiV human infections were observed in the State of Kerala with patients showing acute respiratory syndrome and encephalitis [20,21]. The human NiV strains showed 99.7–100% sequence homology with bat NiV strains, suggesting that bats were the source of outbreak [22]. These fruit bats could infect humans or pigs through the consumption of bat-bitten fruits amplifying the virus diffusion.

2.2. Pigs

It is commonly accepted that pigs contract NiV infection by eating food contaminated with urine or saliva from bats and can then infect humans. The main symptom in pigs infected with NiV is a respiratory disease with about 5% lethality. Pig farmers and abattoir workers who are in direct contact with infected animals are the main group at risk for NiV infection [13,23]. For example, an outbreak among slaughterhouse workers who looked after pigs from Malaysia occurred in Singapore in 1999. The epidemic was limited with a ban on the import of pigs from Malaysia [24]. NiV was isolated from the nose and oropharynx swabs of pigs, which are animals that can act as intermediate hosts [6,12,25,26].

2.3. Other Hosts

Serological studies have demonstrated that other animals such as cattle, horses, dogs, cats, or goats can also be exposed to NiV and can develop specific antibodies against NiV, although transmission to humans in this way is yet to be reported [6,23,27].

3. Modes of Transmission

After the Malaysia outbreak in 1998–1999, later outbreaks of NiV confirmed that the main routes of virus transmission to humans are direct contact with the respiratory secretions or body fluids of infected animals such as bats and pigs or by the consumption of contaminated fruit palm or its derivatives (sap and alcohol). Indeed, several human outbreaks originated from drinking contaminated raw palm sap or climbing the trees coated with contaminated excrement, urine, or saliva from fruit bats [28,29]. Human outbreaks can occur sporadically (as observed in India and The Philippines) or in more specific times of the year, such as in the winter season in Bangladesh during the date palm harvest [30]. In the 2014 Philippines outbreak, close contact with horses or horse meat consumption was reported in 10 of the 17 confirmed cases. In the same period, 10 horses died, 9 of which were reported to have neurological disorders, although no test for NiV was performed on the horse samples [31]. In this study, fruit bats were considered as the most probable source of infection for horses [31].

Human-to-human transmission has also been reported [12,32,33]. For example, in India in 2007, an outbreak originated from one person who contracted the disease due the consumption of alcohol obtained from date palm. The infection was transmitted to other members of the family and to the One Health worker who collected blood and performed a tomography scan of the brain of the initial disease case, suggesting that close contacts are required to transmit the disease [34]. Indeed, contact with the body fluids of NiV positive patients increases the risk of virus transmission [35]. The Bangladesh strain of NiV is usually associated with human-to-human transmission, with no intermediary host and with a case fatality rate of 75% [12,36].

4. Symptoms and Pathogenesis in Humans

In humans, the infection mainly occurs via the oronasal site, and different incubation periods are reported in the literature, mainly from 1 to 2 weeks, but ranging from 4 to 21 days [37–40]. According to the World Health Organization (WHO), the period of incubation for NiV could vary from 4 to 45 days, but mainly from 4 to 14 days [41]. In the outbreak in Malaysia and Singapore in 1998–1999, 3.3 days was the meantime between fever onset and hospitalization, whereas death occurred after 9.5 days [42].

Different clinical courses have been described, and they vary according to the NiV strain. Differences in symptoms and lethality rate have been observed between the Malaysian (NiV-M) and Bangladesh strains (NiV-B). NiV-B is often associated with severe respiratory disease, whereas NiV-M is more often associated with severe encephalitis. Furthermore, in 2018, during the outbreak of Kerala caused by NiV-B, 83% of patients showed respiratory distress syndrome (ARDS) [36,38]. The course of the disease also depends on the strain: the mean disease duration from symptom onset to death was 16 days in Malaysia and only 4–6 days in Bangladesh and India [43,44]. The variability of interval time between symptom onset and death, which has been observed in different outbreaks, is probably linked to the different virulence of the strains.

Generally, the beginning of NiV disease is characterized by flu-like symptoms, with fever, myalgia, cough, vomiting, and headaches [12,39,45].

Respiratory complications characterized by cold, respiratory distress, shortness of breath, and atypical pneumonia can be observed. The poor prognosis is linked to age, thrombocytopenia, or the presence of other comorbidities [39]. In the late stage, NiV-

M infection can also reach the Central Nervous System (CNS) with several degrees of encephalopathy. Neurological symptoms include convulsions, altered functionality of cerebellum, and reduction in consciousness. Different levels of consciousness reductions have been reported, 50% in Malaysia versus 90% in Bangladesh, with NiV-B inducing more severe respiratory troubles than NiV-M [4]. Problems of mental disorientation can also appear [46]. The most common symptoms of encephalitis that could arise after a week are hypotonia, segmental myoclonus, areflexia, limb weakness, and gaze palsy. Coma and death can then happen within a few days. Fatigue, neurological deficits, and depression can persist in about 20% of patients who recover from acute infection [47], and some cases of relapsing or late-onset encephalitis have been described [48]. Furthermore, latent infections have been described after acute infection with a persistence that can last months or even years [49].

Subclinical infections have also been described in NiV epidemics. The rate of asymptomatic infections among confirmed cases varies according to the outbreak, ranging from 8 to 17% in Malaysia to more than 45% in Singapore [4,13,24,50,51]. Asymptomatic infections were absent or rarely observed with NiV-B in Bangladesh and India [52,53].

NiV infection leads to local changes in blood vessels with the appearance of vasculitis in the small vessels in humans. Brain parenchyma often presents necrotic plaques and sometimes syncytia of multinucleated giant endothelial cells. Moreover, the infection spreads to major organs [6,32,45,54].

5. Diagnostics
5.1. Molecular Diagnostics

In suspected cases, an early diagnosis of NiV infection is crucial to limit the outbreak and to provide appropriate care for the patient. NiV infection can be confirmed in several ways using direct detection methods, such as virus isolation, immunohistochemistry or immunofluorescence assays, nucleic acid amplification, or sequencing, but also using indirect detection methods of anti-NiV IgM or IgG antibodies, such as enzyme-linked immunosorbent assays (ELISA) or virus neutralization tests. Due to high specificity and sensitivity, qRT-PCR is currently considered as the first-choice method for the diagnosis of acute NiV infection, and it can reliably diagnose NiV infection within a few hours. As described above, NiV has a period of incubation which mainly ranges from 4 to 14 days, even though it can be longer [55]. It is yet to be established whether or not patients can transmit the virus during the incubation period, but such transmission has been demonstrated in pigs [56]. Nasal swabs, cerebrospinal fluid, urine, and blood are usually used in RT-PCR tests [57].

Originally, heminested or nested RT-PCR were the main methods used to detect NiV RNA. Real-time RT-PCR was set up later and, compared to conventional RT-PCR, provides a better sensitivity, an ease of use, and a reduced risk of contamination, so it has become the gold standard diagnostic method. Real-time PCR is also indicated by the CDC as the reference method to be used for the diagnosis of acute infection [58].

Different kinds of RT-PCR have been set up for NiV detection, most of them being in-house assays (Table 1). Some RT-PCRs are able to detect several paramyxoviruses including NiV. For example, a nested RT-PCR described by Tong [59] is still used for known or new paramyxovirus identification. It consists of a broad range RT-PCR that uses a conserved region of L gene as a target. The sensitivity ranged between 500 and 1000 copies of template RNA [59]. Another duplex RT-PCR was set up for RNA detection from the urine of bats. This RT-PCR included an internal control (IC) consisting of an RNA plasmid containing a 1.2 kb fragment of Kanamycin gene. The primers for NiV were specific for a conserved region of N gene, as described previously [6]. The lower limit of detection was 0.37 pg/μL of total RNA. This method was also applicable to other biological samples such as bat's saliva and blood [60,61]. One of the most frequently used Real-Time PCR was developed by Guillaume [62] and is based on the amplification of a conserved region of the N gene. This assay has a linearity between 10^3 and 10^9 equivalent copies of NiV. It was set up with the NiV-M strain, but the authors did not report whether this sensitivity was confirmed for the NiV-B strain. This assay is specific for NiV and does not amplify Hendra nor measles virus genome. A comparison of TaqMan Real-Time and SYBR Green assay using primers designed in different regions of NiV virus (for N, M, L, and P genes) showed different sensitivities according to the Real-Time PCR platform [63]. Since it was not possible to set up a TaqMan Real-Time PCR assay for L gene, a heminested RT-PCR was carried out for this gene. In this study, one-step RT-PCR appeared to be more sensitive than two-step RT-PCR. Assays based on N and P genes showed the same sensitivity in SYBRGreen format, while P gene was found to be the most sensitive in TaqMan platform. The SYBR Green method can distinguish NiV and HeV on the basis of melting curve analysis. The TaqMan assays for P and N regions, N SYBR Green protocols, and L heminested RT-PCR showed a sensitivity of 20 equivalent genomes per reaction, while M gene TaqMan assay was associated with a sensitivity of 2000 equivalent genomes per reaction. In this case, the limited sensitivity of M gene RT-PCR could be linked to the use of degenerate primers in the assays. M and P gene primer pairs were found to be unsuitable for SYBR Green assays, because the potential primer dimers probably interfere with the performance of the assay [63].

N gene was also used as a target in a SYBR Green I-based qRT-PCR amplification assay used to analyze the kinetics of viral replication in vitro, with a sensitivity of about 100 pfu/μL [64]. For a precise molecular quantification of the viral burden, it is important to note that in most of the real-time assays, target sequences are present both in genomes and in mRNA transcripts. Thus, the viral burden estimation can be influenced by the amount of mRNA transcripts. For example, N transcripts are the most abundant, and L transcripts are the least abundant. A real-time approach to calculate the real burden of genomic NiV RNA excluding amplification of the viral mRNA was developed by Jensen [65]. This one-step RT-PCR targets only viral genomic RNA in the untranslated intergenic region separating the F and G viral proteins and does not amplify mRNA transcripts, making it also possible to detect both NiV-M or NiV-B isolates in the same sample. Detection is linear from 1×10^2 to 1×10^9 copies/mL with a coefficient of correlation of $r^2 = 0.998$. Different LOD were reached with NiV-B and NiV-M strains with 1.63×10^4 genomes/mL compared to 5.82×10^3 genomes/mL, respectively. No cross reactivity was found with Hendra virus (HeV), Respiratory Syncytial Virus (RSV), Ebola virus (EBOV), or Lassa virus (LASV).

Moreover, in order to improve the rapidity of the diagnostics and to test different pathogens at the same time, multiple pathogen card arrays are often used. Different types of platforms exist on the market, and one of the main advantages of these cards is the relatively easy use and storage as well as the stability at 4 °C for two years with shipment at room temperature. For example, one TaqMan array card (TAC) based on N region is able to detect several pathogens simultaneously in cerebrospinal fluid: *Balamuthia mandrillaris* and *Acanthamoeba* (parasites), *Streptococcus pneumoniae*, *Haemophilus influenzae*, *Neisseria meningitidis*, *Mycoplasma pneumoniae*, *Mycobacterium tuberculosis*, and *Bartonella* (bacteria), and 13 viruses (parechovirus, dengue virus, Nipah virus, varicella-zoster virus, mumps virus, measles virus, lyssavirus, herpes simplex 1 and 2, Epstein Barr virus, enterovirus, cytomegalovirus, and chikungunya virus). The target sequence for NiV is the N gene, and the estimated LOD is 54 equivalent copies/well [66].

Another TAC is based on quantitative reverse transcription PCR (qRT-PCR) for the simultaneous detection of 15 viruses including NiV (Chikungunya, Crimean-Congo hemorrhagic fever (CCHF) virus, dengue, EBOV, Bundinbugyo virus, SUDV, hantaviruses (Hantaan and Seoul), Hepatitis E, MARV, Nipah virus, O'nyong-nyoung virus, Rift Valley fever virus, West Nile virus, and YFV), 8 bacteria (*Bartonella* spp., *Brucella* spp., *Coxiella Burneti*, *Leptospira* spp., *Rickettsia* spp., *Salmonella enterica* and *Salmonella enterica* serovar Typhy, and *Yersinya pestis*), and 3 protozoa (*Leishmania* spp., *Plasmodium* spp., and *Trypanosoma brucei*) of particular relevance to Sub-Saharan Africa. TAC exhibited an overall sensitivity of 88% and a specificity of 99%. The LOD for viral genomes is 10^4 copies/mL in blood. No indication is available for cerebrospinal fluid samples. A critical aspect of this assay is the need of a nucleic extraction step that limits its use for NiV surveillance activities in remote areas and could also induce some biosafety problems [67].

Nevertheless, it can be difficult to perform such RT-PCR tests if the outbreak occurs in a remote area with limited facilities, especially for electrical power and working materials. In such cases, other types of PCR techniques could be implemented. For example, N gene is also the target for real-time reverse transcription-loop-mediated isothermal amplification (RT-LAMP), which is able to detect all NiV strains [68]. The results are obtained within 45 min, and the sensitivity is close to 100 pg of total M and B genotypes NiV pseudovirus RNA, corresponding to 10^7 RNA copies. This is ten fold higher than the sensitivity obtained with a conventional N gene RT-PCR [69]. No cross reactivity was found with Influenza A virus and Hendra virus. This assay was able to detect NiV in different biological samples: urine, blood, feces, and throat swabs [68].

Recently, three rapid molecular diagnostic tests were developed. They are based on reverse transcription recombinase-based isothermal amplification coupled with lateral flow detection. For these three assays, a region of N gene is targeted and reaches an analytical sensitivity of 1000 copies/µL, corresponding to 100–200 RNA copies/reaction, for both NiV-B and NiV-M genotypes. These tests provide the results within 30 min without any extraction step allowing their use in low-resource settings [69].

Finally, NiV can also be detected in biological samples using a metagenomic approach based on high-throughput sequencing (HTS). These metagenomic approaches are used frequently nowadays and could become the new reference technique in the future. A combination of RNA baits specific for 35 epizootic and zoonotic viruses made it possible to enrich the samples from about 10- to 10,000-fold, reaching a considerable sensitivity. For NiV, this assay showed a sensitivity of 21 genomes/reaction [70].

5.2. Serological Diagnosis

RT-PCR is the best option to diagnose NiV in humans, especially during the acute phase of infection, but indirect antibody detection methods may be used to diagnose later NiV infections or as a complementary diagnostic method. IgG and IgM are indeed key markers for Nipah seroprevalence studies, but data on kinetics or persistence during human convalescence are limited. ELISA-based serological tests can be used to detect IgM and IgG antibodies in the serum and cerebrospinal fluid (CSF), and it seems that for most patients, the IgM and IgG antibodies appear in the serum during the first week after onset (Figure 2) [71]. IgG antibodies seem to persist up to 8 months in symptomatic patients, while IgM antibodies seem to persist from 3 to 7 months. According to Ramasundram et al., IgM positivity is observed on the first day after the appearance of symptoms in 50% of patients and reaches 100% after three days, while IgG are detected two days after the onset of symptoms in 31% of patients, with 100% positivity reached by day 17 [72]. In a small study performed in India [73], contact cases of NiV-positive patients were tested regularly for a number of months to detect IgM and IgG antibodies. Antibodies were detected among three symptomatic and two asymptomatic contact cases. For symptomatic patients, IgM was detectable from the 5th to 27th days following disease onset, while IgG can persist for more than one year. Comparable IgM and IgG immune responses against NiV infection were observed in the presence or the absence of clinical symptoms.

Due to their more complicated use compared to RT-PCR, need of specific reagents, and longer implementation time, serological tests are more commonly used in epidemiological studies and surveillance activities with no ongoing NiV acute infections. Several tests have thus been implemented over the years to detect NiV IgG and IgM. These tests can generally differentiate NiV from HeV [74–77].

Table 1. Molecular diagnostic assay of Nipah virus.

Test	Test Type	Virus Detected	Target Gene	Biological Matrix	Sensivity	Specificity	Author
Home made	RT-PCR broad range	Paramyxoviruses	L gene of NiV	Paramyxovirus viral strains (henipavirus, Morbilliviruses, respiroviruses, and Rubulaviruses)	10–100 copies/reaction	No cross reactivity with Influenza A and B viruses, Rhinoviruses, adenoviruses, coronaviruses 229E, and OC-43. *Chlamidia pneumoniae*, *Haemophilus influenzae*, *Streptococcus pneumoniae*, and *Mycoplasma pneumoniae*	Tong et al., 2008 [59]
Home made	Duplex RT-PCR	NiV	N gene of NiV	Urine, saliva, and blood	0.37 pg/μL	ND	Chua et al., 2000 [6]
Home made	Real-time RT-PCR with fluorescent reporter dye detected at each PCR cycle	NiV	N gene of NiV M strain	Viral stock	Whole blood: 10^3 copies/mL	No cross-reactivity with measles virus was declared	Guillaume et al., 2004 [62]
Home made	Real-time RT-PCR with fluorescent reporter dye detected at each PCR cycle	NiV	N, M, P genes	In vitro transcribed RNA	20 copies/reaction	Whole blood: 100% [95.9–100].	Feldman et al., 2009 [63]
Home made	SYBR Green RT-PCR	NiV	N gene	Whole blood and urine	20 copies/reaction	Whole blood: 100% for Nipah virus	Feldman et al., 2009 [63]
Home made	Transcription-loop-mediated isothermal amplification (RT-LAMP)	NIV	N gene of NiV	Whole blood sample, fecal sample, throat swab sample, and urine sample	10^7 copies/ reaction	No cross reactivity with HeV, Newcastle disease virus, Japanese encephalitis virus, and Influenza A virus.	Ma et al., 2019 [68]
Home made	SYBR Green RT-PCR	NiV	N gene of NiV	Viral replication in vitro	100 pfu/μL	ND	Chang et al., 2006 [64]
Home made	Real-time RT-PCR with fluorescent dye-labelled probes to detect PCR amplicons	NiV	Intergenic region separating the viral F and G protein coding regions	Serum, body fluid, and urine	LOD: NiV-B assay 1.63×10^4 genomes/mL; NiV-M 5.82×10^3 genomes/mL	No cross reactivity with HeV, RSV, EBOV, and LASV	Jensen et al., 2018 [65]
Home made	NGS	35 epizootic and zoonctic viruses	Full genome	Whole blood, serum, plasma, and urine	21 genomes/reaction	ND	Wylezich et al., 2021 [70]
EZ1 test (DOD)	Real-time TaqMan RT-PCR with fluorescent reporter dye detected at each PCR cycle	Several pathogens	N gene	Whole blood and plasma	Whole blood: 54 copies/well	100%; no cross-reactivity with other Viral Haemorrhagic Fever	Onyango et al., 2017 [66]
Home made	TaqMan array CARD	15 viruses, 8 bacteria and 3 protozoa			Blood samples 10^4 copies/mL		Liu et al., 2016 [67]
Home made	Reverse transcription recombinase-based isothermal amplification coupled with lateral flow detection		N gene	Viral stock produced in vitro	10^3 copies/μL (analytical sensitivity)	No cross reactivity with Hendra virus	Pollak et al., 2023 [69]

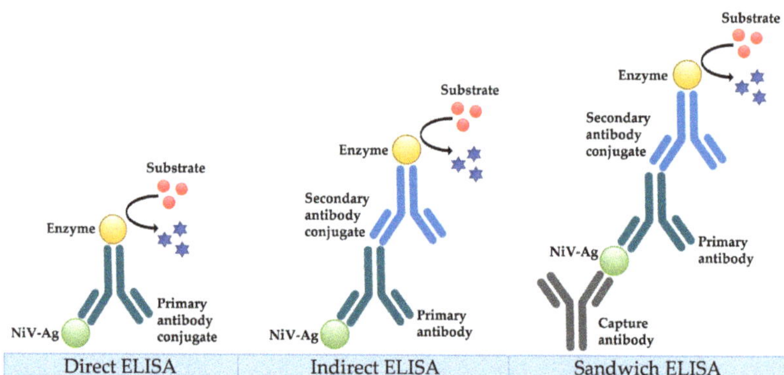

Figure 2. Different types of enzyme-linked immunosorbent assay (ELISA).

Initially, ELISA investigations used a gamma-irradiated NiV antigen [78,79]. However, in this assay, the antigen derived from NiV-infected cells was difficult to standardize due to the variability of different culture parameters such as virus strains, cell line, or multiplicity of virus infection, and an antigen produced in different culture conditions could have different antigenic properties causing potential troubleshooting in the interpretation of the tests. The main advantage of these tests is their ability to detect a very broad spectrum of human antibodies directed against the whole virus and not only restricted to several proteins. On the other hand, more recent serological tests using NiV recombinant proteins allow a greater standardization but cannot detect a large spectrum of antibodies directed against the whole virus.

N recombinant protein is most frequently used in ELISA assays because of its immunodominance and conservation among different NiV strains. This structural protein is the most abundant produced during the infection by NiV, and amino acidic sequences are similar in humans and animal reservoirs [6,80]. NiV-N recombinant protein has been often expressed in *E. coli*. Yu et al. [81] developed an ELISA with the NiV-N recombinant protein that showed a sensitivity and specificity of 91.7% and 91.8% respectively, using human sera and the CDC IgM ELISA assay as a reference test. The NiV antibody detection in swine sera showed a 100% concordance with the CDC IgG ELISA assay. These findings confirmed that N protein is a highly immunogenic protein that must be considered as an excellent target for serodiagnosis assays.

In another assay, the NiV-N protein expressed in *E. coli* was tested in 1709 swine serum samples and showed a sensitivity and specificity of 100% and 98.7%, respectively [82].

In addition to direct ELISA assays, antigen capture ELISA could also be implemented to detect HeV and NiV antibodies. The anti-N antibody 1a11c1 captures proteins from HeV and both NiV-M and NiV-B strains with a high sensitivity (LOD 400 pfu/well) and can detect NiV antigen from a frozen pig lung specimen. No cross-reactivity was observed with Marburg hemorrhagic fever (MHF) virus. The 1a11c1 antibody detects the NiV-M SPB199901924 Malaysia strain [83–85] but a low background signal was observed for uninfected Vero cell lysate, Lassa virus, Marburg, and Crimean-Congo hemorrhagic fever (CCHF), indicating a lack of specificity of assays performed with this antibody. Indeed, a problem linked to NiV-N protein expression is the non-specific binding due to residual bacterial *E. coli* proteins, which are not totally removed during the purification process [86]. This aspect has been addressed by Chen [87] by calculating the background binding due to non-specific interactions using incubation of the sera with excess free antigen to block specific binding. The sample was considered positive only when its total reactivity signal was higher than a pre-determined cut-off value, and the ratio of the total reactivity to the background signal was A450 R sample reactivity/A450 B background reactivity > 2.0. This method eliminates 35% of reactive but non-specific sera, and reached a specificity of 95.8%, which is no higher than those reported by other ELISA assays.

Another ELISA, named solid-phase blocking ELISA, was developed to detect anti-NiV antibodies [88]. The ELISA plates were coated with NiV virus cell infected lysate, and the serum incubated with solid phase was washed after one hour. Anti-NiV monoclonal antibodies (mAb) conjugated to peroxidase were then added to detect the remaining free antigen. A sample was considered positive when the inhibition due to mAb reaction against NiV antigen was over 20%. The relative sensitivity and specificity against the serum neutralization test were >70% and >95%, respectively, with human sera. Considering its lower sensitivity compared to other ELISAs, this test should only be used for anti-NiV screening. Luminex platform is also used for antibody detection. This type of assay uses very small sample volumes, can be used in multiplex, and is much cheaper than traditional ELISA assays and therefore is of relevance to low- and middle-income settings [89,90]. Serum neutralization assay as a confirmatory test should be then performed to confirm the results for doubtful values and to avoid false positive results. This assay was employed to test pig and bat sera but could also be used for other animal species.

5.3. Neutralization Assays

Considering its high specificity, serum neutralization testing is today considered as the reference standard for the confirmatory diagnosis of NiV by the World organization for Animal Health (OIE). Moreover, these tests are useful to check the protective immunity of a serum that is correlated to the level of neutralizing antibodies. First neutralization assays were initially developed with live virus. The sera are serially diluted and incubated with a well-defined number of viral particles, before being added to Vero cells. After 24 h, the syncytia specific of NiV infection can be observed [91]. Such an assay with infectious virus has two main limitations as it must be conducted in a BSL4 laboratory, and the plaque reduction observed could be linked to serum cytotoxicity instead of being a direct effect produced by the virus, even though the serial dilutions used during the test limit this risk.

One alternative to standard plaque reduction neutralization tests (PRNT) is a neutralization assay based on pseudovirions that can be used to detect antibodies against viral surface glycoproteins. Since pseudovirions are able to produce non-infectious viruses, they do not require BSL4 facilities and could be easy to implement. The envelope attachment (G) and fusion (F) glycoproteins can be expressed in different systems: Moloney murine leukemia virus (MuLV), vesicular stomatitis virus (VSV), or human immunodeficiency virus lacking envelope protein [92–94]. The reporter gene can be the green fluorescent protein (GFP) or luciferase [95,96]. All these assays based on pseudovirions are quantitative, but some limitations exist as the sensitivity of each assay has not always been compared with the sensitivity obtained using PNRT, which is considered as the gold-standard neutralization assay. Some aspecific positivity has also been observed. For example, a study using pig vaccinated serum compared the results obtained with PRNT assays and tests with pseudovirions produced by VSV expressing the F and G proteins of NiV as target antigens (pVSV-NiV-F/G) [96]. In this study, there is a good correlation in the results obtained with the two assays, except to detect low levels of antibodies where PRNT results were better than pseudovirion test results. The specificity was high (94–100%). Therefore, at least for the moment, neutralization assays performed with pseudovirions cannot totally replace the assays with live viruses.

5.4. Virus Isolation

Before RT-PCR, virus isolation was the reference method to confirm the diagnostics of NiV infection, because it could characterize the strain precisely. Indeed, through viral isolation, one can obtain enough material to sequence the whole genome or to set up neutralization assays. Nevertheless, viral isolation requires BSL4 facilities, it is time-consuming and less sensitive than RT-PCR because it is not always possible to isolate a virus from a sample. It is therefore no longer used for primary diagnosis but only for confirmatory diagnosis and to work further with the isolated viral strain. NiV can be isolated from throat swabs, urine, CSF, nasal swab samples, and brain and lung tissues using the Vero E6 cell line. A cytopathic effect usually appears within 3 days. It is characterized by the formation of syncytia that may contain 20 or more nuclei. Subsequently, syncytia detach from the substrate leaving holes in the monolayer surface [78]. This technique can be used to differentiate NiV from HeV. Indeed, for NiV, the nuclei and nucleocapsid are localized in the periphery of the infected cells, while in the case of HeV, both structures tend to localize centrally [97].

Moreover, different studies report a correlation between a positive virus isolation in the CSF and a high lethality in patients [98].

6. One Health Concept for Preparedness Applied to Nipah Virus

NiV outbreaks occur regularly in India and Southeast Asia. Surveillance of NiV circulation in endemic countries is therefore crucial for pandemic preparedness but must be performed with adapted diagnostic tests and using the One Health approach.

As indicated on the WHO website, the One Health concept is an integrated, unifying approach to balance and optimize the health of people and animals and the environment [99]. This approach mobilizes multiple sectors, disciplines, and communities at varying levels of society to work together to prevent, predict, detect, and respond to global health threats such as pandemics. One Health involves the public health, veterinary health, and environmental approach and thus can be implemented to improve the surveillance of this pathogen and to minimize the risk of large NiV outbreaks. Indeed, due to the following characteristics, NiV represents a large threat for both animal and human health due to the reservoir of NiV is *Pteropodidae* bats, which are widely distributed in Southeast Asia in densely populated areas; direct transmission to humans could occur via bats or domesticated animals; human-to-human transmission is possible; emergence is often described in densely and connected areas; the mortality rate is high; and there is no effective vaccine or treatment. All these reasons contribute to making NiV a high-risk pathogen priority for the WHO, and using the One Health method seems to be of great importance for NiV preparedness. Since the SARS-CoV-2 pandemic, the One Health approach for infectious diseases has indeed become more significant because prevention and preparedness is the key to better understand the risks of emergence and to limit pandemics.

Due to the mutual dependence between humans and animals, wild animal trade or hunting, deforestation, climate change, intensive agriculture or unlimited urbanization increase contacts between humans and animals and thus also the risk of viral emergence. For example, very recently in Vietnam, the possibility of the emergence of novel viruses with zoonotic potential in bats has been described, due to close human–animal contacts [100]. The spillover surveillance is the key for new viral emergence preparedness, for which it is necessary to conduct risk assessments, to monitor wildlife or farms (pigs, poultry, etc.), and to link all these data to human surveillance data. This approach could be useful to set up appropriate containment measures as quickly as possible [101]. In Asia, animal markets are well-known hotspots for viral emergences and, in particular, the bats that are the reservoir for NiV are sold in many street markets [102]. The analysis of NiV genome in bats using PCR and seroprevalence studies can therefore be useful to follow the circulation of strains, including new ones, and to better understand how human outbreaks begin. Indeed, four factors must be present to initiate an epidemic: the transmission intensity in bats, the dynamic of transmission, the shedding of the virus, and the contact between bats and humans via food consumption [30]. Surveillance of all activities that bring animal reservoirs and humans in contact should be implemented in all at-risk countries for NiV, but in the field, there are multiple organizational, financial, logistical, human, and technical challenges. The successful strategy implemented in Vietnam, based on upstream preparedness in the sanitary surveillance of wildlife, domestic animals, and humans, should serve as an example and favor the identification of the areas with the highest risk of emergence [100]. In the state of Kerala in India, a team composed of public health experts, microbiologists, and other infectious disease experts coordinate a multidisciplinary response team to verify the diagnosis, control outbreaks, and identify the source of infection. They collect and identify bats and ensure the quick availability of adequate PPE and other logistics in the area concerned in order to rapidly limit person-to-person transmission in cases of spillover [103]. To be a successful strategy, like in India, this preparedness must be organized before there is a real emergency associated with NiV, by creating concrete collaborations between human health and animal health institutions, thereby making it possible to detect early signals and then develop a more appropriate and faster public health response at the onset of an emergency [103]. The One Health holistic approach applied to NiV thus appears particularly relevant to prevent outbreaks or limit their consequences.

7. Discussion

Since 2016, the WHO has added NiV on the Blueprint priority disease list. Thanks to the CEPI efforts, at least 13 vaccine candidates are in the preclinical stages of development, but none of them are licensed, and no therapy is available for this infection. As described above, surveillance activities and the rapid identification of positive cases are still essential to prevent or limit the impact of human outbreaks, and diagnosis is a key factor for this success. In endemic countries, the diagnosis of NiV is still hampered by the few BSL4 or BSL3 laboratories available to safely manage suspicious samples, which means additional time is often required to transport the samples to these facilities. Classical diagnosis in these facilities used mainly expensive and not so easy-to-use RT-PCR assays, which must be implemented on specific and expensive machines by personnel trained for this purpose. Rapid point-of-care tests that can be used at a patient's bedside could represent an alternative that do not require the extraction of nucleic acids from the samples. They must be quite easy to use but compatible with biosafety, cheap, and not require to well-trained personnel, rendering the diagnosis more rapid and available in remote areas of these countries. Nevertheless, this kind of assay is often less sensitive than other classical diagnosis assays, limiting their use in diagnoses such as those of RG4 pathogens like Nipah virus. For highly pathogenic viruses, it is fundamental to detect the first cases as soon as possible. It is essential not to miss any positive cases because of a significantly low sensitivity assay because that would have serious consequences. These tests are not yet available for NiV except for some CARDs assays that include a large panel of pathogens and that are described in Section 5.1. They remain expensive, and their sensitivity and specificity must be precisely assessed before being recommended for NiV diagnosis in the field. Nevertheless, once the pathogen responsible for the outbreak is well known, and as it has been observed in the field during the Ebola outbreak in West Africa, rapid tests are very useful to quickly identify some positive cases and allow their isolation as soon as possible to minimize the spread of the virus in the population.

RT-PCR methods therefore remain the technique of choice to diagnose NiV, even though an international standard for NiV is yet to be established, and it is difficult to compare the sensitivity of the different assays. Some RT-PCRs have also been validated only on NiV-M or NiV-B but not on both the strains, impacting their use for diagnosis. Indeed, RT-PCR recommended for diagnosis must recognize different members of the *Henipavirus* genus, in order to not miss the emergence of a new viral strains. Moreover, the sensitivity of molecular tests should still be improved due to the false-negative rate (0.7%) observed by an external quality assessment program in China [104].

However, sensitivity of serological assays in the early phase of infection is too low [72] for use in contact tracing during human outbreaks, and these assays are more often used for surveillance. Serology remains an important tool for the surveillance of viral strain circulation in bats and other animals as long as the assays are interpreted correctly. Cross-reaction with *Henipavirus*-like viruses cannot be excluded, and positive detections should be followed by virological studies to detect NiV shedding in animals [105]. With the same limitations, serology is also used to perform seroprevalence studies in humans. These studies are particularly useful in the field for predicting where the next outbreak may occur and to implement preparedness in advance. Nevertheless, the standardization of these tests is currently lacking, and a precise comparison should be performed before recommending the use of some tests in this context. Finally, recent studies suggest that several bats could act as reservoirs for NiV such as *Rosettus aegypticus*, *Taphozous longimanus*, *Taphozous melanopagon*, *Rhinolophus luctus*, *Chaerophon plicatus*, and *Macroglossus minimus*, indicating that the list of NiV reservoirs can never be considered as definitive [21]. Reliable serological tests will thus be necessary in the future to evaluate the virus circulation in an area that could be larger than initially described.

8. Conclusions

For better NiV surveillance and preparedness, all gaps in diagnostic methods should be filled in the near future. Molecular diagnostics for NiV already includes various tests based on different target genes. It is now crucial to implement comparative studies to assess their sensitivity and specificity in order to make some recommendations for future use. Fewer serological tests are available, but the current lack of standardization prevents them from being recommended for diagnostic use.

Author Contributions: Conceptualization, writing, draft, and reviewing, A.R.G., D.P. and D.L. Data Curation: S.P. and D.L.; Resources: A.R.G.; Writing—Original draft preparation: D.P., D.L., A.R.G. and S.P; supervision: H.R. All authors have read and agreed to the published version of the manuscript.

Funding: This research was funded by European Project SHARP JA (Grant N. 848096), Ricerca Corrente Linea 1, funded by the Italian Ministry of Health.

Institutional Review Board Statement: Not applicable.

Informed Consent Statement: Not applicable.

Data Availability Statement: Not applicable.

Conflicts of Interest: The authors declare no conflict of interest.

References

1. Shoemaker, T.; Choi, M.J. Henipaviruses. In *Centers for Disease Control and Prevention CDC Yellow Book 2020: Health Information for International Travel*; Oxford University Press: Oxford, UK, 2017.
2. Bruno, L.; Nappo, M.A.; Ferrari, L.; Di Lecce, R.; Guarnieri, C.; Cantoni, A.M.; Corradi, A. Nipah Virus Disease: Epidemiological, Clinical, Diagnostic and Legislative Aspects of This Unpredictable Emerging Zoonosis. *Animals* **2022**, *13*, 159. [CrossRef]
3. Kummer, S.; Kranz, D.-C. Henipaviruses—A Constant Threat to Livestock and Humans. *PLoS Negl. Trop. Dis.* **2022**, *16*, e0010157. [CrossRef]
4. Sharma, V.; Kaushik, S.; Kumar, R.; Yadav, J.P.; Kaushik, S. Emerging Trends of Nipah Virus: A Review. *Rev. Med. Virol.* **2019**, *29*, e2010. [CrossRef]
5. Uchida, S.; Horie, R.; Sato, H.; Kai, C.; Yoneda, M. Possible Role of the Nipah Virus V Protein in the Regulation of the Interferon Beta Induction by Interacting with UBX Domain-Containing Protein1. *Sci. Rep.* **2018**, *8*, 7682. [CrossRef]
6. Chua, K.B.; Bellini, W.J.; Rota, P.A.; Harcourt, B.H.; Tamin, A.; Lam, S.K.; Ksiazek, T.G.; Rollin, P.E.; Zaki, S.R.; Shieh, W.-J.; et al. Nipah Virus: A Recently Emergent Deadly Paramyxovirus. *Science* **2000**, *288*, 1432–1435. [CrossRef]
7. Guillaume, V.; Aslan, H.; Ainouze, M.; Guerbois, M.; Fabian Wild, T.; Buckland, R.; Langedijk, J.P.M. Evidence of a Potential Receptor-Binding Site on the Nipah Virus G Protein (NiV-G): Identification of Globular Head Residues with a Role in Fusion Promotion and Their Localization on an NiV-G Structural Model. *J. Virol.* **2006**, *80*, 7546–7554. [CrossRef] [PubMed]
8. Bonaparte, M.I.; Dimitrov, A.S.; Bossart, K.N.; Crameri, G.; Mungall, B.A.; Bishop, K.A.; Choudhry, V.; Dimitrov, D.S.; Wang, L.-F.; Eaton, B.T.; et al. Ephrin-B2 Ligand Is a Functional Receptor for Hendra Virus and Nipah Virus. *Proc. Natl. Acad. Sci. USA* **2005**, *102*, 10652–10657. [CrossRef] [PubMed]
9. Negrete, O.A.; Levroney, E.L.; Aguilar, H.C.; Bertolotti-Ciarlet, A.; Nazarian, R.; Tajyar, S.; Lee, B. EphrinB2 Is the Entry Receptor for Nipah Virus, an Emergent Deadly Paramyxovirus. *Nature* **2005**, *436*, 401–405. [CrossRef] [PubMed]
10. Aguilar, H.C.; Matreyek, K.A.; Filone, C.M.; Hashimi, S.T.; Levroney, E.L.; Negrete, O.A.; Bertolotti-Ciarlet, A.; Choi, D.Y.; McHardy, I.; Fulcher, J.A.; et al. N-Glycans on Nipah Virus Fusion Protein Protect against Neutralization but Reduce Membrane Fusion and Viral Entry. *J. Virol.* **2006**, *80*, 4878–4889. [CrossRef]
11. Patch, J.R.; Crameri, G.; Wang, L.-F.; Eaton, B.T.; Broder, C.C. Quantitative Analysis of Nipah Virus Proteins Released as Virus-like Particles Reveals Central Role for the Matrix Protein. *Virol. J.* **2007**, *4*, 1. [CrossRef] [PubMed]
12. Paton, N.I.; Leo, Y.S.; Zaki, S.R.; Auchus, A.P.; Lee, K.E.; Ling, A.E.; Chew, S.K.; Ang, B.; Rollin, P.E.; Umapathi, T.; et al. Outbreak of Nipah-Virus Infection among Abattoir Workers in Singapore. *Lancet* **1999**, *354*, 1253–1256. [CrossRef]

13. Parashar, U.D.; Sunn, L.M.; Ong, F.; Mounts, A.W.; Arif, M.T.; Ksiazek, T.G.; Kamaluddin, M.A.; Mustafa, A.N.; Kaur, H.; Ding, L.M.; et al. Case-Control Study of Risk Factors for Human Infection with a New Zoonotic Paramyxovirus, Nipah Virus, during a 1998–1999 Outbreak of Severe Encephalitis in Malaysia. *J. Infect. Dis.* **2000**, *181*, 1755–1759. [CrossRef]
14. Rahman, S.A.; Hassan, S.S.; Olival, K.J.; Mohamed, M.; Chang, L.-Y.; Hassan, L.; Saad, N.M.; Shohaimi, S.A.; Mamat, Z.C.; Naim, M.S.; et al. Characterization of Nipah Virus from Naturally Infected Pteropus vampyrus Bats, Malaysia. *Emerg. Infect. Dis.* **2010**, *16*, 1990–1993. [CrossRef]
15. Anderson, D.E.; Islam, A.; Crameri, G.; Todd, S.; Islam, A.; Khan, S.U.; Foord, A.; Rahman, M.Z.; Mendenhall, I.H.; Luby, S.P.; et al. Isolation and Full-Genome Characterization of Nipah Viruses from Bats, Bangladesh. *Emerg. Infect. Dis.* **2019**, *25*, 166–170. [CrossRef] [PubMed]
16. Yob, J.M.; Field, H.; Rashdi, A.M.; Morrissy, C.; Van Der Heide, B.; Rota, P.; Bin Adzhar, A.; White, J.; Daniels, P.; Jamaluddin, A.; et al. Nipah Virus Infection in Bats (Order Chiroptera) in Peninsular Malaysia. *Emerg. Infect. Dis.* **2001**, *7*, 439–441. [CrossRef] [PubMed]
17. Yadav, P.D.; Towner, J.S.; Raut, C.G.; Shete, A.M.; Nichol, S.T.; Mourya, D.T.; Mishra, A.C. Detection of Nipah Virus RNA in Fruit Bat (Pteropus giganteus) from India. *Am. J. Trop. Med. Hyg.* **2012**, *87*, 576–578. [CrossRef] [PubMed]
18. Halpin, K.; Hyatt, A.D.; Fogarty, R.; Middleton, D.; Bingham, J.; Epstein, J.H.; Rahman, S.A.; Hughes, T.; Smith, C.; Field, H.E.; et al. Pteropid Bats Are Confirmed as the Reservoir Hosts of Henipaviruses: A Comprehensive Experimental Study of Virus Transmission. *Am. J. Trop. Med. Hyg.* **2011**, *85*, 946–951. [CrossRef]
19. Chow, V.T.K.; Tambyah, P.A.; Yeo, W.M.; Phoon, M.C.; Howe, J. Diagnosis of Nipah Virus Encephalitis by Electron Microscopy of Cerebrospinal Fluid. *J. Clin. Virol.* **2000**, *19*, 143–147. [CrossRef]
20. Ambat, A.S.; Zubair, S.M.; Prasad, N.; Pundir, P.; Rajwar, E.; Patil, D.S.; Mangad, P. Nipah Virus: A Review on Epidemiological Characteristics and Outbreaks to Inform Public Health Decision Making. *J. Infect. Public Health* **2019**, *12*, 634–639. [CrossRef]
21. Plowright, R.K.; Becker, D.J.; Crowley, D.E.; Washburne, A.D.; Huang, T.; Nameer, P.O.; Gurley, E.S.; Han, B.A. Prioritizing Surveillance of Nipah Virus in India. *PLoS Negl. Trop. Dis.* **2019**, *13*, e0007393. [CrossRef]
22. Yadav, P.D.; Shete, A.M.; Kumar, G.A.; Sarkale, P.; Sahay, R.R.; Radhakrishnan, C.; Lakra, R.; Pardeshi, P.; Gupta, N.; Gangakhedkar, R.R.; et al. Nipah Virus Sequences from Humans and Bats during Nipah Outbreak, Kerala, India, 2018. *Emerg. Infect. Dis.* **2019**, *25*, 1003–1006. [CrossRef] [PubMed]
23. Mhod Nor, M.N.; Gan, C.H.; Ong, B.L. Nipah Virus Infection of Pigs in Peninsular Malaysia. *Rev. Sci. Tech. OIE* **2000**, *19*, 160–165. [CrossRef] [PubMed]
24. Chua, K.B. Nipah Virus Outbreak in Malaysia. *J. Clin. Virol.* **2003**, *26*, 265–275. [CrossRef]
25. Enserink, M. New Virus Fingered in Malaysian Epidemic. *Science* **1999**, *284*, 407–410. [CrossRef]
26. Centers for Disease Control and Prevention (CDC). Update: Outbreak of Nipah Virus—Malaysia and Singapore, 1999. *MMWR Morb. Mortal. Wkly. Rep.* **1999**, *48*, 335–337.
27. Chowdhury, S.; Khan, S.U.; Crameri, G.; Epstein, J.H.; Broder, C.C.; Islam, A.; Peel, A.J.; Barr, J.; Daszak, P.; Wang, L.-F.; et al. Serological Evidence of Henipavirus Exposure in Cattle, Goats and Pigs in Bangladesh. *PLoS Negl. Trop. Dis.* **2014**, *8*, e3302. [CrossRef]
28. Hegde, S.T.; Sazzad, H.M.S.; Hossain, M.J.; Alam, M.-U.; Kenah, E.; Daszak, P.; Rollin, P.; Rahman, M.; Luby, S.P.; Gurley, E.S. Investigating Rare Risk Factors for Nipah Virus in Bangladesh: 2001–2012. *EcoHealth* **2016**, *13*, 720–728. [CrossRef]
29. Rahman, M.; Chakraborty, A. Nipah Virus Outbreaks in Bangladesh: A Deadly Infectious Disease. *WHO South-East Asia J. Public Health* **2012**, *1*, 208–212. [CrossRef]
30. Epstein, J.H.; Anthony, S.J.; Islam, A.; Kilpatrick, A.M.; Ali Khan, S.; Balkey, M.D.; Ross, N.; Smith, I.; Zambrana-Torrelio, C.; Tao, Y.; et al. Nipah Virus Dynamics in Bats and Implications for Spillover to Humans. *Proc. Natl. Acad. Sci. USA* **2020**, *117*, 29190–29201. [CrossRef]
31. Ching, P.K.G.; De Los Reyes, V.C.; Sucaldito, M.N.; Tayag, E.; Columna-Vingno, A.B.; Malbas, F.F.; Bolo, G.C.; Sejvar, J.J.; Eagles, D.; Playford, G.; et al. Outbreak of Henipavirus Infection, Philippines, 2014. *Emerg. Infect. Dis.* **2015**, *21*, 328–331. [CrossRef]
32. Chua, K.B.; Goh, K.J.; Wong, K.T.; Kamarulzaman, A.; Tan, P.S.K.; Ksiazek, T.G.; Zaki, S.R.; Paul, G.; Lam, S.K.; Tan, C.T. Fatal Encephalitis Due to Nipah Virus among Pig-Farmers in Malaysia. *Lancet* **1999**, *354*, 1257–1259. [CrossRef] [PubMed]
33. Gurley, E.S.; Spiropoulou, C.F.; De Wit, E. Twenty Years of Nipah Virus Research: Where Do We Go From Here? *J. Infect. Dis.* **2020**, *221*, S359–S362. [CrossRef] [PubMed]
34. Arankalle, V.A.; Bandyopadhyay, B.T.; Ramdasi, A.Y.; Jadi, R.; Patil, D.R.; Rahman, M.; Majumdar, M.; Banerjee, P.S.; Hati, A.K.; Goswami, R.P.; et al. Genomic Characterization of Nipah Virus, West Bengal, India. *Emerg. Infect. Dis.* **2011**, *17*, 907–909. [CrossRef] [PubMed]
35. Kumar, C.P.G.; Sugunan, A.P.; Yadav, P.; Kurup, K.K.; Aarathee, R.; Manickam, P.; Bhatnagar, T.; Radhakrishnan, C.; Thomas, B.; Kumar, A.; et al. Infections among Contacts of Patients with Nipah Virus, India. *Emerg. Infect. Dis.* **2019**, *25*, 1007–1010. [CrossRef] [PubMed]
36. Hossain, M.J.; Gurley, E.S.; Montgomery, J.M.; Bell, M.; Carroll, D.S.; Hsu, V.P.; Formenty, P.; Croisier, A.; Bertherat, E.; Faiz, M.A.; et al. Clinical Presentation of Nipah Virus Infection in Bangladesh. *Clin. Infect. Dis.* **2008**, *46*, 977–984. [CrossRef] [PubMed]
37. Playford, E.G.; McCall, B.; Smith, G.; Slinko, V.; Allen, G.; Smith, I.; Moore, F.; Taylor, C.; Kung, Y.-H.; Field, H. Human Hendra Virus Encephalitis Associated with Equine Outbreak, Australia, 2008. *Emerg. Infect. Dis.* **2010**, *16*, 219–223. [CrossRef]

38. Arunkumar, G.; Chandni, R.; Mourya, D.T.; Singh, S.K.; Sadanandan, R.; Sudan, P.; Bhargava, B.; Nipah Investigators People and Health Study Group; Gangakhedkar, R.R.; Gupta, N.; et al. Outbreak Investigation of Nipah Virus Disease in Kerala, India, 2018. *J. Infect. Dis.* **2019**, *219*, 1867–1878. [CrossRef] [PubMed]

39. Goh, K.J.; Tan, C.T.; Chew, N.K.; Tan, P.S.K.; Kamarulzaman, A.; Sarji, S.A.; Wong, K.T.; Abdullah, B.J.J.; Chua, K.B.; Lam, S.K. Clinical Features of Nipah Virus Encephalitis among Pig Farmers in Malaysia. *N. Engl. J. Med.* **2000**, *342*, 1229–1235. [CrossRef]

40. Aditi; Shariff, M. Aditi; Shariff, M. Nipah Virus Infection: A Review. *Epidemiol. Infect.* **2019**, *147*, e95. [CrossRef]

41. World Health Organization. Nipah Virus. Available online: https://www.who.int/news-room/fact-sheets/detail/nipah-virus (accessed on 8 May 2023).

42. Hauser, N.; Gushiken, A.C.; Narayanan, S.; Kottilil, S.; Chua, J.V. Evolution of Nipah Virus Infection: Past, Present, and Future Considerations. *Trop. Med. Infect. Dis.* **2021**, *6*, 24. [CrossRef]

43. Ang, B.S.P.; Lim, T.C.C.; Wang, L. Nipah Virus Infection. *J. Clin. Microbiol.* **2018**, *56*, e01875-17. [CrossRef] [PubMed]

44. Pallivalappil, B.; Ali, A.; Thulaseedharan, N.; Karadan, U.; Chellenton, J.; Dipu, K.; Anoop Kumar, A.; Sajeeth Kumar, K.; Rajagopal, T.; Suraj, K.; et al. Dissecting an Outbreak: A Clinico-Epidemiological Study of Nipah Virus Infection in Kerala, India, 2018. *J. Glob. Infect. Dis.* **2020**, *12*, 21–27. [CrossRef] [PubMed]

45. Wong, K.T.; Shieh, W.-J.; Kumar, S.; Norain, K.; Abdullah, W.; Guarner, J.; Goldsmith, C.S.; Chua, K.B.; Lam, S.K.; Tan, C.T.; et al. Nipah Virus Infection: Pathology and Pathogenesis of an Emerging Paramyxoviral Zoonosis. *Am. J. Pathol.* **2002**, *161*, 2153–2167. [CrossRef] [PubMed]

46. Singh, R.K.; Dhama, K.; Chakraborty, S.; Tiwari, R.; Natesan, S.; Khandia, R.; Munjal, A.; Vora, K.S.; Latheef, S.K.; Karthik, K.; et al. Nipah Virus: Epidemiology, Pathology, Immunobiology and Advances in Diagnosis, Vaccine Designing and Control Strategies—A Comprehensive Review. *Vet. Q.* **2019**, *39*, 26–55. [CrossRef]

47. Siva, S.R.; Chong, H.T.; Tan, C.T. Ten Year Clinical and Serological Outcomes of Nipah Virus Infection. *Neurol. Asia* **2009**, *14*, 53–58.

48. Sejvar, J.J.; Hossain, J.; Saha, S.K.; Gurley, E.S.; Banu, S.; Hamadani, J.D.; Faiz, M.A.; Siddiqui, F.M.; Mohammad, Q.D.; Mollah, A.H.; et al. Long-Term Neurological and Functional Outcome in Nipah Virus Infection. *Ann. Neurol.* **2007**, *62*, 235–242. [CrossRef]

49. Thakur, N.; Bailey, D. Advances in Diagnostics, Vaccines and Therapeutics for Nipah Virus. *Microbes Infect.* **2019**, *21*, 278–286. [CrossRef]

50. Tan, K.S.; Tan, C.T.; Goh, K.J. Epidemiological Aspects of Nipah Virus Infection. *Neurol. J. Southeast Asia* **1999**, *4*, 77–81.

51. Chan, K.P.; Rollin, P.E.; Ksiazek, T.G.; Leo, Y.S.; Goh, K.T.; Paton, N.I.; Sng, E.H.; Ling, A.E. A Survey of Nipah Virus Infection among Various Risk Groups in Singapore. *Epidemiol. Infect.* **2002**, *128*, 93–98. [CrossRef]

52. Hsu, V.P.; Hossain, M.J.; Parashar, U.D.; Ali, M.M.; Ksiazek, T.G.; Kuzmin, I.; Niezgoda, M.; Rupprecht, C.; Bresee, J.; Breiman, R.F. Nipah Virus Encephalitis Reemergence, Bangladesh. *Emerg. Infect. Dis.* **2004**, *10*, 2082–2087. [CrossRef]

53. Banerjee, S.; Gupta, N.; Kodan, P.; Mittal, A.; Ray, Y.; Nischal, N.; Soneja, M.; Biswas, A.; Wig, N. Nipah Virus Disease: A Rare and Intractable Disease. *Intractable Rare Dis. Res.* **2019**, *8*, 1–8. [CrossRef] [PubMed]

54. Wong, K.T.; Robertson, T.; Ong, B.B.; Chong, J.W.; Yaiw, K.C.; Wang, L.F.; Ansford, A.J.; Tannenberg, A. Human Hendra Virus Infection Causes Acute and Relapsing Encephalitis. *Neuropathol. Appl. Neurobiol.* **2009**, *35*, 296–305. [CrossRef] [PubMed]

55. Chong, H.T.; Kunjapan, S.R.; Thayaparan, T.; Geok Tong, J.M.; Petharunam, V.; Jusoh, M.R.; Tan, C.T. Nipah Encephalitis Outbreak in Malaysia, Clinical Features in Patients from Seremban. *Can. J. Neurol. Sci.* **2002**, *29*, 83–87. [CrossRef]

56. Middleton, D.J.; Westbury, H.A.; Morrissy, C.J.; Van Der Heide, B.M.; Russell, G.M.; Braun, M.A.; Hyatt, A.D. Experimental Nipah Virus Infection in Pigs and Cats. *J. Comp. Pathol.* **2002**, *126*, 124–136. [CrossRef] [PubMed]

57. Centers for Disease Control and Prevention (CDC). Nipah Virus (NiV). Available online: https://www.cdc.gov/vhf/nipah/index.html (accessed on 12 June 2023).

58. Centers for Disease Control and Prevention. Nipah Virus (NiV): Diagnosis. Available online: https://www.cdc.gov/vhf/nipah/diagnosis/index.html (accessed on 10 June 2023).

59. Tong, S.; Chern, S.-W.W.; Li, Y.; Pallansch, M.A.; Anderson, L.J. Sensitive and Broadly Reactive Reverse Transcription-PCR Assays To Detect Novel Paramyxoviruses. *J. Clin. Microbiol.* **2008**, *46*, 2652–2658. [CrossRef]

60. Wacharapluesadee, S.; Lumlertdacha, B.; Boongird, K.; Wanghongsa, S.; Chanhome, L.; Rollin, P.; Stockton, P.; Rupprecht, C.E.; Ksiazek, T.G.; Hemachudha, T. Bat Nipah Virus, Thailand. *Emerg. Infect. Dis.* **2005**, *11*, 1949–1951. [CrossRef]

61. Wacharapluesadee, S.; Boongird, K.; Wanghongsa, S.; Phumesin, P.; Hemachudha, T. Drinking Bat Blood May Be Hazardous to Your Health. *Clin. Infect. Dis.* **2006**, *43*, 269. [CrossRef]

62. Guillaume, V.; Lefeuvre, A.; Faure, C.; Marianneau, P.; Buckland, R.; Lam, S.K.; Wild, T.F.; Deubel, V. Specific Detection of Nipah Virus Using Real-Time RT-PCR (TaqMan). *J. Virol. Methods* **2004**, *120*, 229–237. [CrossRef]

63. Feldman, K.S.; Foord, A.; Heine, H.G.; Smith, I.L.; Boyd, V.; Marsh, G.A.; Wood, J.L.N.; Cunningham, A.A.; Wang, L.-F. Design and Evaluation of Consensus PCR Assays for Henipaviruses. *J. Virol. Methods* **2009**, *161*, 52–57. [CrossRef]

64. Chang, L.-Y.; Mohd Ali, A.; Hassan, S.S.; AbuBakar, S. Quantitative Estimation of Nipah Virus Replication Kinetics in Vitro. *Virol. J.* **2006**, *3*, 47. [CrossRef]

65. Jensen, K.S.; Adams, R.; Bennett, R.S.; Bernbaum, J.; Jahrling, P.B.; Holbrook, M.R. Development of a Novel Real-Time Polymerase Chain Reaction Assay for the Quantitative Detection of Nipah Virus Replicative Viral RNA. *PLoS ONE* **2018**, *13*, e0199534. [CrossRef] [PubMed]

66. Onyango, C.O.; Loparev, V.; Lidechi, S.; Bhullar, V.; Schmid, D.S.; Radford, K.; Lo, M.K.; Rota, P.; Johnson, B.W.; Munoz, J.; et al. Evaluation of a TaqMan Array Card for Detection of Central Nervous System Infections. *J. Clin. Microbiol.* **2017**, *55*, 2035–2044. [CrossRef] [PubMed]

67. Liu, J.; Ochieng, C.; Wiersma, S.; Ströher, U.; Towner, J.S.; Whitmer, S.; Nichol, S.T.; Moore, C.C.; Kersh, G.J.; Kato, C.; et al. Development of a TaqMan Array Card for Acute-Febrile-Illness Outbreak Investigation and Surveillance of Emerging Pathogens, Including Ebola Virus. *J. Clin. Microbiol.* **2016**, *54*, 49–58. [CrossRef]

68. Ma, L.; Chen, Z.; Guan, W.; Chen, Q.; Liu, D. Rapid and Specific Detection of All Known Nipah Virus Strains' Sequences With Reverse Transcription-Loop-Mediated Isothermal Amplification. *Front. Microbiol.* **2019**, *10*, 418. [CrossRef]

69. Pollak, N.M.; Olsson, M.; Marsh, G.A.; Macdonald, J.; McMillan, D. Evaluation of Three Rapid Low-Resource Molecular Tests for Nipah Virus. *Front. Microbiol.* **2023**, *13*, 1101914. [CrossRef]

70. Wylezich, C.; Calvelage, S.; Schlottau, K.; Ziegler, U.; Pohlmann, A.; Höper, D.; Beer, M. Next-Generation Diagnostics: Virus Capture Facilitates a Sensitive Viral Diagnosis for Epizootic and Zoonotic Pathogens Including SARS-CoV-2. *Microbiome* **2021**, *9*, 51. [CrossRef]

71. Arunkumar, G.; Devadiga, S.; McElroy, A.K.; Prabhu, S.; Sheik, S.; Abdulmajeed, J.; Robin, S.; Sushama, A.; Jayaram, A.; Nittur, S.; et al. Adaptive Immune Responses in Humans During Nipah Virus Acute and Convalescent Phases of Infection. *Clin. Infect. Dis.* **2019**, *69*, 1752–1756. [CrossRef]

72. Ramasundram, V.; Tan, C.T.; Chua, K.B.; Chong, H.T.; Goh, K.J.; Chew, N.K.; Tan, K.S.; Thayaparan, T.; Kunjapan, S.R.; Petharunam, V.; et al. Kinetics of IgM and IgG Seroconversion in Nipah Virus Infection. *Neurol. J. Southeast Asia* **2000**, *5*, 23–28.

73. Shete, A.; Radhakrishnan, C.; Pardeshi, P.; Yadav, P.; Jain, R.; Sahay, R.; Sugunan, A. Antibody Response in Symptomatic & Asymptomatic Nipah Virus Cases from Kerala, India. *Indian J. Med. Res.* **2021**, *154*, 533–535. [CrossRef]

74. Kulkarni, D.D.; Tosh, C.; Venkatesh, G.; Senthil Kumar, D. Nipah Virus Infection: Current Scenario. *Indian J. Virol.* **2013**, *24*, 398–408. [CrossRef]

75. Abdullah, S.; Tan, C.T. Henipavirus Encephalitis. In *Handbook of Clinical Neurology*; Elsevier: Amsterdam, The Netherlands, 2014; Volume 123, pp. 663–670. ISBN 978-0-444-53488-0.

76. Satterfield, B.A.; Dawes, B.E.; Milligan, G.N. Status of Vaccine Research and Development of Vaccines for Nipah Virus. *Vaccine* **2016**, *34*, 2971–2975. [CrossRef] [PubMed]

77. Chattu, V.; Kumar, R.; Kumary, S.; Kajal, F.; David, J. Nipah Virus Epidemic in Southern India and Emphasizing "One Health" Approach to Ensure Global Health Security. *J. Fam. Med. Prim. Care* **2018**, *7*, 275–283. [CrossRef] [PubMed]

78. Daniels, P.; Ksiazek, T.; Eaton, B.T. Laboratory Diagnosis of Nipah and Hendra Virus Infections. *Microbes Infect.* **2001**, *3*, 289–295. [CrossRef] [PubMed]

79. Field, H.; Young, P.; Yob, J.M.; Mills, J.; Hall, L.; Mackenzie, J. The Natural History of Hendra and Nipah Viruses. *Microbes Infect.* **2001**, *3*, 307–314. [CrossRef]

80. Harcourt, B.H.; Tamin, A.; Ksiazek, T.G.; Rollin, P.E.; Anderson, L.J.; Bellini, W.J.; Rota, P.A. Molecular Characterization of Nipah Virus, a Newly Emergent Paramyxovirus. *Virology* **2000**, *271*, 334–349. [CrossRef]

81. Yu, F.; Khairullah, N.S.; Inoue, S.; Balasubramaniam, V.; Berendam, S.J.; Teh, L.K.; Ibrahim, N.S.W.; Abdul Rahman, S.; Hassan, S.S.; Hasebe, F.; et al. Serodiagnosis Using Recombinant Nipah Virus Nucleocapsid Protein Expressed in Escherichia coli. *J. Clin. Microbiol.* **2006**, *44*, 3134–3138. [CrossRef]

82. Kulkarni, D.D.; Venkatesh, G.; Tosh, C.; Patel, P.; Mashoria, A.; Gupta, V.; Gupta, S.; Senthilkumar, D. Development and Evaluation of Recombinant Nucleocapsid Protein Based Diagnostic ELISA for Detection of Nipah Virus Infection in Pigs. *J. Immunoassay Immunochem.* **2016**, *37*, 154–166. [CrossRef]

83. Chiang, C.-F.; Lo, M.K.; Rota, P.A.; Spiropoulou, C.F.; Rollin, P.E. Use of Monoclonal Antibodies against Hendra and Nipah Viruses in an Antigen Capture ELISA. *Virol. J.* **2010**, *7*, 115. [CrossRef]

84. Ksiazek, T.G.; Rollin, P.E.; Jahrling, P.B.; Johnson, E.; Dalgard, D.W.; Peters, C.J. Enzyme Immunosorbent Assay for Ebola Virus Antigens in Tissues of Infected Primates. *J. Clin. Microbiol.* **1992**, *30*, 947–950. [CrossRef]

85. Saijo, M.; Georges-Courbot, M.-C.; Fukushi, S.; Mizutani, T.; Philippe, M.; Georges, A.-J.; Kurane, I.; Morikawa, S. Marburgvirus Nucleoprotein-Capture Enzyme-Linked Immunosorbent Assay Using Monoclonal Antibodies to Recombinant Nucleoprotein: Detection of Authentic Marburgvirus. *Jpn. J. Infect. Dis.* **2006**, *59*, 323–325.

86. Warnes, A.; Fooks, A.R.; Stephenson, J.R. Design and Preparation of Recombinant Antigens as Diagnostic Reagents in Solid-Phase Immunosorbent Assays. *Methods Mol. Med.* **2004**, *94*, 373–391. [CrossRef] [PubMed]

87. Chen, J.-M.; Yu, M.; Morrissy, C.; Zhao, Y.-G.; Meehan, G.; Sun, Y.-X.; Wang, Q.-H.; Zhang, W.; Wang, L.-F.; Wang, Z.-L. A Comparative Indirect ELISA for the Detection of Henipavirus Antibodies Based on a Recombinant Nucleocapsid Protein Expressed in Escherichia Coli. *J. Virol. Methods* **2006**, *136*, 273–276. [CrossRef] [PubMed]

88. Kashiwazaki, Y.; Na, Y.N.; Tanimura, N.; Imada, T. A Solid-Phase Blocking ELISA for Detection of Antibodies to Nipah Virus. *J. Virol. Methods* **2004**, *121*, 259–261. [CrossRef]

89. McNabb, L.; Barr, J.; Crameri, G.; Juzva, S.; Riddell, S.; Colling, A.; Boyd, V.; Broder, C.; Wang, L.-F.; Lunt, R. Henipavirus Microsphere Immuno-Assays for Detection of Antibodies against Hendra Virus. *J. Virol. Methods* **2014**, *200*, 22–28. [CrossRef] [PubMed]

90. Foord, A.J.; White, J.R.; Colling, A.; Heine, H.G. Microsphere Suspension Array Assays for Detection and Differentiation of Hendra and Nipah Viruses. *BioMed Res. Int.* **2013**, *2013*, 289295. [CrossRef] [PubMed]

91. Wang, L.-F.; Daniels, P. Diagnosis of Henipavirus Infection: Current Capabilities and Future Directions. In *Henipavirus*; Lee, B., Rota, P.A., Eds.; Current Topics in Microbiology and Immunology; Springer: Berlin/Heidelberg, Germany, 2012; Volume 359, pp. 179–196. ISBN 978-3-642-29818-9.
92. Bae, S.E.; Kim, S.S.; Moon, S.T.; Cho, Y.D.; Lee, H.; Lee, J.-Y.; Shin, H.Y.; Lee, H.-J.; Kim, Y.B. Construction of the Safe Neutralizing Assay System Using Pseudotyped Nipah Virus and G Protein-Specific Monoclonal Antibody. *Biochem. Biophys. Res. Commun.* **2019**, *513*, 781–786. [CrossRef]
93. Kaku, Y.; Noguchi, A.; Marsh, G.A.; McEachern, J.A.; Okutani, A.; Hotta, K.; Bazartseren, B.; Fukushi, S.; Broder, C.C.; Yamada, A.; et al. A Neutralization Test for Specific Detection of Nipah Virus Antibodies Using Pseudotyped Vesicular Stomatitis Virus Expressing Green Fluorescent Protein. *J. Virol. Methods* **2009**, *160*, 7–13. [CrossRef]
94. Khetawat, D.; Broder, C.C. A Functional Henipavirus Envelope Glycoprotein Pseudotyped Lentivirus Assay System. *Virol. J.* **2010**, *7*, 312. [CrossRef]
95. Kaku, Y.; Noguchi, A.; Marsh, G.A.; Barr, J.A.; Okutani, A.; Hotta, K.; Bazartseren, B.; Fukushi, S.; Broder, C.C.; Yamada, A.; et al. Second Generation of Pseudotype-Based Serum Neutralization Assay for Nipah Virus Antibodies: Sensitive and High-Throughput Analysis Utilizing Secreted Alkaline Phosphatase. *J. Virol. Methods* **2012**, *179*, 226–232. [CrossRef]
96. Tamin, A.; Harcourt, B.H.; Lo, M.K.; Roth, J.A.; Wolf, M.C.; Lee, B.; Weingartl, H.; Audonnet, J.-C.; Bellini, W.J.; Rota, P.A. Development of a Neutralization Assay for Nipah Virus Using Pseudotype Particles. *J. Virol. Methods* **2009**, *160*, 1–6. [CrossRef]
97. Hyatt, A.D.; Zaki, S.R.; Goldsmith, C.S.; Wise, T.G.; Hengstberger, S.G. Ultrastructure of Hendra Virus and Nipah Virus within Cultured Cells and Host Animals. *Microbes Infect.* **2001**, *3*, 297–306. [CrossRef]
98. Chua, K.B.; Lam, S.K.; Tan, C.T.; Hooi, P.S.; Goh, K.J.; Chew, N.K.; Tan, K.S.; Kamarulzaman, A.; Wong, K.T. High Mortality in Nipah Encephalitis Is Associated with Presence of Virus in Cerebrospinal Fluid. *Ann. Neurol.* **2000**, *48*, 802–805. [CrossRef]
99. World Health Organization. One Health. Available online: https://www.who.int/health-topics/one-health#tab=tab_1 (accessed on 10 July 2023).
100. Latinne, A.; Nga, N.T.T.; Long, N.V.; Ngoc, P.T.B.; Thuy, H.B.; PREDICT Consortium; Long, N.V.; Long, P.T.; Phuong, N.T.; Quang, L.T.V.; et al. One Health Surveillance Highlights Circulation of Viruses with Zoonotic Potential in Bats, Pigs, and Humans in Viet Nam. *Viruses* **2023**, *15*, 790. [CrossRef]
101. Charlier, J.; Barkema, H.W.; Becher, P.; De Benedictis, P.; Hansson, I.; Hennig-Pauka, I.; La Ragione, R.; Larsen, L.E.; Madoroba, E.; Maes, D.; et al. Disease Control Tools to Secure Animal and Public Health in a Densely Populated World. *Lancet Planet. Health* **2022**, *6*, e812–e824. [CrossRef] [PubMed]
102. Morcatty, T.Q.; Pereyra, P.E.R.; Ardiansyah, A.; Imron, M.A.; Hedger, K.; Campera, M.; Nekaris, K.A.-I.; Nijman, V. Risk of Viral Infectious Diseases from Live Bats, Primates, Rodents and Carnivores for Sale in Indonesian Wildlife Markets. *Viruses* **2022**, *14*, 2756. [CrossRef]
103. Singhai, M.; Jain, R.; Jain, S.; Bala, M.; Singh, S.; Goyal, R. Nipah Virus Disease: Recent Perspective and One Health Approach. *Ann. Glob. Health* **2021**, *87*, 102. [CrossRef]
104. Zhang, R.; Tan, P.; Feng, L.; Li, R.; Yang, J.; Zhang, R.; Li, J. External Quality Assessment of Molecular Testing of 9 Viral Encephalitis-Related Viruses in China. *Virus Res.* **2021**, *306*, 198598. [CrossRef] [PubMed]
105. Gilbert, A.T.; Fooks, A.R.; Hayman, D.T.S.; Horton, D.L.; Müller, T.; Plowright, R.; Peel, A.J.; Bowen, R.; Wood, J.L.N.; Mills, J.; et al. Deciphering Serology to Understand the Ecology of Infectious Diseases in Wildlife. *EcoHealth* **2013**, *10*, 298–313. [CrossRef] [PubMed]

 *viruses*

MDPI

Article

Identification of the Tembusu Virus in Mosquitoes in Northern Thailand

Rodolphe Hamel [1,2,3,*,†], Ronald Enrique Morales Vargas [4,5,†], Dora Murielle Rajonhson [4], Atsushi Yamanaka [6], Jiraporn Jaroenpool [7,8], Sineewanlaya Wichit [2,3], Dorothée Missé [1], Anamika Kritiyakan [9], Kittipong Chaisiri [10], Serge Morand [1,9,‡] and Julien Pompon [1,‡]

[1] MIVEGEC, Université de Montpellier, IRD, CNRS, 34394 Montpellier, France; dorothee.misse@ird.fr (D.M.); serge.morand@umontpellier.fr (S.M.); julien.pompon@ird.fr (J.P.)

[2] Department of Clinical Microbiology and Applied Technology, Faculty of Medical Technology, Mahidol University, Nakhon Pathom 73170, Thailand; sineewanlaya.wic@mahidol.ac.th

[3] Viral Vector Joint Unit, Join Laboratory, Mahidol University, Nakhon Pathom 73170, Thailand

[4] Department of Medical Entomology, Faculty of Tropical Medicine, Mahidol University, Bangkok 10400, Thailand; ronald.mor@mahidol.ac.th (R.E.M.V.); dorarajonhson@gmail.com (D.M.R.)

[5] Department of Pharmacology, Faculty of Science, Mahidol University, Bangkok 10400, Thailand

[6] Research Institute for Microbial Diseases, Osaka University, Osaka 565-0871, Japan; knmya@biken.osaka-u.ac.jp

[7] Department of Medical Technology, School of Allied Health Sciences, Walailak University, Nakhon Si Thammarat 80160, Thailand; jjirapor@mail.wu.ac.th

[8] Excellent Center for Dengue and Community Public Health, Walailak University, Nakhon Si Thammarat 80160, Thailand

[9] Faculty of Veterinary Technology, Kasetsart University, Bangkok 10900, Thailand; anamikanose.k@gmail.com

[10] Department of Helminthology, Faculty of Tropical Medicine, Mahidol University, Bangkok 10900, Thailand; kittipong.cha@mahidol.ac.th

* Correspondence: rodolphe.hamel@ird.fr

† These authors contributed equally to this work.

‡ These authors contributed equally to this work.

Citation: Hamel, R.; Vargas, R.E.M.; Rajonhson, D.M.; Yamanaka, A.; Jaroenpool, J.; Wichit, S.; Missé, D.; Kritiyakan, A.; Chaisiri, K.; Morand, S.; et al. Identification of the Tembusu Virus in Mosquitoes in Northern Thailand. *Viruses* **2023**, *15*, 1447. https://doi.org/10.3390/v15071447

Academic Editor: Norbert Nowotny

Received: 26 May 2023
Revised: 20 June 2023
Accepted: 23 June 2023
Published: 27 June 2023

Abstract: Among emerging zoonotic pathogens, mosquito-borne viruses (MBVs) circulate between vertebrate animals and mosquitoes and represent a serious threat to humans via spillover from enzootic cycles to the human community. Active surveillance of MBVs in their vectors is therefore essential to better understand and prevent spillover and emergence, especially at the human–animal interface. In this study, we assessed the presence of MBVs using molecular and phylogenetic methods in mosquitoes collected along an ecological gradient ranging from rural urbanized areas to highland forest areas in northern Thailand. We have detected the presence of insect specific flaviviruses in our samples, and the presence of the emerging zoonotic Tembusu virus (TMUV). Reported for the first time in 1955 in Malaysia, TMUV remained for a long time in the shadow of other flaviviruses such as dengue virus or the Japanese encephalitis virus. In this study, we identified two new TMUV strains belonging to cluster 3, which seems to be endemic in rural areas of Thailand and highlighted the genetic specificities of this Thai cluster. Our results show the active circulation of this emerging flavivirus in Thailand and the need for continuous investigation on this poorly known but threatening virus in Asia.

Keywords: mosquito-borne viruses; Tembusu virus; Culex mosquito; emergent arboviruses

1. Introduction

Mosquito-borne viruses (MBVs) of zoonotic origins are responsible for multiple animal and human diseases worldwide and represent a large reservoir of viruses with emergence potential via spillover from their enzootic cycles. Tembusu virus (TMUV)

is an emerging mosquito-borne flavivirus that belongs to the Ntaya serocomplex, including Ntaya virus, Bagaza virus and Israel Turkey virus (refer to the ICTV database—https://ictv.global/report/chapter/flaviviridae/flaviviridae/orthoflavivirus (accessed on 26 June 2023)). Similar to other flaviviruses, including DENV and JEV, TMUV is an enveloped, positive-sense single-stranded RNA virus with an approximately 11-kb genome (reviewed in [1]). Despite recent sporadic outbreaks, knowledge on TMUV ecology and biology remains fragmented, precluding a thorough evaluation of its emergence potential.

TMUV was first isolated in Malaysia in 1955 [2], before being reported in different surveys in Asia and Southeast Asia (SEA) including China, Malaysia, Taiwan and Thailand [3–6]. TMUV has been intermittently reported in wild and domestic birds and in trapped mosquitoes [7,8]. TMUV infects a wide variety of avian species such as ducks, geese, chickens, sparrows and pigeons [1]. Since 2000, new variants of TMUV have been reported to cause several outbreaks in poultry and birds. Symptoms include dramatic decreases in egg production, severe neurologic disorders and retarded growth [5,7]. Migration of wild birds close to poultry farms could allow transmission to domestic ducks, while retention of the virus in high-density duck-producing areas could facilitate the rapid spread of the disease. Because of the symptom severity in ducks and the economic importance of ducks, some reports named the new viral variant Duck-TMUV (DTMUV) [3]. Nonetheless, hereafter, we will use TMUV as a generic term to refer to all viruses belonging to the TMUV phylogenetic group. TMUVs are phylogenetically divided into two lineages: the "TMUV lineage" including the original viruses, and the "DTMUV lineage". The DTMUV lineage is divided into three different clusters, named "TMUV cluster-1", "TMUV cluster-2" (with sub-cluster a and b) and "TMUV cluster-3" [9]. *Culex* mosquitoes are likely the main vector of transmission, as TMUVs have been isolated from several *Culex* species such as *Culex tritaeniorhynchus*, *Cx. Vishnui*, and *Cx. Gellidus* [1,10]. In addition to vector transmission, vertical transmission and non-vector transmission in birds (by air droplet exposure or by close contact) are suspected [11–13].

Located in the heart of South East Asia, Thailand has tight interactions with surrounding countries, including China and Laos. Endemic transmission of numerous mosquito-borne flaviviruses such as JEV and DENV occurs in Thailand, and other arboviruses associated with diseases in humans [14,15]. Thailand is largely covered with forests and rural areas, with an increasing entanglement of rural and urban territories and a densification of urban areas. In Thailand, TMUV was isolated in mosquitoes from the rural parts of the country, including the provinces of Kamphaengphet [8], Chiang Mai [16] and Kanchanaburi [5,17]. TMUV strains were also detected in broiler and layer ducks from the provinces of Chonburi, Nakhom Pathom, Nakhon Ratchasima, Prachinburi and Signburi [5]. Such recurrent detection indicates a wide distribution of TMUV in Thailand. Accordingly, in 2013, TMUV outbreaks occurred throughout the year (August 2013–September 2014) and many farms were affected, leading to losses in the poultry industry. Alarmingly, seroconversion was detected in humans, irrespective of contact with ducks, suggesting a zoonotic emergence of TMUV [18]. However, the serological survey in humans was conducted with a limited number of samples, and the survey lacked methodological details. In this context, TMUV surveillance in animals and vectors is essential to prevent agro-economical losses and evaluate emergence in humans.

In this study, we conducted agnostic arbovirus surveillance in mosquitoes along an ecological gradient in the Thai northern province of Nan, which shares a border with Lao PDR. The sub-district of Saenthong in the province of Nan is divided into two geographical landscape types: an agricultural lowland, including low-density urbanized villages, and a highland with sparse villages and an agricultural zone embedded in the forest zone located close to the protected Nanthaburi National Park. This contrasted area is separated by a transition zone including agricultural areas with rice paddy fields and dwellings. These landscapes provided an ideal study area with low and high levels of human-impacted habitats to study the ecology of MBVs and their mosquito vectors. In the different landscapes we sampled, we detected TMUV and insect-specific flaviviruses in the transition area, and in the lowland as well as the forest. We further characterized the phylogeny of the new TMUV strains, revealing a potential endemic cluster in Thailand.

2. Materials and Methods

2.1. Mosquito Collections

Mosquitoes were collected in the province of Nan in the northern part of Thailand (Figure 1a). The survey area was located in the Saen Thong sub-district of the Tha Wang Pha district, a rural area localized in an ecological gradient between forests, paddy fields lowlands and peridomestic urban areas. The eight collection points were distributed along a transect covering eight villages and a forested area (three sessions) (Figure 1a,b). The research proposal, involving specimen collection in Nan province, was approved by the Faculty of Tropical Medicine, Mahidol University, under agreement number FTM ECF-033-00.

(a)

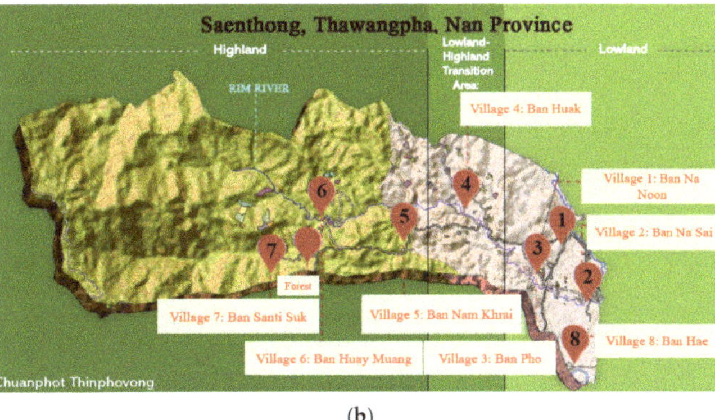

(b)

Figure 1. Geographic localization of the study area. (**a**) Localization of Nan province in Thailand. (map created with mapchart.net) and (**b**) localization of the collection sites in the Saen Thong sub-district, Tha Wang Pha district (design credit to Chuanphot Thinphovong) on a schematic flat map reflecting the different areas.

Samples were collected using BG-sentinel traps combined with BG Lure (Biogents AG, Regensburg, Germany). BG-sentinel traps operated for 48 h, for day- and night-time sessions.

Mosquito specimens were transported in cold boxes containing frozen cold packs to the field laboratory, for sorting up to the species identification level. In cases in which species could not be determined, specimens were grouped according to their genus only, and noted *"Genus"* sp. Mosquitoes were sorted and pooled in a 15 mL tube according to genus, sex and collection site, and then stored frozen. The samples were then transported in a liquid nitrogen tank to the department of Medical Entomology, Faculty of Tropical Medicine, Mahidol University, where samples were identified up to the species level following the morphological identification keys outlined by Rattanarithikul, R. et al. [19–21]. Mosquitoes were identified on a chilled table set to $-4\,^{\circ}\text{C}$, then mosquitoes were pooled according to species, sex and collection site and stored in a $-80\,^{\circ}\text{C}$ freezer.

Mosquito pools (1 to 15 specimens per pool) were made according to mosquito species, sex and collection location. Stainless steel beads (5 mm diameter) were added to tubes containing mosquitoes before homogenizing using a TissueLyzer (Qiagen, Hilden, Germany) at 50 cycles/s for 5 min in 500 µL of DMEM medium (Gibco, Waltham, MA, USA), complemented with 1% of penicillin/streptomycin solution (Gibco, Waltham, MA, USA) and $1\times$ of Fungizone solution (Gibco, Waltham, MA, USA). After homogenization, an additional 1 mL of DMEM medium was added. Tubes were clarified using centrifugation at $13,000\times g$ at $4\,^{\circ}\text{C}$ for 10 min. The supernatants were collected and filtered, using an

0.2 μm syringe filter (Sartorius, Bangkok, Thailand), into 1.5 mL tube and stored at −80 °C before RNA extraction.

2.2. RNA Extraction and Reverse Transcription

Viral RNA was extracted from the supernatant of mosquito homogenate using a NucleoSpin® virus kit (Macherey-Nagel, Düren, Germany) according to the manufacturer's protocol. Briefly, a 200 μL of homogenized sample was lysed in 5 μL of proteinase K and 200 μL lysis buffer containing guanidine hydrochloride. Carrier RNA was then added to the mixture, and the viral nucleic acid was then extracted and collected in an elution volume of 30 μL of RNase-free water. Purified RNA extracts were stored at −80 °C until virus screening using RT-PCR. Reverse transcription (RT) was performed using an M-MLV reverse transcriptase kit (Promega, Madison, WI, USA) on 14 μL of an RNA sample, following the manufacturer's instructions. cDNA was stored at −20 °C until subsequent analyses.

2.3. Detection of Flaviviruses and Alphaviruses Using PCR

The pan-flavivirus primers [22] PFlav-fAAR (5′-TACAACATGATGGGAAAGAGAG AGAARAA-3′) and PFlav-rKR (5′-GTGTCCCAKCCRGCTGTGTCATC-3′) were used to amplify a 256 base pair(bp) region of the NS5 gene of Flaviviruses. PCR was performed using GoTaq G2 Master Mix (Promega, Charbonnières-les-Bains, France) and 2 μL of cDNA with the following parameters: 95 °C for 3 min, 45 cycles of 95 °C 15 s, 56 °C 15 s, 72 °C 20 s and 72 °C for 2 min. PCR products were visualized on 1.8% agarose gel. Amplicons were purified from gel using a PureLink Gel extraction kit (Thermo Fisher Scientific, Illkirch-Graffenstaden, France) and stored at −20 °C.

The pan-alphavirus primers [23] PanAlpha F2A forward primer (5′-ATGATGAARTCI GGIATGTTYYT-3′), and reverse primers R2A (5′-ATYTTIACTTCCATGTTCATCCA-3′), R3A (5′-ATYTTIACTTCCATRTTCARCCA-3′), R4A (5′-ATYTTIACTTCCATGTTGACCCA-3′) were used to amplify a 200-pb region of the nsP4 gene in the alphavirus genome. PCR was performed using GoTaq G2 Master Mix (Promega, Charbonnières-les-Bains, France) and 2 μL of cDNA with the following parameters: 95 °C for 3 min, 45 cycles of 95 °C 15 s, 54 °C 15 s, 72 °C 20 s and 72 °C for 2 min. PCR products were visualized on 1.8% agarose gel.

All purified amplicons obtained with pan-flavivirus- or pan-alphavirus-PCR were characterized using Sanger sequencing in both forward and reverse directions (Eurofins, Vergèze, France). Sequence identities were determined via BLAST alignment (https://blast. ncbi.nlm.nih.gov/Blast.cgi, accessed on 26 June 2023).

2.4. TMUV Envelope Sequencing

A 1503-bp amplicon covering the entire TMUV envelope gene was amplified using the primers TMUV-E_F (5′-TTCAGCTGTCTGGGGATGCA-3′) and TMUV-E_R (5′-GGCATTGACATTTACTGCCA-3′). PCR amplification was conducted from 2 μL of cDNA using Q5 High Fidelity DNA Polymerase (NEB, Évry-Courcouronnes, France) and the following parameters: 98 °C for 1 min, 40 cycles of 98 °C 10 s, 60 °C 15 s, 72 °C 60 s and 72 °C for 2 min. PCR products were visualized on 1.5% agarose gel, and amplicons were gel-purified using a PureLink Gel extraction kit (Thermo Fisher Scientific, Illkirch-Graffenstaden, France) and stored at −20 °C. Purified amplicons were sequenced via Sanger sequencing in both forward and reverse directions (Eurofins, Vergèze, France).

2.5. Phylogenic Analysis

To characterize TMUV isolated from mosquito homogenates, sequences encoding TMUV Envelope were subjected to phylogenetic analysis, along with representative TMUV sequences obtained from the NCBI GenBank database (Table 1).

Table 1. TMUV sequences used in the phylogenetic tree, including the two strains identified in this study.

Virus	GenBank Accession n°	Year	Country
Tembusu virus strains	JX477685	1955	Malaysia
	AB110495	1992	Thailand
	JX477686	2000	Malaysia
	KC810847	2002	Thailand
	KC810846	2002	Thailand
	MF621927	2007	Thailand
	JX273153	2010	China
	JF270480	2010	China
	MN649260	2010	China
	JF895923	2010	China
	JF312912	2010	China
	JF459991	2010	China
	KX686578	2011	China
	KF557893	2012	China
	KF826767	2012	China
	KX097989	2012	Malaysia
	AB917090	2012	China
	KX097990	2012	Malaysia
	KR061333	2013	Thailand
	KJ740748	2013	China
	KF573582	2013	Thailand
	KX686577	2013	China
	MH748542	2014	China
	MN649267	2014	China
	KU323595	2014	China
	KP742476	2015	China
	KX686572	2015	China
	KT824876	2015	China
	MN649261	2015	China
	MK276420	2015	Thailand
	MH460536	2015	Thailand
	MK276427	2016	Thailand
	MK276442	2016	Thailand
	MN649266	2016	China
	MK276459	2017	Thailand
	MK907880	2018	China
	MK542820	2019	China
	MN747003	2019	Taiwan
Ntaya virus	JX236040	2013	-
JEV	NC001437	1989	Japan
WNV	NC009942	1999	USA
ZIKV	KY766069	2013	French Polynesia
Usutu virus	AY453411	2001	Austria
Israel Turkey virus	KC734553	2010	Israel
Bagaza virus	AY632545	2010	Central African Republic
P49_TH_2019 *	ON254216	2019	Thailand
P73_TH_2019 *	OQ543571	2019	Thailand

Year: year of isolation; Country: country of isolation. *: sequence obtained in this study.

Reference sequences were selected to cover all the diversity of TMUV strains and a broad range of geographical origins. All sequences were referenced into the phylogenetic tree in a format consisting of "accession number_country_year of isolation". Multiple sequence alignments and edits were carried out using MEGA 11. Sequences were edited and sites that could not be unambiguously aligned were excluded from the analyses. Maximum likelihood trees were constructed using PhyML software [24,25] with the best-fit nucleotide substitution model (GTR+G) identified by Akaike's information criterion (AIC).

The bootstrap method was used to estimate the robustness of nodes with 1000 iterations. Phylogenetic trees were edited using FigTree v1.4.4 software. Sequences of Zika virus (ZIKV) (GenBank number: KY766069), JEV (GenBank number: NC001437) and West Nile virus (WNV) (GenBank number: NC009942) were used as the outgroup to provide a relative framework for analyzing the phylogenetic differences between viruses within the Ntaya complex. All sequences from this study have been deposited in the GenBank database, and their accession numbers are shown in Table 1 Amino acid sequences were aligned using the MEGA 11 program to identify specific amino acid variations in the envelop protein sequences of TMUV strains.

3. Results

3.1. Collection of Mosquitoes

Mosquitoes were collected in eight villages and in one forested area (over three sessions) from the 19th to 26th of July 2019 in the Saenthong sub-district of the province of Nan, Thailand (Figure 1b). The Saenthong sub-district is located in the rural district of Thawangpha. Samples were collected along an ecological gradient from lowland areas, including villages, farms, and agricultural areas (village 1, 2, 3 and 8), to highland areas wherein three villages (village 5, 6 and 7) are located, with small agricultural zones surrounded by a vast forest area. Village 4 is located in a transition zone between lowland and highland (Figure 1b).

A total of 596 mosquitoes were collected and homogenized in 116 pools (Table 2). genera total of 5 genera of *Culicidae* were reported, including 12 species. Some specimens were damaged during collection, making it impossible to identify the species. In these cases, specimens have been listed as *"Genus"* sp. (Table 2). Overall, the *Culex* genus was the most represented (75%, n = 436), with *Cx. vishnui* representing the most prevalent species (39.8%, n = 237). The *Aedes* genus represented 15.4% (n = 92), and *Armigeres* 8.9% (n = 53), while *Mansonia* and *Toxorhynchites* both represented 0.17% (n = 1). However, the mosquito genera largely varied depending on location. All *Toxorhynchites* (n = 1), 78.3% of *Aedes* (n = 72) and 50.9% of *Armigeres* (n = 27) mosquitoes were caught in the forested area, whereas only 1.1% of *Culex* (n = 5) were collected in this area. In contrast, *Culex* mosquitoes represented 74% (n = 431) of all mosquitoes collected in the villages. Furthermore, the majority of *Culex* mosquitoes (89%, n = 387) were collected in the lowland villages and in the transition zone to the highland (villages 1–4 and 8, Figure 1b).

3.2. Detection of Flaviviruses and Alphaviruses

A total of 116 pools of mosquito homogenates were tested for flaviviruses and alphaviruses (Table 2). Six pools were positive for flaviviruses, and none for alphaviruses. Positive PCR products were sequenced and sequences were identified using the NCBI BLAST® website. Out of six samples, one *Aedes* and one *Culex* genus pool contained sequences related to the *Yunnan Culex flavivirus* (YNCxFV), with a maximum identity of 80.68% and 81.31%, respectively (Table 3). The sequences of one pool of *Aedes aegypti* and one pool of *Culex vishnui* contained the sequence of *Phlebotomus-associated flavivirus* (PAFV) with a maximum identity of 98.07% and 98.78%, respectively (Table 3). The pools P#73 with *Cx. vishnui* and P#49 with *Cx* sp. were both collected in "Ban Huak" village 4, and contained a TMUV-like sequence, which we noted as P73_TH_2019 and P49_TH_2019, respectively. The identity score for the first two hits for P49_TH_2019 sequence was 97.36% (coverage = 99%) and 96.04% (coverage = 99%) to TMUV KAN2016 (GenBank access number: KX184310) and TMUV HNU-NX2-2019 (GenBank access number: OP186478), respectively (Tables 3 and S1). The P73_TH_2019 sequence was similar, at 98.83% (coverage = 100%) with TMUV_ GX2021 (GenBank access number: OM240641) and 98.44% (coverage = 100%), with TMUV_ SD2021 (GenBank access number: OM240640) (Tables 3 and S1). The first ten hits of each sample are visualized in the Supplementary Materials (Table S1). The alignment of the P73_TH_2019 with P49_TH_2019 shows a very high similarity (96%), suggesting close phylogenic history between the two TMUVs collected in two different mosquito-pools.

Table 2. Summary of mosquito species, number of mosquitoes, and pools tested in the study. Mosquitoes for which we only identified the *Genus* were recorded in the column indicated as "*Genus* sp.".

Collection Site	Ae. albopictus		Ae. aegypti		Aedes sp.		Cx. quinquefasciatus		Cx trevipalpis		Cx. hutchinsoni		Cx. nigropunctatus		Culex sp.		Cx. vishnui		Cx. tritaeniorhynchus		Armigeres sp.		Arm. kesseli		Arm. subalbatus		Anopheles sp.		An. subpictus		Mansonia sp.		Toxorhynchites sp.		Total
	♂	♀	♂	♀	♂	♀	♂	♀	♂	♀	♂	♀	♂	♀	♂	♀	♂	♀	♂	♀	♂	♀	♂	♀	♂	♀	♂	♀	♂	♀	♂	♀	♂	♀	
Village 1	0	0	0	0	0	0	2	6	0	0	0	0	0	0	0	10	0	17	0	0	0	0	0	1	0	0	0	0	0	0	0	0	0	0	36
Village 2	0	2	0	0	0	0	1	5	0	1	0	1	0	0	1	8	1	8	0	0	0	1	0	2	0	0	0	0	0	1	0	0	0	0	32
Village 3	0	2	0	0	0	1	0	0	0	0	0	0	0	0	0	10	1	5	0	0	0	0	0	0	0	0	0	1	0	0	0	0	0	0	20
Village 4	0	0	0	0	0	0	0	1	0	3	0	0	0	0	0	96★	1	181★	0	7	0	7	0	5	0	0	0	3	0	3	0	0	0	0	307
Village 5	0	2	0	0	0	0	0	0	0	0	0	1	0	0	0	19	0	19	0	0	0	1	0	4	0	1	0	0	0	0	0	0	0	0	47
Village 6	0	3	0	0	0	1	0	0	0	0	0	0	0	0	1	4	0	0	0	0	0	0	0	0	0	0	0	0	0	0	0	0	0	0	9
Village 7	0	1	0	0	0	2	0	0	0	0	0	0	0	0	0			0	0	0	0	1	0	0	0	0	0	0	0	0	0	0	0	0	4
Village 8	1	3	0	2★	0	0	0	1	0	0	0	0	0	1	2	14	1	2	0	0	0	3	0	0	0	0	0	4	0	2	0	1	0	0	37
Forest session 1	0	15	0	0	0	3	0	0	1	0	0	0	0	0	0	1	0	1	0	0	0	1	0	6	0	1	0	0	0	0	0	0	0	0	29
Forest session 2	0	18	0	0	0	1	0	0	0	0	0	0	0	0	0	2★	0	0	0	0	0	5	0	4	0	0	0	0	0	0	0	0	0	0	30
Forest session 3	0	8	0	0	0	27★	0	0	0	0	0	0	0	0	0	0	0	0	0	0	0	6	0	4	0	0	0	0	0	0	0	0	0	1	46
Individuals	1	54	0	2	0	35	3	13	0	4	0	2	0	1	4	164	4	233	0	7	0	25	0	26	0	2	0	8	0	6	0	1	0	1	596
positive pools/Nbr of pool for each species	0/1	0/11	0	1/1	0	1/8	0/2	0/6	0/1	0/2	0	0/2	0	0/1	0/3	2/24	0/4	2/24	0	0/1	0	0/8	0	0/7	0	0/2	0	0/3	0	0/3	0	0/1	0	0/1	6/116

★ mosquito collection presenting positive pools for flavivirus family detection.

Table 3. Detection and identification of viruses in the mosquito pools.

Mosquito Pool ID	Mosquito Species Identification	Number of Mosquitoes per Pool	First Hit with BLAST® Alignment			
			Collection Site	Viral Identification	Coverage Score	Identity Score
P#13	*Ades aegypti*	2 ♀	Village 8	PAFV	97%	98.07%
P#20	*Aedes* sp.	9 ♀	Forest	YNCxFV	97%	80.68%
P#49	*Culex* sp.	10 ♀	Village 4	TMUV	99%	97.36%
P#60	*Culex* sp.	1 ♀	Forest	YNCxFV	98%	81.31%
P#73	*Culex vishnui*	15 ♀	Village 4	TMUV	100%	98.83%
P#77	*Culex vishnui*	15 ♀	Village 4	PAFV	98%	98.78%

PAFV: Phlebotomus-associated flavivirus; YNCxFV: Yunnan Culex flavivirus; TMUV: Tembusu virus, ♀: female.

3.3. Phylogenetic Analysis of the TMUV Isolates

To phylogenetically characterize the two virus isolates from the pools P49_TH_2019 and P73_TH_2019, we analyzed the envelope sequences. Alignment with representatives of TMUVs' phylogenetic diversity (Table 1) illustrated the distribution of TMUV strains in five distinctives clusters, and their relation to the Ntaya virus (Figure 2a,b) [1].

A first group, named "TMUV", includes the original strain isolated in 1955 in Malaysia and the MN747003 strain isolated in Taiwan in 2019. The other viruses are divided into four different clusters, named "cluster 1", comprising strains isolated in Malaysia and Thailand, "cluster 2.a", including strains isolated in Thailand and China, "cluster 2.b", corresponding to strains isolated only in China, and "cluster 3", including strains isolated in Thailand and China. P49_TH_2019 and P73_TH_2019 isolates from this study form a monophyletic lineage closely related to strains belonging to the cluster 3. The strains in cluster 3 were isolated in Thailand in 2016 and in China in 2014. Furthermore, we generated another phylogenetic tree by including the partial sequences of the envelope gene of three other TMUV strains isolated in Thailand in 1992 and 2002 [10,16]. All three Thai TMUV strains were clustered into the "cluster 3" with the isolates from this study (Figure 2b).

3.4. Identification of Envelope Amino Acid Modifications Specific to the TMUV Isolates from Nan Province

The genomic sequences of the P49_TH_2019 and P73_TH_2019 samples cover the entire coding sequence of the envelope protein (E protein), and were used to reveal amino acid differences from other strains belonging to every cluster of TMUV. The positions of amino acids with characteristic substitutions are shown in Table 4. All TMUV clusters have specific amino acid variations at certain positions, with unique patterns for every cluster (Table 4). The strains of cluster 1 carry specific amino acids on position 52 and 83, except for the strain KX097989, isolated in Malaysia in 2012, which has unique substitutions at position 373, 390 and 394. The strains of cluster 2 present unique amino acids on position 89, 180,185, 312, 332 and 451.

The strains of cluster 3 have nine amino acids unique to this cluster, except for MH748542_CH2014 (Chinese isolate from 2014), and five other amino acids in common with the ancestral cluster "TMUV". Positions 69, 91, 135, 149, 150, 365, 371, 391 and 394 have common amino acids for the strains MK276427-TH-2016 and P49_TH_2019, and P73_TH_2019 isolates. These amino acids are positioned in the DI (position 135; 149 and 150), DII (position 69) and DIII (position 365; 371; 391 and 394) domains. All TMUV strains present an Asn residue at the position 154; in addition, the strains MK276427_TH_2016, P49_TH_2019 and P73_TH_2019 present an S150N substitution. Finally, both P49_TH_2019 and P73_TH_2019 isolates have two unique substitutions at position 358 (V358I) of the domain III of the envelope protein.

a)

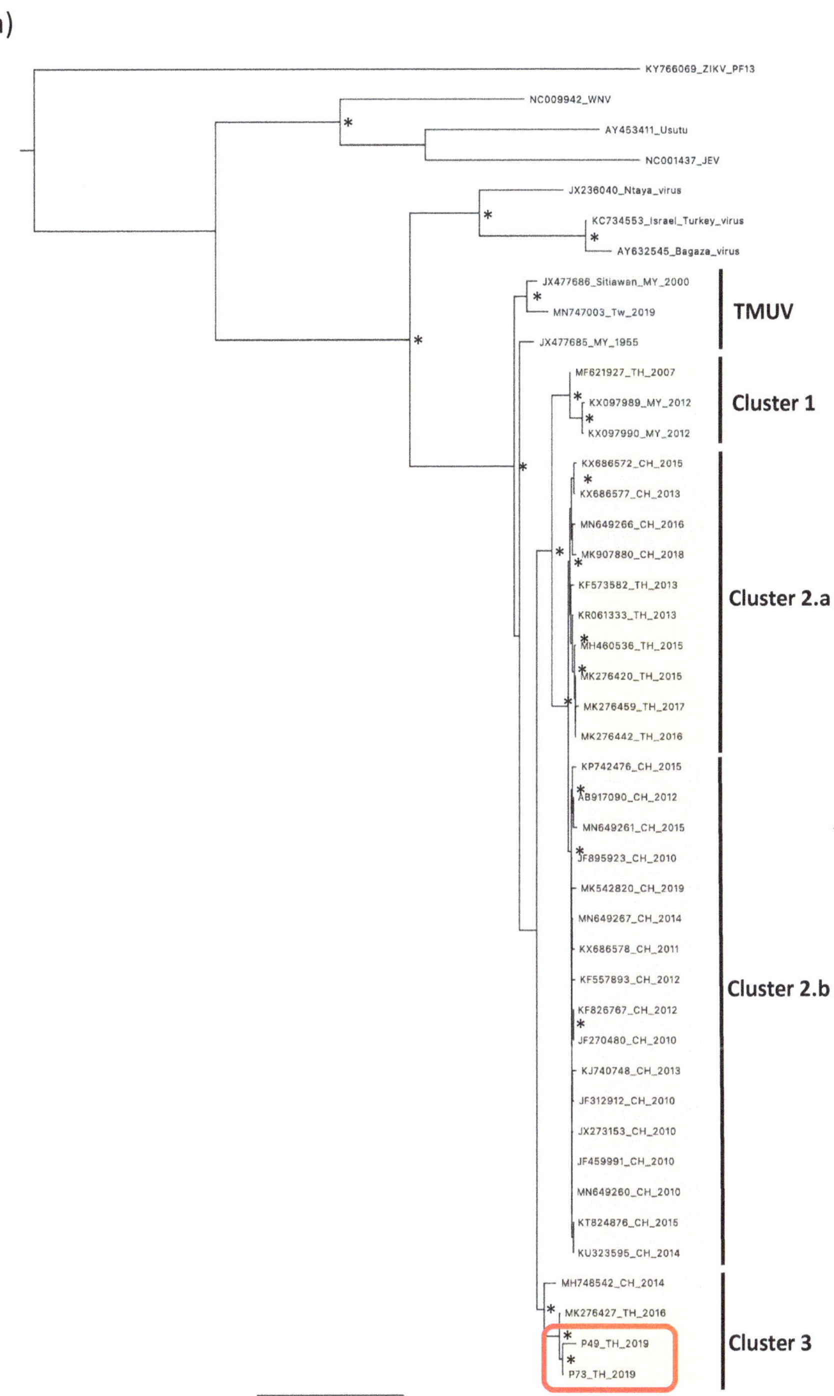

Figure 2. *Cont.*

b)

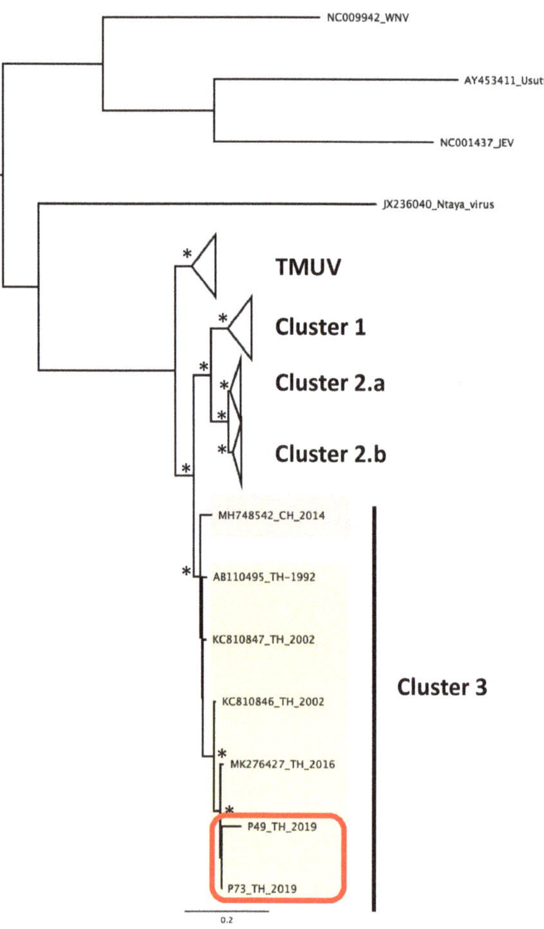

Figure 2. Phylogenetic tree of TMUV isolates from Nan province. Maximum likelihood tree of TMUV envelope sequences, generated using the GTR+G substitution model. Bootstrap values higher than 0.85 are shown on branch nodes by an asterisk. Samples collected in this study are indicated by a red-colored rectangle. (**a**) Phylogenetic tree based on the full envelope gene sequence's alignment. (**b**) Phylogenetic tree based on the partial envelope gene sequence's alignment. Country acronyms are abbreviated as CH for China, MY for Malaysia, TW for Taiwan, and TH for Thailand. Strains from this study are marked in red.

Table 4. Cluster-specific amino acid substitution for the envelope among TMUV strains. The different domains of the envelope are indicated as DI to DIII.

Strain Identification	Cluster Name	Amino Acid Substitution in Envelope Protein																											
		DI 1–50			DII 51–132						DI 133–197						DII 197–278		DIII 301–505											
Amino Acid position		2	21	38	52	69	72	83	89	91	135	149	150	157	180	185	236	277	312	332	358	365	373	371	390	391	394	451	467	487
MN747003_TW_2019	TMUV	N	I	R	E	T	P	P	D	I	V	V	S	V	L	S	R	S	A	S	V	S	V	I	K	G	R	K	I	A
JX477685_MY_1955	TMUV	-	-	-	-	-	-	-	-	-	-	-	-	-	-	A	-	-	-	-	-	-	-	-	-	-	-	-	V	-
MF621927_TH_2007	Cluster 1	S	I	R	D	-	S	S	D	-	-	A	-	A	L	S	K	S	A	S	-	-	S	-	K	-	R	K	A	V
KX097989_MY_2012	Cluster 1	-	-	-	-	-	-	-	-	-	-	-	-	-	-	-	-	-	-	-	-	-	I	-	N	-	K	-	-	A
KX686572_CH_2015	Cluster 2.a	-	I	K	E	-	-	P	E	-	-	-	-	-	M	T	-	N	V	T	-	-	V	-	K	-	R	K	-	V
MK907880_CH_2018	Cluster 2.a	-	-	R	-	-	-	-	-	-	-	-	-	-	-	-	-	-	-	-	-	-	-	-	-	-	-	R	-	-
KF573582_TH_2013	Cluster 2.a	-	-	-	-	-	-	-	-	-	-	-	-	-	-	-	-	-	-	-	-	-	-	-	-	-	-	-	-	-
MK276458_TH_2017	Cluster 2.a	-	-	-	-	-	-	-	-	-	-	-	-	-	-	-	-	-	-	-	-	-	-	-	-	-	-	-	-	-
MK542820_CH_2019	Cluster 2.b	-	-	K	-	-	-	-	-	-	-	-	-	-	-	-	-	-	-	-	-	-	-	-	-	-	-	-	-	A
JF270480_CH_2010	Cluster 2.b	-	T	R	-	-	-	-	-	-	-	-	-	-	-	-	-	-	-	-	-	-	-	-	-	-	-	-	-	-
KT824876_CH_2015	Cluster 2.b	-	I	R	-	-	-	-	-	-	-	-	-	-	-	-	-	-	-	-	-	-	-	-	-	-	-	-	-	-
KJ740748_CH_2013	Cluster 2.b	-	-	K	-	-	-	-	-	-	-	-	-	-	-	-	-	-	-	-	-	-	-	-	-	-	-	-	-	-
MH748542_CH_2014	Cluster 3	N	V	K	-	T	P	-	D	I	V	A	S	V	L	T	R	-	A	S	V	K	V	I	K	G	R	R	A	A
MK276427_TH_2016	Cluster 3	N	I	R	-	S	P	-	-	V	I	T	N	V	L	A	R	-	A	S	V	P	-	V	-	E	K	K	V	-
P49_TH_2019	Cluster 3	N	I	R	-	S	P	-	-	V	I	T	N	V	L	A	R	-	G	S	I	P	-	V	-	E	K	K	V	-
P73_TH_2019	Cluster 3	N	I	R	-	S	P	-	-	V	I	T	N	V	L	A	R	-	A	S	I	P	-	V	-	E	K	K	V	-

Legend:
- ☐ Amino acid unique to cluster TMUV and cluster 3
- ☐ Amino acid unique to cluster-1 and -2
- ■ Amino acid unique to cluster 1 (blue)
- ■ Amino acid unique to cluster 3 (yellow)
- ☐ Amino acid unique to cluster-2
- ■ Amino acid unique to strain of this study (orange)
- ■ Amino acid unique to strain KX087989_MY_2012 (brown)
- ▨ Amino acid shared by MH748542_CH_2014 with Cluster-TMUV, -1 and -2 (hatched)

Strain identification: GenBank accession n°#_Country_Year of isolation. The cluster of each strain was determined using the phylogenetic tree in this study. DI: domain I of the envelope protein; DII: domain II of the envelope protein; DIII: domain III of the envelope protein.

4. Discussion

In this study, we report the investigation of more than 596 mosquitoes collected in 2019 in the Nan province of northern Thailand. Samples were collected along a transect covering an ecological gradient ranging from a sparsely urbanized rural area to dwellings clustered in a few villages surrounded by forests and highland. Although we screened for flaviviruses and alphaviruses, we only detected three different flaviviruses in six different mosquito pools. These flaviviruses included two insect-specific flaviviruses and two TMUV strains in two pools of *Culex* mosquitoes.

Composition of mosquito species varied along the ecological gradient depending on the type of landscape. *Aedes albopictus* was abundantly collected in the forest, although a few specimens were captured inside the villages. Our observations were as expected based on its reported peridomestic distribution [26], and were in accordance with previous *Ae. albopictus* collections in the rural and forested habitats of Thailand [27,28]. Additionally, *Ae. albopictus* was the most prevalent species in the forest sample site, although we could not identify all *Aedes* sp. due to sample damage. *Aedes aegypti* was strictly found in village 8, which corresponds to an urbanized rural area in the lowland. *Aedes aegypti* prefers urban zones in part because of its use of human-made containers as breeding sites [29]. *Culex* spp. were the most prevalent genera in the villages, likely in relation to the breeding conditions made available by agricultural activities. We identified *Cx vishnui* in all habitats (from forested to urban areas), as previously reported [30]. In addition to variations in landscape, environmental conditions including altitude can influence mosquito species' composition and abundance [31]. Accordingly, we reported that 90% of *Culex* spp. were collected in the lowland villages and at the intersection of lowland and highland. Our study provides important information about mosquito species' distribution in different ecological settings, which will be important when conducting spatial risk analyses for arbovirus circulation.

We were able to identify three different flaviviruses, including two insect-specific flaviviruses: the *Phlebotomus-associated flavivirus* (PAFV) and the *Yunnan Culex flavivirus* (YNCxFV). Interestingly, both viruses were detected in Culex and Aedes spp. These viruses were previously identified in other regions in *Cx. gellidus*, *Cx. tritaeniorhynchus*, *Cx. vishnui* and *Cx. quinquefasciatus* [32,33]. There is a growing interest in insect flaviviruses, as two of these have been shown to increase transmission of pathogenic flaviviruses [34], although the mechanism remains elusive. The wide distribution of insect-specific flaviviruses and their potential role in pathogenic virus transmission warrants further studies [35,36].

Importantly, we identified two isolates as belonging to the TMUV group. The isolate P73_TH_2019 was isolated from a pool of *Cx. Vishnui*, and the isolate P49_TH_2019 was isolated from *Culex* spp. in the same village located in the transition zone between the highlands and lowlands. Although the species of the pool of *Culex* spp. could not be identified at the species level, it is likely *Cx. Vishnui*, since this species was overwhelmingly present among the other identified mosquitoes at the same site. Previous studies looked at the ability of mosquito vectors to transmit TMUV, and observed that mosquitoes of the genus *Culex* were very competent [10]. Accordingly, in Malaysia, China, Thailand, and Taiwan, TMUV was detected in *Cx. tritaeniorhynchus*, *Cx. vishnui*, *Cx. quinquefasciatus*, *Cx. annulus* and *Cx. pipiens* [1]. Although the transmission capacity of *Culex* mosquitoes seems variable, and a source of discussion [17,37], *Cx. tritaeniorhynchus* was proposed as the principal vector [38]. In Thailand, TMUV infection in *Cx. tritaeniorhynchus* collected in paddy fields was reported in the vicinity of Kamphaeng Phet province both in 1982 and 2002 [8,10]. *Culex vishnui* and *Cx. tritaeniorhynchus* belong to the same subgroup, and are also considered major vectors of JEV.

After its first identification in 1955 in Malaysia, TMUV has only been reported in four different countries in Asia. In Thailand, TMUV has been detected mostly in association with large duck farms in the center of the country [9,39]. In both China and Thailand and in a wetland habitat for waterbirds in a suburban area of Taipei city in Taiwan, the strains isolated were related to cluster 2. In contrast, strains belonging to the cluster TMUV and cluster 1 were found in rural or forest areas in Malaysia and Taiwan [6]. In our study, we identified TMUV strains that belong to cluster 3 in a rural area in northern Thailand. The collection site was in a village of 360 inhabitants, surrounded by rice fields and forested areas downstream to a dam. Prior to the large outbreaks of the 2010s, associated with TMUV cluster 2, in China and Thailand, TMUV cluster 3 had already been identified in Thailand, in rural provinces in the west of the country [2,16]. Although Ninvilai et al. previously reported a cluster 3 TMUV being present on a large duck farm in central Thailand [9], it would seem that viruses affiliated with clusters TMUV, 1 and 3 are found preferentially in rural and forested areas. These locations, including wetlands or rice fields, are particularly favorable to the development of mosquitoes of the genus *Culex*, which are described as major vectors of TMUV, and are also areas with domestic and wild birdlife. Thus, these areas could be favorable for the maintenance and spread of TMUV. Further investigation should be carried out to elucidate the possible relationships between the type of viral strain and the kind of ecological area.

It is interesting to note that, similar to the strain P73_TH_2019 in our study, the strains isolated from mosquitoes in 1982 and 2002 in Thailand also belong to cluster 3. Although TMUV has been detected in large parts of the Thai territory, cluster 3 strains were mostly found in rural areas in the north-west and north parts of the country. Our phylogenetic analysis showed that the two new TMUV strains form a monophyletic group closely related to TMUV cluster 3. While they do display the specificities of recent TMUV strains, such as the presence of the S156P substitution in «loop 150» region of the envelope protein [40], Thai cluster 3 strains are phylogenetically closer to the TMUV cluster than to clusters 1 and 2. Previous studies have shown the important effect of amino acid substitution sequences in the viral envelope on the virulence and pathogenicity of TMUV [40–42]. Recently, Nivilai et al. showed the presence of unique residues in the envelope protein for strains belonging to cluster 3 [9]. The cluster 3 strains identified in our study possess unique amino acid substitutions that are partially different from those described by Ninvilai et al. We found the same residues in position 149, 150, 391 and 394, as previously described by Ninvilai et al., but we also found other amino acid substitutions in positions 69, 91, 135, 365 and 371. As with most flaviviruses, we found an Asn residue at position 154 for the two TMUV isolates identified in our study. The position of this Asn residue forms an N-x-(S/T) glycosylation motif that plays an important role in pathogenicity of flaviviruses [43,44]. In the cluster 3 strains, except for the strain MH748542_2014_Ch, we also found the presence of an Asn residue in position 150, resulting from an S150N substitution. However, this substitution does not lead to the formation of an N-x-(S/T) motif, thus suggesting the absence of glycosylation in this region. Moreover, a unique V358I substitution appears in both strains P49_TH_2019 and P73_TH_2019, whereas the substitution was not present in the cluster 3 strains MK276427_TH_2016 and MH748542_2014_Ch, or in strains isolated from other clusters. Differences in the envelope sequence may stem from selection in either the host or mosquito. Finally, the whole envelope sequence of the strains isolated in 1992 and 2002 is not available, and therefore we are unable to evaluate if the amino acid signature of the cluster 3 strains isolated in Thailand is recent or not.

We would like to propose that the cluster 3 strains circulate at a regional scale between Thailand and the south of China. The local maintenance of these strains could be explained by the presence of the virus in more isolated areas of Thailand, such as the province of Nan, which carry out less trading than the central regions of the country; these central regions are more densely populated and carry out more economic exchanges with other Asian countries. However, the low number of strains from cluster 3 does not provide sufficient hindsight on the evolution of TMUV in Thailand, and further studies should be conducted to evaluate the impact of the different TMUV clusters on its dissemination, the evolution of pathogenicity and TMUV emergence risk in humans.

In conclusion, we report the detection of TMUV in *Culex* mosquito populations in northern Thailand. Our phylogenetic analysis classified these isolates into TMUV cluster 3, and highlighted the genetic specificities of this cluster, providing insights into the diversity and evolution of TMUV. TMUV surveillance is particularly important in a context of global changes and the intensification of trade between China, Laos, and Thailand. The opening of new economic corridors and new trade routes in the Indo-Pacific region will have a major impact on the emergence of pathogens that were previously restricted to small geographical areas. It is therefore essential to maintain active surveillance for arbovirus emergence and spillover in areas in which there is strong interaction between wildlife, domestic animals, and human communities.

Supplementary Materials: The following supporting information can be downloaded at: https://www.mdpi.com/article/10.3390/v15071447/s1, Table S1: The first ten hits alignment of sample P49_TH_2019 and P73_TH_2019 were obtained using the NCBI BLAST® website.

Author Contributions: Conceptualization, R.H., R.E.M.V. and S.M.; methodology, R.H., R.E.M.V. and S.M.; formal analysis, R.H., D.M.R. and R.E.M.V.; investigation, R.H., D.M.R., R.E.M.V., A.K., K.C. and S.W.; resources, R.H. and A.Y.; writing—original draft preparation, R.H. and J.J.; writing—review and editing, R.H., R.E.M.V., S.M. and J.P.; visualization, R.H.; supervision, R.H., S.M. and R.E.M.V.; project administration, R.H., A.K., S.M. and K.C.; funding acquisition, R.H., R.E.M.V., A.Y., J.P., J.J., S.W., D.M. and D.M.R. All authors have read and agreed to the published version of the manuscript.

Funding: This research was funded through (i) the French project INGENIOUS, supported by the Labex CEMEB and the I-SITE Excellence Program of the University of Montpellier, under the Investissements France 2030, granted to Dr. Rodolphe Hamel; (ii) the French project BILAO supported by FSPI OHSEA «One Health (Une seule Santé) en pratique en Asie du Sud-Est», a program supervised by IRD, CIRAD and CNRS and granted to Dr. Rodolphe Hamel; (iii) the French ANR Project FutureHealthSEA (grant number: ANR-17-CE35-0003-02) "Predictive scenarios of health in Southeast Asia: linking land use and climate changes to infectious diseases", and supported by the Thailand International Cooperation Agency for the project "Innovative Animal Health" and National Research Council of Thailand, granted to Dr. Serge Morand; and (iiii) the ANR project V-DOSAGE (grant number: ANR-22-CE35-0012-03) granted to Dr. Julien Pompon. The funders had no role in study design, data collection and analysis, decision to publish, or preparation of the manuscript. D.M. Rajonhson is a student enrolled in the international PhD program at the faculty of Tropical Medicine, Mahidol University, supported by the Organization for Women in Science for the Developing World (OWSD) and the Swedish International Development Cooperation Agency (SIDA).

Data Availability Statement: Not applicable.

Acknowledgments: We would like to express our gratitude to the National Research Council of Thailand (NRCT) and the local administration in Nan province, including the Nan Provincial Public Health Office, Tha Wang Pha District Public Health Office, Tha Wang Pha Hospital and Saen Thong Subdistrict Health Promoting Hospital for supporting and facilitating research in the field. We thank Malee Tanita (Saen Thong Health Promoting Hospital), Chuanphot Thinphovong, Yossapong Paladsing and Phurin Makaew, the health village volunteers of Saen Thong, for their great help in carrying out the field study. Special thanks also go to Chuanphot Thinphovong of the Faculty of Veterinary Technology from Kasetsart University for the illustrative map of the field location. We also thank the International Join Unit (LMI) PRESTO (Protect-Detect-Stop), the representation of IRD in Thailand and the Japan Agency for Medical Research and Development (AMED) (JP22wm0325004s0503) for their help in the smooth running of this study in the field and in the lab.

Conflicts of Interest: The authors declare no conflict of interest. The funders had no role in the design of the study; in the collection, analyses, or interpretation of data; in the writing of the manuscript; or in the decision to publish the results.

References

1. Hamel, R.; Phanitchat, T.; Wichit, S.; Morales Vargas, R.E.; Jaroenpool, J.; Diagne, C.T.; Pompon, J.; Missé, D. New Insights into the Biology of the Emerging Tembusu Virus. *Pathogens* **2021**, *10*, 1010. [CrossRef] [PubMed]
2. US Army Medical Research Unit (Malaya), Institute for Medical Research. *Annual Report*; Federation of Malaya: Kuala Lumpur, Malaysia, 1957; pp. 100–103.
3. Cao, Z.; Zhang, C.; Liu, Y.; Ye, W.; Han, J.; Ma, G.; Zhang, D.; Xu, F.; Gao, X.; Tang, Y.; et al. Tembusu virus in ducks, China. *Emerg. Infect. Dis.* **2011**, *17*, 1873–1875. [CrossRef] [PubMed]
4. Homonnay, Z.G.; Kovács, E.W.; Bányai, K.; Albert, M.; Fehér, E.; Mató, T.; Tatár-Kis, T.; Palya, V. Tembusu-like flavivirus (Perak virus) as the cause of neurological disease outbreaks in young Pekin ducks. *Avian Pathol.* **2014**, *43*, 552–560. [CrossRef]

5.   Thontiravong, A.; Ninvilai, P.; Tunterak, W.; Nonthabenjawan, N.; Chaiyavong, S.; Angkabkingkaew, K.; Mungkundar, C.; Phuengpho, W.; Oraveerakul, K.; Amonsin, A. Tembusu-Related Flavivirus in Ducks, Thailand. *Emerg. Infect. Dis.* **2015**, *21*, 2164–2167. [CrossRef]

6.   Peng, S.H.; Su, C.L.; Chang, M.C.; Hu, H.C.; Yang, S.L.; Shu, P.Y. Genome Analysis of a Novel Tembusu Virus in Taiwan. *Viruses* **2020**, *12*, 567. [CrossRef]

7.   Kono, Y.; Tsukamoto, K.; Abd Hamid, M.; Darus, A.; Lian, T.C.; Sam, L.S.; Yok, C.N.; Di, K.B.; Lim, K.T.; Yamaguchi, S.; et al. Encephalitis and retarded growth of chicks caused by Sitiawan virus, a new isolate belonging to the genus Flavivirus. *Am. J. Trop. Med. Hyg.* **2000**, *63*, 94–101. [CrossRef]

8.   Leake, C.J.; Ussery, M.A.; Nisalak, A.; Hoke, C.H.; Andre, R.G.; Burke, D.S. Virus isolations from mosquitoes collected during the 1982 Japanese encephalitis epidemic in northern Thailand. *Trans. R. Soc. Trop. Med. Hyg.* **1986**, *80*, 831–837. [CrossRef]

9.   Ninvilai, P.; Tunterak, W.; Oraveerakul, K.; Amonsin, A.; Thontiravong, A. Genetic characterization of duck Tembusu virus in Thailand, 2015–2017: Identification of a novel cluster. *Transbound. Emerg. Dis.* **2019**, *66*, 1982–1992. [CrossRef] [PubMed]

10.   O'Guinn, M.L.; Turell, M.J.; Kengluecha, A.; Jaichapor, B.; Kankaew, P.; Miller, R.S.; Endy, T.P.; Jones, J.W.; Coleman, R.E.; Lee, J.S. Field detection of Tembusu virus in western Thailand by rt-PCR and vector competence determination of select culex mosquitoes for transmission of the virus. *Am. J. Trop. Med. Hyg.* **2013**, *89*, 1023–1028. [CrossRef]

11.   Tunterak, W.; Prakairungnamthip, D.; Ninvilai, P.; Tiawsirisup, S.; Oraveerakul, K.; Sasipreeyajan, J.; Amonsin, A.; Thontiravong, A. Patterns of duck Tembusu virus infection in ducks, Thailand: A serological study. *Poult. Sci.* **2021**, *100*, 537–542. [CrossRef]

12.   Ninvilai, P.; Limcharoen, B.; Tunterak, W.; Prakairungnamthip, D.; Oraveerakul, K.; Banlunara, W.; Thontiravong, A. Pathogenesis of Thai duck Tembusu virus in Cherry Valley ducks: The effect of age on susceptibility to infection. *Vet. Microbiol.* **2020**, *243*, 108636. [CrossRef] [PubMed]

13.   Li, X.; Shi, Y.; Liu, Q.; Wang, Y.; Li, G.; Teng, Q.; Zhang, Y.; Liu, S.; Li, Z. Airborne Transmission of a Novel Tembusu Virus in Ducks. *J. Clin. Microbiol.* **2015**, *53*, 2734–2736. [CrossRef] [PubMed]

14.   Hamel, R.; Surasombatpattana, P.; Wichit, S.; Dauvé, A.; Donato, C.; Pompon, J.; Vijaykrishna, D.; Liegeois, F.; Vargas, R.M.; Luplertlop, N.; et al. Phylogenetic analysis revealed the co-circulation of four dengue virus serotypes in Southern Thailand. *PLoS ONE* **2019**, *14*, e0221179. [CrossRef] [PubMed]

15.   Raksakoon, C.; Potiwat, R. Current Arboviral Threats and Their Potential Vectors in Thailand. *Pathogens* **2021**, *10*, 80. [CrossRef]

16.   Pandey, B.D.; Karabatsos, N.; Cropp, B.; Tagaki, M.; Tsuda, Y.; Ichinose, A.; Igarashi, A. Identification of a flavivirus isolated from mosquitos in Chiang Mai Thailand. *Southeast Asian J. Trop. Med. Public Health* **1999**, *30*, 161–165.

17.   Nitatpattana, N.; Apiwatanason, C.; Nakgoi, K.; Sungvornyothin, S.; Pumchompol, J.; Wanlayaporn, D.; Chaiyo, K.; Siripholvat, V.; Yoksan, S.; Gonzalez, J.-P. Isolation of Tembusu virus from *Culex quinquefasciatus* in Kanchanaburi Province, Thailand. *Southeast Asian J. Trop. Med. Public Health* **2017**, *48*, 546–551.

18.   Pulmanausahakul, R.; Ketsuwan, K.; Jaimipuk, T.; Smith, D.R.; Auewarakul, P.; Songserm, T. Detection of antibodies to duck tembusu virus in human population with or without the history of contact with ducks. *Transbound. Emerg. Dis.* **2021**, *69*, 870–873. [CrossRef]

19.   Rattanarithikul, R.; Harbach, R.E.; Harrison, B.A.; Panthusiri, P.; Jones, J.W.; Coleman, R.E. Illustrated keys to the mosquitoes of Thailand. II. Genera Culex and Lutzia. *Southeast Asian J. Trop. Med. Public Health* **2005**, *36* (Suppl. S2), 1–97.

20.   Rattanarithikul, R.; Harbach, R.E.; Harrison, B.A.; Panthusiri, P.; Coleman, R.E.; Richardson, J.H. Illustrated keys to the mosquitoes of Thailand. VI. Tribe Aedini. *Southeast Asian J. Trop. Med. Public Health* **2010**, *41* (Suppl. S1), 1–225.

21.   Rattanarithikul, R.; Harrison, B.A.; Panthusiri, P.; Peyton, E.L.; Coleman, R.E. Illustrated keys to the mosquitoes of Thailand III. Genera Aedeomyia, Ficalbia, Mimomyia, Hodgesia, Coquillettidia, Mansonia, and Uranotaenia. *Southeast Asian J. Trop. Med. Public Health* **2006**, *37* (Suppl. S1), 1–85.

22.   Vina-Rodriguez, A.; Sachse, K.; Ziegler, U.; Chaintoutis, S.C.; Keller, M.; Groschup, M.H.; Eiden, M. A Novel Pan-*Flavivirus* Detection and Identification Assay Based on RT-qPCR and Microarray. *BioMed Res. Int.* **2017**, *2017*, 4248756. [CrossRef]

23.   Giry, C.; Roquebert, B.; Li-Pat-Yuen, G.; Gasque, P.; Jaffar-Bandjee, M.C. Improved detection of genus-specific Alphavirus using a generic TaqMan® assay. *BMC Microbiol.* **2017**, *17*, 164. [CrossRef] [PubMed]

24.   Guindon, S.; Dufayard, J.F.; Lefort, V.; Anisimova, M.; Hordijk, W.; Gascuel, O. New algorithms and methods to estimate maximum-likelihood phylogenies: Assessing the performance of PhyML 3.0. *Syst. Biol.* **2010**, *59*, 307–321. [CrossRef] [PubMed]

25.   Dereeper, A.; Guignon, V.; Blanc, G.; Audic, S.; Buffet, S.; Chevenet, F.; Dufayard, J.F.; Guindon, S.; Lefort, V.; Lescot, M.; et al. Phylogeny.fr: Robust phylogenetic analysis for the non-specialist. *Nucleic Acids Res.* **2008**, *36*, W465–W469. [CrossRef]

26.   Mendenhall, I.H.; Manuel, M.; Moorthy, M.; Lee, T.T.M.; Low, D.H.W.; Missé, D.; Gubler, D.J.; Ellis, B.R.; Ooi, E.E.; Pompon, J. Peridomestic Aedes malayensis and Aedes albopictus are capable vectors of arboviruses in cities. *PLoS Negl. Trop. Dis.* **2017**, *11*, e0005667. [CrossRef]

27.   Sumodan, P.K.; Vargas, R.M.; Pothikasikorn, J.; Sumanrote, A.; Lefait-Robin, R.; Dujardin, J.-P. Rubber plantations as a mosquito box amplification in South and Southeast Asia. In *Socio-Ecological Dimensions of Infectious Diseases in Southeast Asia*; Springer: Singapore, 2015; pp. 155–167.

28.   Tsuda, Y.; Suwonkerd, W.; Chawprom, S.; Prajakwong, S.; Takagi, M. Different spatial distribution of Aedes aegypti and Aedes albopictus along an urban–rural gradient and the relating environmental factors examined in three villages in northern Thailand. *J. Am. Mosq. Control Assoc.* **2006**, *22*, 222–228. [CrossRef]

29. Chareonviriyaphap, T.; Akratanakul, P.; Nettanomsak, S.; Huntamai, S. Larval habitats and distribution patterns of Aedes aegypti (Linnaeus) and Aedes albopictus (Skuse), in Thailand. *Southeast Asian J. Trop. Med. Public Health* **2003**, *34*, 529–535.
30. Maquart, P.O.; Chann, L.; Boyer, S. Culex vishnui (Diptera: Culicidae): An Overlooked Vector of Arboviruses in South-East Asia. *J. Med. Entomol.* **2022**, *59*, 1144–1153. [CrossRef]
31. Eisen, L.; Bolling, B.G.; Blair, C.D.; Beaty, B.J.; Moore, C.G. Mosquito species richness, composition, and abundance along habitat-climate-elevation gradients in the northern Colorado Front Range. *J. Med. Entomol.* **2008**, *45*, 800–811. [CrossRef]
32. Fang, Y.; Tambo, E.; Xue, J.B.; Zhang, Y.; Zhou, X.N.; Khater, E.I.M. Detection of DENV-2 and Insect-Specific Flaviviruses in Mosquitoes Collected From Jeddah, Saudi Arabia. *Front. Cell. Infect. Microbiol.* **2021**, *11*, 626368. [CrossRef] [PubMed]
33. Fang, Y.; Zhang, W.; Xue, J.B.; Zhang, Y. Monitoring Mosquito-Borne Arbovirus in Various Insect Regions in China in 2018. *Front. Cell. Infect. Microbiol.* **2021**, *11*, 640993. [CrossRef]
34. Olmo, R.P.; Todjro, Y.M.H.; Aguiar, E.R.G.R.; de Almeida, J.P.P.; Ferreira, F.V.; Armache, J.N.; de Faria, I.J.S.; Ferreira, A.G.A.; Amadou, S.C.G.; Silva, A.T.S.; et al. Mosquito vector competence for dengue is modulated by insect-specific viruses. *Nat. Microbiol.* **2023**, *8*, 135–149. [CrossRef]
35. Xia, H.; Wang, Y.; Atoni, E.; Zhang, B.; Yuan, Z. Mosquito-Associated Viruses in China. *Virol. Sin.* **2018**, *33*, 5–20. [CrossRef]
36. Agboli, E.; Zahouli, J.B.Z.; Badolo, A.; Jöst, H. Mosquito-Associated Viruses and Their Related Mosquitoes in West Africa. *Viruses* **2021**, *13*, 891. [CrossRef]
37. Guo, X.; Jiang, T.; Jiang, Y.; Zhao, T.; Li, C.; Dong, Y.; Xing, D.; Qin, C. Potential Vector Competence of Mosquitoes to Transmit Baiyangdian Virus, a New Tembusu-Related Virus in China. *Vector Borne Zoonotic Dis.* **2020**, *20*, 541–546. [CrossRef] [PubMed]
38. Sanisuriwong, J.; Yurayart, N.; Thontiravong, A.; Tiawsirisup, S. Vector competence of Culex tritaeniorhynchus and Culex quinquefasciatus (Diptera: Culicidae) for duck Tembusu virus transmission. *Acta Trop.* **2021**, *214*, 105785. [CrossRef] [PubMed]
39. Sanisuriwong, J.; Yurayart, N.; Thontiravong, A.; Tiawsirisup, S. Duck Tembusu virus detection and characterization from mosquitoes in duck farms, Thailand. *Transbound. Emerg. Dis.* **2020**, *67*, 1082–1088. [CrossRef] [PubMed]
40. Yan, D.; Shi, Y.; Wang, H.; Li, G.; Li, X.; Wang, B.; Su, X.; Wang, J.; Teng, Q.; Yang, J.; et al. A Single Mutation at Position 156 in the Envelope Protein of Tembusu Virus Is Responsible for Virus Tissue Tropism and Transmissibility in Ducks. *J. Virol.* **2018**, *92*, e00427-18. [CrossRef] [PubMed]
41. Sun, M.; Zhang, L.; Cao, Y.; Wang, J.; Yu, Z.; Sun, X.; Liu, F.; Li, Z.; Liu, P.; Su, J. Basic Amino Acid Substitution at Residue 367 of the Envelope Protein of Tembusu Virus Plays a Critical Role in Pathogenesis. *J. Virol.* **2020**, *94*, e02011-19. [CrossRef]
42. Sun, X.; Sun, M.; Zhang, L.; Yu, Z.; Li, J.; Xie, W.; Su, J. Amino Acid Substitutions in NS5 Contribute Differentially to Tembusu Virus Attenuation in Ducklings and Cell Cultures. *Viruses* **2021**, *13*, 921. [CrossRef]
43. Carbaugh, D.L.; Lazear, H.M. Flavivirus Envelope Protein Glycosylation: Impacts on Viral Infection and Pathogenesis. *J. Virol.* **2020**, *94*, 94. [CrossRef] [PubMed]
44. Wen, D.; Li, S.; Dong, F.; Zhang, Y.; Lin, Y.; Wang, J.; Zou, Z.; Zheng, A. N-glycosylation of Viral E Protein Is the Determinant for Vector Midgut Invasion by Flaviviruses. *mBio* **2018**, *9*, e00046-18. [CrossRef] [PubMed]

viruses

Article

Interspecies Transmission from Pigs to Ferrets of Antigenically Distinct Swine H1 Influenza A Viruses with Reduced Reactivity to Candidate Vaccine Virus Antisera as Measures of Relative Zoonotic Risk

J. Brian Kimble [1,†], Carine K. Souza [1], Tavis K. Anderson [1], Zebulun W. Arendsee [1], David E. Hufnagel [1], Katharine M. Young [1], Nicola S. Lewis [2], C. Todd Davis [3], Sharmi Thor [3] and Amy L. Vincent Baker [1,*]

[1] Virus Prion Research Unit, National Animal Disease Center, United States Department of Agriculture-ARS, Ames, IA 50010, USA

[2] Department of Pathology and Population Sciences, Royal Veterinary College, University of London, London NW1 0TU, UK

[3] Influenza Division, National Center for Immunization and Respiratory Diseases, Centers for Disease Control and Prevention, Atlanta, GA 30333, USA

[*] Correspondence: amy.vincent@usda.gov

[†] Current Address: Foreign Arthropod-Borne Animal Disease Research Unit, National Bio and Agro-Defense Facility, USDA-ARS, Manhattan, KS 66502, USA.

Citation: Kimble, J.B.; Souza, C.K.; Anderson, T.K.; Arendsee, Z.W.; Hufnagel, D.E.; Young, K.M.; Lewis, N.S.; Davis, C.T.; Thor, S.; Vincent Baker, A.L. Interspecies Transmission from Pigs to Ferrets of Antigenically Distinct Swine H1 Influenza A Viruses with Reduced Reactivity to Candidate Vaccine Virus Antisera as Measures of Relative Zoonotic Risk. *Viruses* 2022, 14, 2398. https://doi.org/10.3390/v14112398

Academic Editors: Myriam Ermonval and Serge Morand

Received: 28 September 2022
Accepted: 26 October 2022
Published: 29 October 2022

Publisher's Note: MDPI stays neutral with regard to jurisdictional claims in published maps and institutional affiliations.

Abstract: During the last decade, endemic swine H1 influenza A viruses (IAV) from six different genetic clades of the hemagglutinin gene caused zoonotic infections in humans. The majority of zoonotic events with swine IAV were restricted to a single case with no subsequent transmission. However, repeated introduction of human-seasonal H1N1, continual reassortment between endemic swine IAV, and subsequent drift in the swine host resulted in highly diverse swine IAV with human-origin genes that may become a risk to the human population. To prepare for the potential of a future swine-origin IAV pandemic in humans, public health laboratories selected candidate vaccine viruses (CVV) for use as vaccine seed strains. To assess the pandemic risk of contemporary US swine H1N1 or H1N2 strains, we quantified the genetic diversity of swine H1 HA genes, and identified representative strains from each circulating clade. We then characterized the representative swine IAV against human seasonal vaccine and CVV strains using ferret antisera in hemagglutination inhibition assays (HI). HI assays revealed that 1A.3.3.2 (pdm09) and 1B.2.1 (delta-2) demonstrated strong cross reactivity to human seasonal vaccines or CVVs. However, swine IAV from three clades that represent more than 50% of the detected swine IAVs in the USA showed significant reduction in cross-reactivity compared to the closest CVV virus: 1A.1.1.3 (alpha-deletion), 1A.3.3.3-clade 3 (gamma), and 1B.2.2.1 (delta-1a). Representative viruses from these three clades were further characterized in a pig-to-ferret transmission model and shown to exhibit variable transmission efficiency. Our data prioritize specific genotypes of swine H1N1 and H1N2 to further investigate in the risk they pose to the human population.

Keywords: influenza A virus; pandemic preparedness; zoonosis; risk assessment; variant; antigenic drift

1. Introduction

Influenza A viruses (IAV) infect a broad range of wild and domestic animal species and humans, and result in disease states ranging from asymptomatic to severe pneumonia and death. Wild waterfowl act as the natural reservoir for IAV, but various subtypes and lineages are endemic in populations such as domestic poultry, swine, and human. IAVs cause significant economic impact on swine productions systems [1–3]. In the United States (US), three primary subtypes of IAV, H1N1, H1N2, and H3N2, circulate endemically in swine populations with multiple hemagglutinin (HA) genetic clades present within each subtype [4]. Within the H1 subtype, which accounted for approximately 68% of all IAVs isolated from US pigs in 2020, 8 HA phylogenetic clades and 9 neuraminidase (NA) clades were detected [4,5]. This considerable diversity is driven by a number of factors, including intra-species genetic evolution, reassortment, and the repeated introduction of human-origin IAV strains to pig populations [6–8]. Additionally, a large diverse swine IAV population may have significant impact on human health, where swine-origin IAV may zoonotically transmit sporadically, termed "variant" in humans, or cause pandemics infecting millions of people, as seen during the 2009 H1N1 pandemic (H1N1pdm09) [9].

Swine H1 IAV in the US are classified by hemagglutinin (HA), and are either the 1A lineage that evolved from the 1918 H1N1 pandemic, or the 1B lineage that resulted from introduction and subsequent persistence of pre-2009 human seasonal H1N1 [10]. The 1B lineage has 3 genetic clades that are currently circulating in the US: 1B.2.1, 1B.2.2.1, and 1B.2.2.2 [10]. The 1A lineage viruses include 5 genetic clades that are currently circulating: 1A.1.1.3, 1A.2, 1A.2-3-like, 1A.3.3.2, and 1A.3.3.3 [10,11]. Within the 1A lineage is the H1N1pdm09, with the HA assigned the global nomenclature of clade 1A.3.3.2. This clade of viruses emerged in swine, zoonotically infected humans and has since become endemic, replacing the existing seasonal H1N1 in humans [12,13]. Thus, the swine-origin pandemic 1A.3.3.2 clade gained sustained transmission, evolution, and adaptation in the human population [13]. Over the years since 2009, the 1A.3.3.2 human viruses were repeatedly reintroduced into swine herds [12], and have increased diversification of other swine HA clades by constantly adding human-origin internal genes to endemic swine H1 via reassortment [14]. Thus, reverse-zoonoses alters and enhances viral diversity in swine, and potentially impacts the likelihood of zoonotic infection through the pairing of human-origin genes to antigenically unique swine surface proteins.

Since 2009, efforts increased to prepare for the next potential IAV pandemic of swine origin. Applying the lessons learned from generating the H1N1pdm09 human vaccine, a selection of variant IAV from human zoonotic isolates have been used to create candidate vaccine viruses (CVV) [15,16]. If a swine-origin variant IAV emerged in the human population, a CVV could be used as seed stock to rapidly initiate vaccine production, provided there was antigenic cross-reactivity between the CVV and the variant IAV. Upon initial selection and generation, CVVs typically exhibit high cross reactivity to genetically similar viruses in swine [16]. However, evolution of IAV in the swine host can result in antigenic change that will reduce the efficacy of CVVs. Further, of the eight swine H1 clades currently circulating in the US, only five have an available CVV, and there is limited understanding of how well those CVVs react with the diverse array of contemporary swine viruses.

In a previous study, we demonstrated that swine H1 lineage strains from 2012–2019 were significantly different from human seasonal vaccine strains and this antigenic dissimilarity increased over time as the viruses evolved in swine [11,17]. Pandemic preparedness CVV strains also demonstrated a loss in within-clade cross-reactivity with tested swine strains. Human sera revealed a range of responses to swine H1 IAV, including two lineages of viruses with little to no immunity, 1A.1.1.3 and 1B.2.1 [11]. In this study, to further assess these swine H1 viruses, we identified contemporary, representative swine IAVs collected from 2019–2020. Selected viruses were tested against ferret antisera as a proxy for predicting the efficacy of available seasonal vaccines and CVV against current circulating swine IAVs. Of the tested strains, three swine H1 IAVs demonstrated reduced cross-reactivity to relevant CVVs and were derived from genetic clades that are frequently detected in surveillance. These strains were used in a pig-to-ferret transmission model to assess zoonotic transmission potential. This work uses in silico, in vitro, and in vivo approaches and identified gaps in current pandemic preparedness vaccine strategies by identifying three swine-origin H1 IAVs of zoonotic concern with a natural host species-based risk assessment.

2. Material and Methods

2.1. Genetic Analysis and Strain Selection

Human IAV vaccine composition and pandemic preparedness CVV assessments occur biannually at the WHO Vaccine Composition Meeting. In these meetings, animal influenza activity data are presented with 6-month windows along with human seasonal influenza activity data. Consequently, we downloaded all available swine HA H1 sequences that were collected and/or deposited in GISAID between 1 January 2020 and 30 June 2020 [18]. These sequences were aligned alongside CVV strains and human seasonal vaccine strains with MAFFT v7.453 [19], and each HA gene was classified to genetic clade within the octoFLU pipeline [20]; if whole genome data were available for a strain, each gene was similarly classified to evolutionary lineage to determine genome constellation. Following classification, sequences were translated to amino acid, and a consensus HA1 for each identified clade was generated using flutile (https://github.com/flu-crew/flutile accessed on 28 October 2022). A pairwise distance matrix was generated in Geneious Prime, and a wildtype field strain that was the best match to the HA1 clade consensus and that was available in the USDA IAV in swine virus repository was selected for additional characterization by hemagglutinin inhibition (HI) assay. Amino acid differences between CVVs or human seasonal vaccine strains and characterized swine IAV and clade consensus HA1s were generated using flutile (https://github.com/flu-crew/flutile accessed on 28 October 2022). These data were visualized through the inference of a maximum-likelihood phylogeny for the HA nucleotide alignment using IQ-TREE v2 implementing automatic model selection [21]. After selecting strains to represent contemporary swine H1 clades, selection criteria were expanded to include representative neuraminidase (NA) and internal gene constellations, the predominant evolutionary lineages were identified using octoFLUshow [5], and representative strains were tested against human seasonal vaccine and CVV ferret anti-sera.

2.2. Viruses and Ferret Antisera

Selected H1N1 and H1N2 isolates were obtained from the National Veterinary Services Laboratories (NVSL) through the U.S. Department of Agriculture (USDA) IAV swine surveillance system in conjunction with the USDA-National Animal Health Laboratory Network (NAHLN). Viruses used in this study were: 1A.1.1.3 A/swine/North Carolina/A02245416/2020 (sw/NC/20) and A/swine/Texas/A02245420/2020 (sw/TX/20); 1A.3.3.2 A/swine/Utah/A02432386/2019 (sw/UT/19); 1A.3.3.3 A/swine/Minnesota/A02245409/2020 (sw/MN/20); 1B.2.2.1 A/swine/Iowa/A02478968/2020 (sw/IA/20); 1B.2.2.2 A/swine/Colorado/A02245414/2020 (sw/CO/20); and 1B.2.1 A/swine/Illinois/A02139356/2018 (sw/IL/18). Vaccine viruses were provided by the Centers for Disease Control and Prevention (CDC), Atlanta, Georgia, USA and included 1A.1.1.3. IDCDC-RG59 A/Ohio/24/2017-CVV (OH/24/17), 1A.3.3.3 A/Ohio/9/2015 (OH/15), 1A.3.3.2 A/Idaho/7/2018 (ID/18), 1B.2.2.1 A/Iowa/32/2016 (IA/16), 1B.2.1 A/Ohio/35/2017 (OH/35/17), and 1B.2.1 A/Michigan/383/2018 (MI/18). Viruses were grown in Madin-Darby canine kidney (MDCK) cells in Opti-MEM (Life Technologies, Waltham, MA) with 10% fetal calf serum and antibiotics/antimycotics supplemented with 1g/mL tosyl phenylalanyl chloromethyl ketone (TPCK)-trypsin (Worthington Biochemical Corp., Lakewood, NJ, USA) at BSL2 containment.

Ferret antisera produced against CVV strains were kindly provided by CDC, Atlanta, Georgia, U.S. Antisera raised in ferrets against the following viruses were used: 1A.1.1.3 IDCDC-RG59 A/Ohio/24/2017-CVV, 1A.3.3.3 IDCDC-RG48 A/Ohio/9/2015-CVV, 1A.3.3.2 A/Idaho/7/2018, 1B.2.2.1 A/Iowa/32/2016, 1B.2.1 A/Ohio/35/2017, and 1B.2.1 A/Michigan/383/2018.

2.3. Hemagglutination Inhibition

Ferret antisera were heat inactivated at 56 °C for 30 min then treated with a 20% Kaolin suspension (Sigma-Aldrich, St. Louis, MO, USA) followed by adsorption with 0.75% guinea pig red blood cells (gpRBC) to remove nonspecific hemagglutination inhibitors as previously described [17]. Treated ferret antisera were used in HI assay with gpRBCs. Briefly, 4 HAU of virus in 25 µL was mixed with 25 µL of two-fold serially diluted serum. After a 30-min incubation at room temperature, 50 µL of 0.75% gpRBCs were added and allowed to settle for 1 h. Wells were observed for hemagglutination activity and the reciprocal of the highest serum dilution factor that prevented hemagglutination was recorded as the HI titer.

2.4. Swine-to-Ferret Transmission Study Design

Twenty 3-week-old piglets of mixed sex were obtained from an IAV- and porcine reproductive and respiratory syndrome virus-free herd. Prophylactic antibiotics (Excede; Zoetis, Florham Park, NJ, USA) were administered upon arrival to prevent potential respiratory bacterial infections. Sixteen 4–6-month-old male and female ferrets were obtained from an influenza-free high health source. Animals were housed under BSL2 containment in compliance with the USDA-ARS NADC institutional animal care and use committee. Serum was collected from each pig and ferret and screened by a commercial enzyme-linked immunosorbent assay (ELISA) (MultiS ELISA; Idexx, Westbrook, ME, USA) prior to experimental manipulations to confirm all animals were free of prior immunity and maternally acquired IAV specific antibodies. Pigs were divided randomly into groups of 5 and placed into separate containment rooms. Three groups received 2mL of sw/TX/20, sw/MN/20, or sw/IL/18 IAV inoculum at 1×10^6 TCID$_{50}$ via intranasal administration, while the fourth group served as an non-inoculated control At two days post inoculation (dpi) four ferrets were placed in the room in separate, open-fronted isolators placed approximately 4 feet from pig decking [22]. All animals received a subcutaneous radio frequency microchip (pigs: Deston Fearing, Dallas, TX, USA; Ferrets: Biomedic Data Systems Inc., Seaford, DE, USA) for identification and body temperature monitoring purposes. Body temperature and weight (ferrets only) were recorded from −3 to 14 dpi, with the readings recorded prior to exposure used for establishing a baseline. Ferrets were provided routine care and handled before pigs, with a change in outer gloves and decontamination of equipment with 70% ethanol between individual ferrets.

Three pigs from each experimental group were euthanized at 5dpi and necropsied to evaluate lung lesions and collect bronchoalveolar lavage fluid (BALF) [23]. All pigs were nasal swabbed at 0, 1, 3, and 5 dpi as previously described. The remaining pigs were swabbed on 7 and 9 dpi and euthanized at 14dpi. Blood samples were collected prior to exposure and at necropsy for all pigs.

Contact ferrets were sampled by nasal wash collection at 0, 1, 3, 5, 7, 9, 11 and 12-days post contact (dpc) [22]. BALF samples from ferrets were collected at necropsy (12 dpc) [22]. Blood samples were collected prior to exposure and at 12 dpc to test for seroconversion by HI assays and [17,24] by a commercial NP-ELISA (MultiS ELISA; Idexx, Westbrook, ME, USA).

2.5. Virus Replication and Shedding

Swine nasal swabs, ferret nasal washes and BALF samples were titrated on MDCK cells to evaluate virus replication in the nose and lungs, as previously described [23]. Inoculated monolayers were evaluated for cytopathic effect (CPE) between 48 and 72 h post-infection, and positive wells were identified by testing supernatant via hemagglutination assay with turkey RBC. A TCID$_{50}$/mL titer was calculated for each sample using the method described by Reed and Muench [25]. Animal samples were processed in BSL2 containment.

2.6. Pathology Examination

Swine lungs were evaluated for lesions at 5 dpi following standard protocols to assess pathogenesis in swine and potential for transmission to ferrets [23]. Tissue samples from the trachea and right middle or affected lung lobe were fixed in 10% buffered formalin for histopathologic examination. Tissues were processed by routine histopathologic procedures and slides stained with hematoxylin and eosin (H&E). The percentage of the lung affected with pneumonic consolidation typical of influenza virus in ferrets was visually estimated at 12 dpc to assess resolution of disease following transmission, following methods of scoring previously described [22].

2.7. Microbiological Assays

Swine BALF samples were cultured for aerobic bacteria on blood agar and Casmin (NAD-enriched) plates to indicate the presence of concurrent bacterial pneumonia. To exclude other causes of pneumonia in pigs, qPCR assays were conducted for porcine circovirus 2 (PCV2) [26], and for *Mycoplasma hyopneumoniae* and North American and European PRRSV (VetMax; Life Technologies, Carlsbad, CA, USA) according to the manufacturer's recommendations. Data not shown as no confounding infections were identified.

2.8. Data Analysis

Results were analyzed with Prism 8 (GraphPad, San Diego, CA, USA) with analysis of variance (ANOVA), with $p < 0.05$ considered significant. Variables with significant effects by treatment group were subjected to pairwise mean comparisons using the Tukey–Kramer test.

3. Results

3.1. Genetic and Phylogenetic Characterization of US Swine H1 Hemagglutinin

Between 1 January 2020 and 30 June 2020, 342 swine IAV isolates with an H1 HA were identified. These viruses represented 8 genetic clades across two evolutionary lineages: 1A.1.1.3 (n = 36, 10.6%), 1A.2 (n = 3, 0.9%), 1A.2-3-like (n = 3, 0.9%), 1A.3.3.2 (n = 53, 15.5%), 1A.3.3.3 (n = 145, 42.4%), 1B.2.1 (n = 80, 23.4%), 1B.2.2.1 (n = 15, 4.4%), 1B.2.2.2 (n = 7, 2%) (Figure 1). For each detected H1 clade, HA1 amino acid sequences were aligned, a consensus sequence generated, and a wildtype virus with highest HA1 similarity to consensus was selected to represent the clade (Table 1). The percent amino acid identity of selected strains ranged from 96.63–99.39% when compared to matching within-clade consensus and ranged from 90.83–98.16% when compared to within-clade CVVs (Table 1). The number of amino acid differences between the representative swine HA gene and the within-clade CVV or human seasonal vaccine ranged from 6 to 29 amino acid differences (Supplemental Tables S1–S6).

3.2. Dominant U.S. Swine H1 Strains Drifted from Human Seasonal H1 or CVV

In addition to current human seasonal vaccines, CVV strains were selected and generated by WHO collaborating centers to mitigate a future potential outbreak of swine IAV in humans. Ferret antisera generated against human vaccine and CVV strains and other variant IAV viruses were tested by HI to determine the relative cross-reactivity to contemporary swine viruses. Within the 1A lineage of HA genes, there were 2 CVVs and the human seasonal H1pdm09 vaccine strain that correspond to the 1A1.1, 1A.3.3.3 and 1A.3.3.2 clades, respectively (Table 2). The consensus 1A1.1 contemporary representative virus, sw/NC/20, had an HI titer of 80 against CVV OH/24/17 antiserum compared to the homologous OH/24/17 titer of 1280, representing a 16-fold reduction in cross-reactivity. Similarly, the 1A.3.3.3 contemporary representative sw/MN/20 virus displayed a 32-fold reduction in HI activity compared to the homologous 1A.3.3.3 CVV OH/15 titer. Conversely, the 1A.3.3.2 selected virus, sw/UT/19, displayed no loss in HI titer as compared to the homologous titer for ID/18 (A/Brisbane/02/2018 (H1N1)-like) human strain.

The 1B lineage of HA genes also had three antisera generated against variant viruses used to generate CVVs: 1B.2.2.1 IA/16 and 1B.2.1 OH/35/17 and MI/18. The 1B.2.2.1 virus, sw/IA/20, displayed an 8-fold reduction in cross-reactivity compared to the 1B.2.2.1 CVV IA/16 virus (Table 3). There is no clade-specific CVV for the 1B.2.2.2 clade of swine viruses and the representative virus, sw/CO/20 had limited reactivity to all potential vaccine sera. Finally, the 1B.2.1 virus, sw/IL/18, was antigenically very similar (2-fold or less reduction) to both 1B.2.1 CVV antisera.

3.3. Swine-to-Ferret Transmission

The antigenic data indicated the three clades of swine IAV with lowest reactivity to vaccine antisera: 1A.1.1.3, 1A.3.3.3, and 1B.2.2.1. These 3 clades accounted for 54.2% of H1 subtype IAV swine isolates during 2020 [5]. Therefore, viruses from 1A.1.1.3, 1A.3.3.3, and 1B.2.2.1 clades were selected to test zoonotic potential using a pig-to-ferret interspecies transmission model. While antigenicity is primarily driven by genetic factors within the HA1 domain of the HA gene, zoonoses is affected by viral factors attributed throughout the genome. To address this, we expanded our selection criteria to include NA and internal gene constellations as defined with the octoFLU tool (20) (n = 225 H1N1 and H1N2 whole genome sequences collected in 2020: Supplemental Figure S1). During 2020, the 1B.2.2.1 HA gene was primarily paired with a N2-2002B gene with a TTTTPT internal gene constellation; the 1A.3.3.3 HA gene was primarily paired with a N1-Classical gene with a TTTPPT internal gene constellation. The 1A.3.3.3 (sw/MN/20) and 1B.2.2.1 (sw/IA/20) viruses selected for the antigenic characterization matched the predominant circulating NA and dominant internal gene constellations and thus remained unchanged. For the 1A.1.1.3 HA clade, the primary NA pairing in 2020 was a N2-2002A gene with detections of TTTTPT, TTTPPT, TTPTPT, and TTPPPT. The sw/NC/20 virus used for HI assays did not match the predominant N2-NA gene, and had a TTTPPT internal gene constellation, and consequently, the strain selected for subsequent in vivo studies was sw/TX/20 that had a N2-2002A gene with a TTTTPT internal gene constellation.

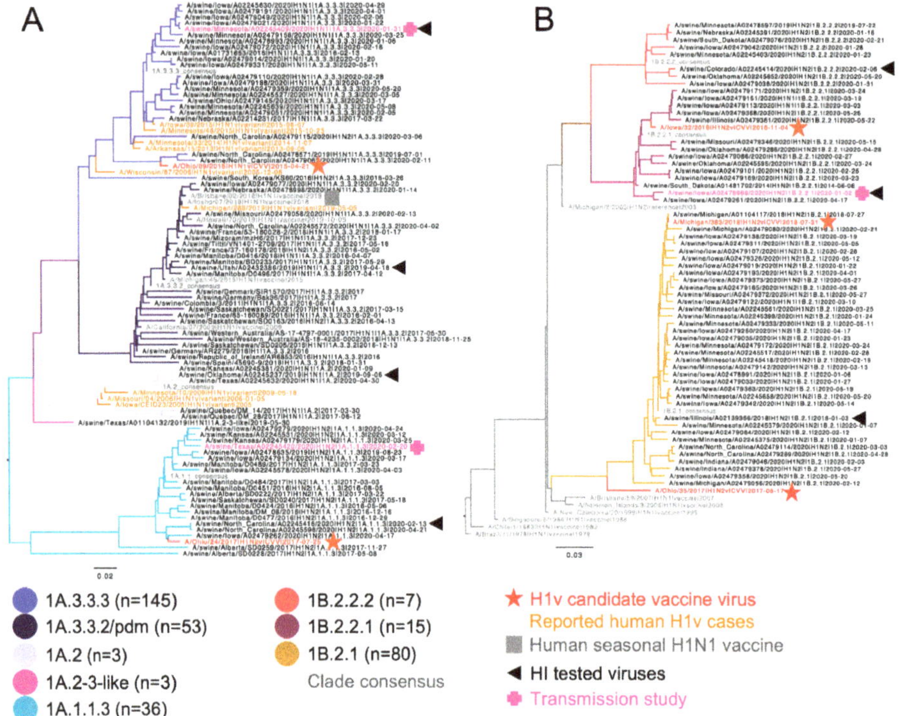

Figure 1. Phylogenetic relationships of North American swine H1 IAV. A representative random sample of (**A**) 1A classical swine lineage and (**B**) 1B human-like lineage swine HA genes from January 2020 through June 2020. Reference human HA genes, CVV, and variant cases are indicated by branch color or shapes. Swine IAV strains tested in hemagglutination inhibition assays are marked by a black triangle and those used in transmission studies by a pink plus (+). The numbers in parentheses in the color key indicate number of each genetic clade detected during the sampling period.

Table 1. Pairwise amino acid sequence similarity of the HA1 domain from swine H1 clade consensus sequences to CVV or seasonal vaccine virus and the swine hemagglutinin clade representative viruses used in this study. Within-clade comparisons are highlighted in grey.

	1A.1.1.3 Cons	A/OH/24/2017 CVV	A/sw/NC/A02245416/2020	A/sw/TX/A02245420/2020	1A.3.3.2 Cons	A/ID/07/2018	A/sw/UT/A02432386/2019	1A.3.3.3 Cons	A/OH/9/2015 CVV	A/sw/MN/A02245409/2020	1B.2.1 Cons	A/OH/35/2017 CVV	A/MI/383/2018	A/sw/IL/A02139356/2018	1B.2.2.1 Cons	A/IA/32/2016 CVV	A/sw/IA/A02478968/2020	1B.2.2.2 Cons	A/sw/CO/A02245414/2020
1A.1.1.3 consensus		98.77	98.77	95.08	82.57	82.57	82.28	81.35	82.57	80.43	70.86	70.86	70.86	71.47	71.78	71.47	71.47	71.78	70.25
A/OH/24/2017	98.77		97.54	94.46	83.49	83.49	82.89	81.96	83.49	81.04	71.47	71.17	71.47	72.09	72.39	72.09	72.09	72.09	70.55
A/sw/NC/A02245416/2020	98.77	97.54		93.85	81.35	81.35	81.05	80.12	81.35	79.21	69.63	69.63	69.63	70.25	70.55	70.25	70.25	70.55	69.63
A/sw/TX/A02245420/2020	95.08	94.46	93.85		81.35	81.35	81.36	79.82	81.35	78.90	69.33	69.33	69.63	69.94	70.86	70.55	70.55	70.55	69.02
1A.3.3.2 consensus	82.57	83.49	81.35	81.35		99.69	97.26	87.77	86.85	86.54	70.03	70.34	70.64	70.64	70.95	70.03	70.64	70.95	69.42
A/ID/07/2018	82.57	83.49	81.35	81.35	99.69		97.57	88.07	87.16	86.85	70.03	70.34	70.64	70.64	70.95	70.03	70.64	70.95	69.42
A/sw/UT/A02432386/2019	82.28	82.89	81.05	81.36	97.26	97.57		88.70	86.56	87.48	70.35	70.66	70.96	70.96	71.27	70.35	70.96	70.66	69.13
1A.3.3.3 consensus	81.35	81.96	80.12	79.82	87.77	88.07	88.70		91.44	98.78	71.56	71.56	71.56	72.17	71.87	71.25	71.56	72.78	71.87
A/OH/9/2015	82.57	83.49	81.35	81.35	86.85	87.16	86.56	91.44		90.83	70.34	69.42	70.95	70.64	70.34	69.73	70.34	70.64	69.11
A/sw/MN/A02245409/2020	80.43	81.04	79.21	78.90	86.54	86.85	87.48	98.78	90.83		71.56	71.87	71.56	71.87	72.17	71.56	71.87	73.09	72.17
1B.2.1 Consensus	70.86	71.47	69.63	69.33	70.03	70.03	70.35	71.56	70.34	71.56		93.56	98.77	99.39	85.58	84.97	85.58	86.50	84.05
A/OH/35/2017	70.86	71.17	69.63	69.33	70.34	70.34	70.66	71.56	69.42	71.87	93.56		93.25	93.56	85.89	85.58	85.28	85.58	84.05
A/MI/383/2018	70.86	71.47	69.63	69.63	70.64	70.64	70.96	71.56	70.95	71.56	98.77	93.25		98.16	85.89	85.28	85.89	86.20	83.74
A/sw/IL/A02139356/2018	71.47	72.09	70.25	69.94	70.64	70.64	70.96	72.17	70.64	71.87	99.39	93.56	98.16		86.20	85.58	86.20	87.12	84.66
1B.2.2.1 consensus	71.78	72.39	70.55	70.86	70.95	70.95	71.27	71.87	70.34	72.17	85.58	85.89	85.89	86.20		98.47	98.16	91.72	88.65
A/IA/32/2016	71.47	72.09	70.25	70.55	70.03	70.03	70.35	71.25	69.73	71.56	84.97	85.58	85.28	85.58	98.47		96.63	91.72	88.65
A/sw/IA/A02478968/2020	71.47	72.09	70.25	70.55	70.64	70.64	70.96	71.56	70.34	71.87	85.58	85.28	85.89	86.20	98.16	96.63		90.49	87.42
1B.2.2.2 consensus	71.78	72.09	70.55	70.55	70.95	70.95	70.66	72.78	70.64	73.09	86.50	85.58	86.20	87.12	91.72	91.72	90.49		96.63
A/sw/CO/A02245414/2020	70.25	70.55	69.63	69.02	69.42	69.42	69.13	71.87	69.11	72.17	84.05	84.05	83.74	84.66	88.65	88.65	87.42	96.63	

Table 2. Antigenic cross-reactivity of 1A viruses and within-clade CVVs. Ferret antisera raised against CVV and vaccine viruses were tested for the ability to inhibit hemagglutination of contemporary swine viruses. Vaccine strains and homologous titers are bolded; grey highlighted cells indicate the within-clade titer of contemporary swine strain.

Strain	Lineage	IDCDC-RG59 A/Ohio/24/2017-CVV	IDCDC-RG48 A/Ohio/9/2015-CVV	A/Idaho/7/2018 *
A/Ohio/24/2017	**1A.1.1.3**	**1280**	<10	640
A/swine/North Carolina/A02245416/2020	1A.1.1.3	80	<10	<10
A/swine/Oklahoma/A02245237/2019	1A.2	20	320	640
A/Ohio/9/2015	**1A.3.3.3**	<10	**2560**	<10
A/swine/Minnesota/102245409/2020	1A.3.3.3	<10	80	40
A/Idaho/7/2018 *	**1A.3.3.2**	<10	<10	**1280**
A/swine/Utah/A02432386/2019	1A.3.3.2	40	160	2560

* A/Brisbane/02/2018 (H1N1)-like.

Table 3. Antigenic cross-reactivity of 1B viruses and within-clade CVVs. Ferret antisera raised against CVV were tested for the ability to inhibit hemagglutination of contemporary swine viruses. Vaccine strains and homologous titers are bolded; grey highlighted cells indicate the within-clade titer of contemporary swine strains.

Strain	Lineage	A/Iowa/32/2016	A/Ohio/5/2017	A/Michigan/383/2018
A/Iowa/32/2016	1B.2.2.1	**640**	20	10
A/swine/Iowa/A02478968/2020	1B.2.2.1	80	10	40
A/swine/Colorado/A02245414/2020	1B.2.2.2	40	10	10
A/Ohio/35/2017	1B.2.1	80	**640**	40
A/Michigan/383/2018	1B.2.1	40	160	**1280**
A/swine/Illinois/A02139356/2018	1B.2.1	20	320	1280

All three selected viruses had similar shedding patterns in pigs in terms of peak and duration (Figure 2). Evaluation of samples collected at the 5dpi necropsy revealed similar levels of macroscopic pathology between the 1A.1.1.3 (alpha) and the 1B.2.2.1 (delta-1a) virus groups at 2.4 and 2.3 percent of affected lung surface respectively. These two groups had a mean BALF titer of 5.6 $\log_{10}$ TCID50/mL and 5.8 $\log_{10}$ TCID$_{50}$/mL respectively. The 1A.3.3.3 (gamma) virus had higher average percentage of lung lesions (8.1%) and higher BALF titers (7.3 $\log_{10}$ TCID$_{50}$/mL) than the other groups. All 6 remaining pigs seroconverted against the respective virus at 14dpi with an average HI titer of 905, 640, and 80 in 1A.1.1.3, 1A.3.3.3, and 1B.2.2.1, respectively. These data demonstrated the propensity for the pigs to seed the room with aerosolized virus to expose the ferrets.

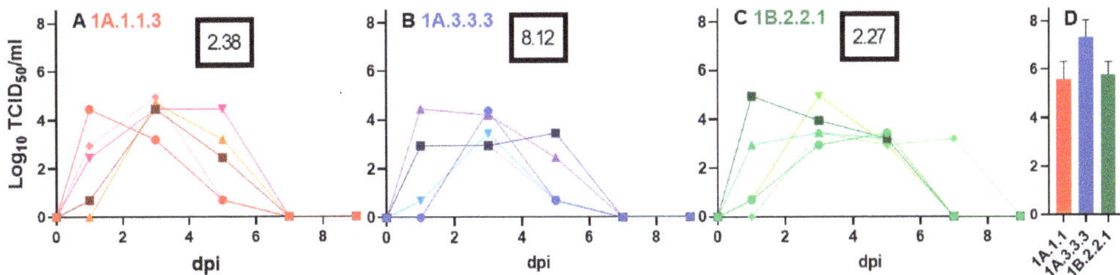

Figure 2. Shedding and replication of clade representative viruses in pigs. Individual pig nasal swabs ((**A–C**), each line represents an individual pig) and group mean bronchoalveolar lavage fluid (BALF, n = 3) (**D**) viral load shown as $\log_{10}$ TCID$_{50}$/mL on MDCK cells. Number in the black box indicates the average (n = 3) percentage of lung surface with visible pneumonic lesions at the 5dpi necropsy.

The three viruses displayed different levels of transmissibility to contact ferrets (Figure 3). All four 1A.1.1.3 contact ferrets shed virus with an average of 2.3 positive samples over the course of the study and an average peak titer of 5.9 $\log_{10}$ TCID$_{50}$/mL in nasal washes. All four ferrets seroconverted with a geometric mean HI titer of 679 at 12 dpc. The four 1A.3.3.3 contact ferrets all seroconverted as well, but with a lower geometric mean titer (231) and only two of the four ferrets had recoverable viral loads in the nasal washes with an average of 2.5 positive samples and an average peak titer of 5.7 $\log_{10}$ TCID$_{50}$/mL in nasal washes. In contrast, only one of four 1B.2.2.1 contact ferrets seroconverted (HI = 160) and had 2 nasal wash positive samples with a peak titer of 4.9 $\log_{10}$ TCID$_{50}$/mL. No ferret had detectable virus in the BALF samples collected at 12 dpc.

Minimal signs of disease were observed in the ferrets. Daily temperature measurements revealed minimal elevation in temperatures and no differences among or between treatment groups (all temperatures were within ± 1.1 °C of baseline temps). Body weight monitoring revealed if a ferret shed virus, regardless of group, it gained significantly less weight (3.5 $\pm$ 2.9%) compared to ferrets that did not shed virus (10.9 $\pm$ 6.5%) (Table 4), but there were no significant differences in change in body weight between virus groups. Necropsy on 12dpc revealed minimal gross pathology, with only two ferrets, one 1A.1.1.3 (4.6%) and one 1B.2.2.1 (3.5%), demonstrating visible lung lesions. No lesions were observed in any 1A.3.3.3 exposed ferrets at 12 dpc.

Table 4. Cumulative clinical, viral, and serological measures of ferret infection.

1	Ferret Number	Change in Bodyweight (%) from 0–12 DPC	DPC with Nasal Shedding	Peak Nasal Titer *	12 DPC HI Titer
1A.1.1.3	1	4.61	5, 7	5.45	160
	2	1.19	3, 5, 7	5.94	1280
	3	4.89	3, 5, 7	5.94	1280
	4	7.71	5	6.2	1280
1A.3.3.3	5	21.36	none	none	80
	6	16.83	none	none	160
	7	2.40	5, 7	5.87	320
	8	−1.13	3, 5, 7	5.53	1280
1B.2.2.1	9	6.74	none	none	<10
	10	9.84	none	none	<10
	11	5.02	5, 7	4.95	160
	12	12.01	none	none	<10
No virus	13	7.25	none	none	<10
	14	2.35	none	none	<10

* $\log_{10}$ TCID$_{50}$/mL.

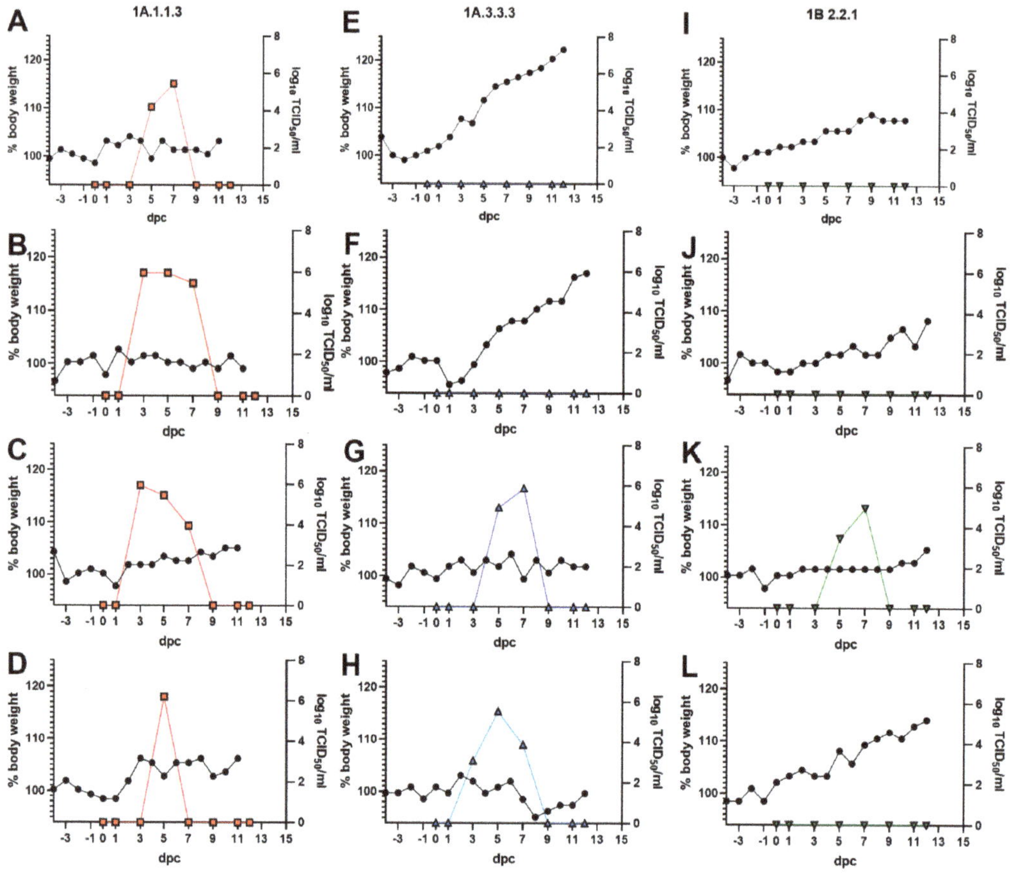

Figure 3. Transmission and replication of swine viruses to ferrets. Individual ferrets exposed to pigs infected with a 1A.1.1.3 ((**A–D**), red squares), 1A.3.3.3 ((**E–H**), blue up-triangles) or 1B.2.2.1((**I–L**), green down-triangles) had body weights recorded daily and converted to a percentage of the 3-day average body weight prior to exposure with 100% being baseline (left axis, black circles). Nasal washes were measured for viral shedding in MDCK cells and recorded as $\log_{10}$ TCID$_{50}$/mL (right axis, color shape).

4. Discussion

Animal origin IAV from avian or swine are a documented source of human IAV zoonotic infections, epidemics, and pandemics. The four most recent IAV pandemics were all driven by either direct zoonosis or by reassortment and zoonosis [9,27]. Regional epidemics and individual infections were also caused by avian and swine viruses (https://www.cdc.gov/flu/weekly/fluviewinteractive.htm accessed on 9 September 2020). In addition to the 2009 pandemic, swine-origin IAV are also responsible for human infections, termed variants, ranging from single cases up to outbreaks of several hundred individual infections without onward transmission. Much progress has been made in preparing for future zoonotic IAV pandemics, with the most proactive efforts centered on the generation of CVV from animal isolates and human isolates of animal origin, including variant viruses of swine origin. These stockpiled CVVs would be used as seed viruses for rapid vaccine generation should an antigenically similar animal origin virus initiate a human pandemic.

Swine IAV in the US are very diverse. In 2020 there were 14 antigenically distinct HA clades isolated from US swine herds, 8 of which were of the H1 subtype [4,5]. Contemporary clade consensus HA1 sequences can have as little as 70% amino acid similarity to divergent HA1 between swine H1 clades and within-clade HA1 sequences can be as much as 15% different. This high level of within- and between-clade genetic diversity makes achieving and maintaining high levels of vaccine coverage difficult and necessitates the continued evaluation of CVV antisera reactivity against contemporary swine IAV isolates.

Of the 8 circulating swine H1 clades, five have an existing within-clade human seasonal or CVV vaccine virus, including the 1A.3.3.2 component of human seasonal flu vaccines [15]. Greater than 95% of 2020 US swine IAV isolates fall within those 5 human vaccine-covered clades [4,5]. Contemporary, clade-representative viruses of two of these clades showed high levels of cross-reactivity to existing vaccines, 1A.3.3.2 and 1B.2.1. Cross-species events involving human-to-swine infection of 1A.3.3.2 viruses in pigs are common in the US [6,12]. This continuous influx of human viruses makes it unsurprising that a representative swine 1A.3.3.2 virus had high levels of cross reactivity with a human 1A.3.3.2 vaccine. The 1B.2.1 A/Michigan/383/2018 is the most recently generated CVV. As such, it follows that the 1B.2.1 clade had not antigenically drifted and high levels of cross-reactivity were expected and observed. Three swine IAV clades have no within-clade vaccine or CVV available: 1A.2, 1A.2-3-like, and 1B.2.2.2. Additionally, these three clades had limited cross-reactivity to CVVs from other genetic clades. However, these three clades only represented 4.5% of 2020 US swine IAV isolates. This relative scarcity may minimize the opportunities for zoonotic transmission and reduced the priority for assessing their pandemic risk posed to humans at this time. However, relative detection frequency of swine HA clades changes over time and these clades may need to be reassessed in the future given frequent interstate movement of pigs and viruses [28–31]. Contemporary clade representative isolates from 1A.1.1.3 (16-fold reduction), 1A.3.3.3 (32-fold reduction) and 1B.2.2.1 (8-fold reduction) exhibited high levels of antigenic drift from relevant CVVs and these three H1 swine clades represented 54.2% of 2020 US swine IAV isolates [5]. These clades have high frequency of detection in US swine herds and have reduced vaccine reactivity to human CVV, indicating a higher potential pandemic risk and requiring further examination of transmission risk factors. The 1A.1.1.3 and 1B.2.2.1 swine H1 clades also showed low detection by human population sera in a previous study [11].

To address zoonotic potential, these viruses were used in a swine-to-ferret interspecies transmission study utilizing an aerosol respiratory contact model. While all three viruses exhibited some level of interspecies transmission, they did so with varying efficiency. The 1A.1.1.3 virus had 100% transmission from pigs to ferrets, indicated by all ferrets shedding virus and seroconverting. All four of the 1A.3.3.3 exposed ferrets also seroconverted, albeit with a lower average HI titer compared to the 1A.1.1.3, but only two of the four ferrets shed virus. Finally, one 1B.2.2.1 ferret seroconverted and shed virus while the other three remained naïve, indicating a reduced propensity for interspecies transmission. The infected ferrets displayed signs of disease measured as a cessation of weight gain compared to noninfected ferrets, but no other overt signs and postmortem evaluation of the lungs revealed minimal pathological damage at 12 dpc. Since this study was focused on transmission rather than pathogenesis in ferrets, further work to determine lung pathology during the active infection phase would be necessary. Nasal titers over the time course and BALF titers on 5 dpi in pigs were similar for all three viruses, indicating that infection and replication kinetics in pigs did not affect transmission to ferrets. The virus and host factors contributing to the lower nasal shedding of the 1A.3.3.3 and the lower transmission of the 1B.2.2.1 swine strains in the contact ferrets are currently unknown, but potentially associated with the diverse gene segment combinations and evolutionary origins of the three viruses.

Results of this study indicate that swine IAV from the US may escape vaccine immunity from CVV or seasonal vaccines as they continue to circulate and evolve in the swine population. Three H1 clades demonstrated antigenic drift away from available CVV antisera. Additionally, contemporary clade representatives showed the ability to transmit from pigs to ferrets, a gold standard for human influenza transmissibility. These data highlight the increased risk to human populations posed by H1 clades of swine IAV, particularly the 1A.1.1.3. Since the conclusion of these experiments in July 2020, there were an additional 15 H1 variant cases in North America with the HA clade determined; an additional 3 variants had insufficient data to identify the HA clade. Of these variant IAVs, 2 came from the 1A.1.1.3 clade, 4 were derived from the 1A.3.3.3 clade, 5 were from the 1A.3.3.2 clade, and 4 were from the 1B.2.1 clade, overlapping with the strains tested here. These data highlight the utility of swine-to-ferret transmission studies as a pandemic risk assessment tool and identifies the gaps in CVV coverage of US H1 swine IAV. These results stress the need to continually assess the within-clade cross-reactivity of existing CVVs to identify and develop more contemporarily relevant pandemic preparedness strains.

Supplementary Materials: The following supporting information can be downloaded at: https://www.mdpi.com/article/10.3390/v14112398/s1, Table S1: Amino acid differences between 1A.1.1.3 clade consensus, the within-clade CVV A/Ohio/24/2017, and the clade representative viruses A/swine/North Caroline/A02245416/2020 and A/swine/Texas/A02245420/2020; Table S2: Amino acid differences between 1A.3.3.2 clade consensus, the within-clade human seasonal vaccine A/Idaho/07/2018, and the clade representative A/swine/Utah/A02432386/2019; Table S3: Amino acid differences between 1A.3.3.3 clade consensus, the within-clade CVV A/Ohio/09/2015, and the clade representative virus A/swine/Minnesota/A02245409/2020; Table S4: Amino acid differences between 1B.2.1 clade consensus, the within-clade CVVs A/Ohio/35/2017 and A/Michigan/383/2018, and the clade representative virus A/swine/Illinois/A02139356/2018; Table S5: Amino acid differences between 1B.2.2.1 clade consensus, the within-clade CVV A/Iowa/32/2016 and the clade representative virus A/swine/Iowa/A02478968/2020; Table S6: Amino acid differences between 1B.2.2.2 clade consensus and the clade representative virus A/swine/Colorado/A02245414/2020; Figure S1: Detection proportions of H1N1 and H1N2 influenza A virus in swine collected in 2020 in the USDA influenza A virus in swine surveillance system.

Author Contributions: Conceptualization, J.B.K., C.K.S., T.K.A. and A.L.V.B.; methodology, J.B.K., C.K.S., T.K.A., N.S.L. and A.L.V.B.; formal analysis, J.B.K., T.K.A., D.E.H., Z.W.A. and T.K.A.; investigation, J.B.K., C.K.S., K.M.Y. and A.L.V.B.; resources, S.T. and C.T.D.; writing—original draft preparation, J.B.K.; writing—review and editing, J.B.K., T.K.A. and A.L.V.B.; supervision, T.K.A. and A.L.V.B.; funding acquisition, T.K.A. and A.L.V.B. All authors have read and agreed to the published version of the manuscript.

Funding: This work was supported in part by: the U.S. Department of Agriculture (USDA) Agricultural Research Service [ARS project number 5030-32000-231-000-D]; the National Institute of Allergy and Infectious Diseases, National Institutes of Health, Department of Health and Human Services [contract numbers HHSN272201400008C and 75N93021C00015]; the Centers for Disease Control and Prevention [21FED2100395IPD]; the USDA Agricultural Research Service Research Participation Program of the Oak Ridge Institute for Science and Education (ORISE) through an interagency agreement between the U.S. Department of Energy (DOE) and USDA Agricultural Research Service [contract number DE-AC05- 06OR23100]; and the SCINet project of the USDA Agricultural Research Service [ARS project number 0500-00093-001-00-D]. The funders had no role in study design, data collection and interpretation, or the decision to submit the work for publication.

Institutional Review Board Statement: The study was approved by the USDA-ARS National Animal Disease Center Institutional Animal Care and Use Committee (protocol code ARS-2017-660; ARS-2019-779 and date of approval are 13 November 2017 and 9 March 2019).

Data Availability Statement: Clinical data associated with this study are available for download from the USDA Ag Data Commons at https://doi.org/10.15482/USDA.ADC/1528065 and the phylogenetic analyses from https://github.com/flu-crew/datasets (Accessed on 28 October 2022).

Acknowledgments: We gratefully acknowledge pork producers, swine veterinarians, and laboratories for participating in the USDA Influenza A Virus in Swine Surveillance System and publicly sharing sequences. We also gratefully acknowledge all data contributors, i.e., the Authors and their Originating laboratories responsible for obtaining the specimens, and their Submitting laboratories for generating the genetic sequence and metadata and sharing via the GISAID Initiative, on which components of this research is based. Mention of trade names or commercial products in this article is solely for the purpose of providing specific information and does not imply recommendation or endorsement by the USDA, CDC, DOE, or ORISE. USDA and CDC are equal opportunity providers and employers. The findings and conclusions in this report are those of the authors and do not necessarily represent the views of the Centers for Disease Control and Prevention or the Agency for Toxic Substances and Disease Registry.

Conflicts of Interest: The authors declare no conflict of interest.

References

1. Lambert, L.C.; Fauci, A.S. Influenza vaccines for the future. *N. Engl. J. Med.* **2010**, *363*, 2036–2044. [CrossRef] [PubMed]
2. Girard, M.P.; Cherian, T.; Pervikov, Y.; Kieny, M.P. A review of vaccine research and development: Human acute respiratory infections. *Vaccine* **2005**, *23*, 5708–5724. [CrossRef] [PubMed]
3. Centers for Disease Control and Prevention. Estimates of deaths associated with seasonal influenza—United States, 1976–2007. *Morb. Mortal. Wkly. Rep.* **2010**, *59*, 1057–1062.
4. Zeller, M.A.; Anderson, T.K.; Walia, R.W.; Vincent, A.L.; Gauger, P.C. ISU FLUture: A veterinary diagnostic laboratory web-based platform to monitor the temporal genetic patterns of Influenza A virus in swine. *BMC Bioinform.* **2018**, *19*, 397. [CrossRef] [PubMed]
5. Arendsee, Z.W.; Chang, J.; Hufnagel, D.E.; Markin, A.; Janas-Martindale, A.; Vincent, A.L.; Anderson, T.K. octoFLUshow: An Interactive Tool Describing Spatial and Temporal Trends in the Genetic Diversity of Influenza A Virus in U.S. Swine. *Microbiol. Resour. Announc.* **2021**, *10*, e01081-21. [CrossRef] [PubMed]
6. Nelson, M.I.; Vincent, A.L. Reverse zoonosis of influenza to swine: New perspectives on the human-animal interface. *Trends Microbiol.* **2015**, *23*, 142–153. [CrossRef]
7. Vijaykrishna, D.; Smith, G.J.; Pybus, O.G.; Zhu, H.; Bhatt, S.; Poon, L.L.; Riley, S.; Bahl, J.; Ma, S.K.; Cheung, C.L.; et al. Long-term evolution and transmission dynamics of swine influenza A virus. *Nature* **2011**, *473*, 519–522. [CrossRef] [PubMed]
8. Rajao, D.S.; Gauger, P.C.; Anderson, T.K.; Lewis, N.S.; Abente, E.J.; Killian, M.L.; Perez, D.R.; Sutton, T.C.; Zhang, J.; Vincent, A.L. Novel Reassortant Human-Like H3N2 and H3N1 Influenza A Viruses Detected in Pigs Are Virulent and Antigenically Distinct from Swine Viruses Endemic to the United States. *J. Virol.* **2015**, *89*, 11213–11222. [CrossRef] [PubMed]
9. Neumann, G.; Noda, T.; Kawaoka, Y. Emergence and pandemic potential of swine-origin H1N1 influenza virus. *Nature* **2009**, *459*, 931–939. [CrossRef]
10. Anderson, T.K.; Macken, C.A.; Lewis, N.S.; Scheuermann, R.H.; Van Reeth, K.; Brown, I.H.; Swenson, S.L.; Simon, G.; Saito, T.; Berhane, Y.; et al. A Phylogeny-Based Global Nomenclature System and Automated Annotation Tool for H1 Hemagglutinin Genes from Swine Influenza A Viruses. *Msphere* **2016**, *1*, e00275-16. [CrossRef] [PubMed]
11. Venkatesh, D.; Anderson, T.K.; Kimble, J.B.; Chang, J.; Lopes, S.; Souza, C.K.; Pekosz, A.; Shaw-Saliba, K.; Rothman, R.E.; Chen, K.-F. Antigenic characterization and pandemic risk assessment of North American H1 influenza A viruses circulating in swine. *Spectrum* **2022**.
12. Nelson, M.I.; Stratton, J.; Killian, M.L.; Janas-Martindale, A.; Vincent, A.L. Continual Reintroduction of Human Pandemic H1N1 Influenza A Viruses into Swine in the United States, 2009 to 2014. *J. Virol.* **2015**, *89*, 6218–6226. [CrossRef] [PubMed]
13. Su, Y.C.F.; Bahl, J.; Joseph, U.; Butt, K.M.; Peck, H.A.; Koay, E.S.C.; Oon, L.L.E.; Barr, I.G.; Vijaykrishna, D.; Smith, G.J.D. Phylodynamics of H1N1/2009 influenza reveals the transition from host adaptation to immune-driven selection. *Nat. Commun.* **2015**, *6*, 7952. [CrossRef] [PubMed]
14. Vijaykrishna, D.; Poon, L.L.; Zhu, H.C.; Ma, S.K.; Li, O.T.; Cheung, C.L.; Smith, G.J.; Peiris, J.S.; Guan, Y. Reassortment of pandemic H1N1/2009 influenza A virus in swine. *Science* **2010**, *328*, 1529. [CrossRef] [PubMed]
15. WHO. Summary of Status of Development and Availability of Variant Influenza A(H1) Candidate Vaccine Viruses and Potency Testing Reagents. 2020. Available online: https://www.who.int/influenza/vaccines/virus/candidates_reagents/summary_a_h1 v_cvv_sh2021_20200930.pdf?ua=1 (accessed on 15 June 2020).
16. Robertson, J.S.; Nicolson, C.; Harvey, R.; Johnson, R.; Major, D.; Guilfoyle, K.; Roseby, S.; Newman, R.; Collin, R.; Wallis, C.; et al. The development of vaccine viruses against pandemic A(H1N1) influenza. *Vaccine* **2011**, *29*, 1836–1843. [CrossRef] [PubMed]
17. Anderson, T.K.; Chang, J.; Arendsee, Z.W.; Venkatesh, D.; Souza, C.K.; Kimble, J.B.; Lewis, N.S.; Davis, C.T.; Vincent, A.L. Swine Influenza A Viruses and the Tangled Relationship with Humans. *Cold Spring Harb Perspect. Med.* **2021**, *11*, a038737. [CrossRef]
18. Shu, Y.; McCauley, J. GISAID: Global initiative on sharing all influenza data—from vision to reality. *Eurosurveill* **2017**, *22*, 30494. [CrossRef]
19. Katoh, K.; Standley, D.M. MAFFT multiple sequence alignment software version 7: Improvements in performance and usability. *Mol. Biol. Evol.* **2013**, *30*, 772–780. [CrossRef]
20. Chang, J.; Anderson, T.K.; Zeller, M.A.; Gauger, P.C.; Vincent, A.L. octoFLU: Automated Classification for the Evolutionary Origin of Influenza A Virus Gene Sequences Detected in US Swine. *Microbiol. Resour. Announc.* **2019**, *8*, e00673-19. [CrossRef]
21. Minh, B.Q.; Schmidt, H.A.; Chernomor, O.; Schrempf, D.; Woodhams, M.D.; von Haeseler, A.; Lanfear, R. IQ-TREE 2: New Models and Efficient Methods for Phylogenetic Inference in the Genomic Era. *Mol. Biol Evol.* **2020**, *37*, 1530–1534. [CrossRef]
22. Kaplan, B.S.; Kimble, J.B.; Chang, J.; Anderson, T.K.; Gauger, P.C.; Janas-Martindale, A.; Killian, M.L.; Bowman, A.S.; Vincent, A.L. Aerosol Transmission from Infected Swine to Ferrets of an H3N2 Virus Collected from an Agricultural Fair and Associated with Human Variant Infections. *J. Virol.* **2020**, *94*, e01009-20. [CrossRef] [PubMed]

23. Gauger, P.C.; Vincent, A.L.; Loving, C.L.; Henningson, J.N.; Lager, K.M.; Janke, B.H.; Kehrli, M.E., Jr.; Roth, J.A. Kinetics of lung lesion development and pro-inflammatory cytokine response in pigs with vaccine-associated enhanced respiratory disease induced by challenge with pandemic (2009) A/H1N1 influenza virus. *Vet. Pathol.* **2012**, *49*, 900–912. [CrossRef] [PubMed]
24. Lin, Y.P.; Gregory, V.; Collins, P.; Kloess, J.; Wharton, S.; Cattle, N.; Lackenby, A.; Daniels, R.; Hay, A. Neuraminidase receptor binding variants of human influenza A(H3N2) viruses resulting from substitution of aspartic acid 151 in the catalytic site: A role in virus attachment? *J. Virol.* **2010**, *84*, 6769–6781. [CrossRef]
25. Reed, L.J.; Muench, H. A simple method of estimating fifty per cent endpoints12. *Am. J. Epidemiol.* **1938**, *27*, 493–497. [CrossRef]
26. Opriessnig, T.; Yu, S.; Gallup, J.M.; Evans, R.B.; Fenaux, M.; Pallares, F.; Thacker, E.L.; Brockus, C.W.; Ackermann, M.R.; Thomas, P.; et al. Effect of vaccination with selective bacterins on conventional pigs infected with type 2 porcine circovirus. *Vet. Pathol.* **2003**, *40*, 521–529. [CrossRef]
27. Kilbourne, E.D. Influenza pandemics of the 20th century. *Emerg. Infect. Dis.* **2006**, *12*, 9–14. [CrossRef] [PubMed]
28. Nelson, M.I.; Lemey, P.; Tan, Y.; Vincent, A.; Lam, T.T.-Y.; Detmer, S.; Viboud, C.; Suchard, M.A.; Rambaut, A.; Holmes, E.C. Spatial dynamics of human-origin H1 influenza A virus in North American swine. *PLoS Pathog.* **2011**, *7*, e1002077. [CrossRef]
29. Neveau, M.N.; Zeller, M.A.; Kaplan, B.S.; Souza, C.K.; Gauger, P.C.; Vincent, A.L.; Anderson, T.K.; Lowen, A.C. Genetic and Antigenic Characterization of an Expanding H3 Influenza A Virus Clade in U.S. Swine Visualized by Nextstrain. *Msphere* **2022**, *7*, e00994-21. [CrossRef]
30. Zeller, M.A.; Chang, J.; Vincent, A.L.; Gauger, P.C.; Anderson, T.K. Spatial and temporal coevolution of N2 neuraminidase and H1 and H3 hemagglutinin genes of influenza A virus in US swine. *Virus Evol.* **2021**, *7*, veab090. [CrossRef]
31. Trovao, N.S.; Nelson, M.I. When Pigs Fly: Pandemic influenza enters the 21st century. *PLoS Pathog.* **2020**, *16*, e1008259. [CrossRef]

Article

Climate Anomalies and Spillover of Bat-Borne Viral Diseases in the Asia–Pacific Region and the Arabian Peninsula

Alice Latinne [1,2,3,4,*] and Serge Morand [3,4,5]

1   Wildlife Conservation Society, Viet Nam Country Program, Ha Noi 100000, Vietnam
2   Wildlife Conservation Society, Global Conservation Program, Bronx, NY 10460, USA
3   MIVEGEC, CNRS—IRD—Montpellier Université, 911 Avenue Agropolis, BP 6450, 34394 Montpellier, France; serge.morand@umontpellier.fr
4   Faculty of Veterinary Technology, University of Kasetsart, Bangkok 10900, Thailand
5   Faculty of Tropical Medicine, University of Mahidol, Bangkok 10400, Thailand
*   Correspondence: alice.latinne@gmail.com

Abstract: Climate variability and anomalies are known drivers of the emergence and outbreaks of infectious diseases. In this study, we investigated the potential association between climate factors and anomalies, including El Niño Southern Oscillation (ENSO) and land surface temperature anomalies, as well as the emergence and spillover events of bat-borne viral diseases in humans and livestock in the Asia–Pacific region and the Arabian Peninsula. Our findings from time series analyses, logistic regression models, and structural equation modelling revealed that the spillover patterns of the Nipah virus in Bangladesh and the Hendra virus in Australia were differently impacted by climate variability and with different time lags. We also used event coincidence analysis to show that the emergence events of most bat-borne viral diseases in the Asia–Pacific region and the Arabian Peninsula were statistically associated with ENSO climate anomalies. Spillover patterns of the Nipah virus in Bangladesh and the Hendra virus in Australia were also significantly associated with these events, although the pattern and co-influence of other climate factors differed. Our results suggest that climate factors and anomalies may create opportunities for virus spillover from bats to livestock and humans. Ongoing climate change and the future intensification of El Niño events will therefore potentially increase the emergence and spillover of bat-borne viral diseases in the Asia–Pacific region and the Arabian Peninsula.

Keywords: bat-borne virus; spillover; SARS-CoV-2; Nipah virus; Hendra virus; climate change; El Niño Southern Oscillation; event coincidence analysis; temporal analysis; structural equation modelling

Citation: Latinne, A.; Morand, S. Climate Anomalies and Spillover of Bat-Borne Viral Diseases in the Asia–Pacific Region and the Arabian Peninsula. *Viruses* **2022**, *14*, 1100. https://doi.org/10.3390/v14051100

Academic Editor: Peng Zhou

Received: 5 April 2022
Accepted: 16 May 2022
Published: 20 May 2022

Publisher's Note: MDPI stays neutral with regard to jurisdictional claims in published maps and institutional affiliations.

1. Introduction

Most zoonotic and vector-borne diseases are climate-sensitive, particularly to temperature or precipitations [1], and several are also sensitive to climate variability and anomalies [2–4]. The El Niño Southern Oscillation (ENSO), with its alternating warming (El Niño), cooling (La Niña), and neutral phases, is one of the most important climate phenomena due to its ability to modify the global atmospheric circulation and the temperature and precipitation patterns across the globe [5]. The ENSO has significant cascade effects on ecosystems [6–8] and agriculture productivity [9]. ENSO-related climate variability is also a known driver of the emergence and outbreaks of infectious diseases [3,10]. Outbreaks of numerous infectious diseases, such as cholera [11], Rift Valley fever [12], visceral leishmaniasis [13], dengue [14,15], Zika virus [16], and malaria [17], among others, have been linked to the ENSO. The emergences of some viral diseases of bat origin, such as the Hendra virus (HeV) in Australia and the Nipah virus (NiV) in Malaysia, have also been associated with El Niño events [18,19]. Several mechanisms by which the ENSO affects and facilitates the transmission of zoonotic diseases have been suggested, including modifications of seasonal cycles, population dynamics, and distribution ranges of vectors and hosts of zoonotic pathogens, as well as alterations of the replication and transmission patterns of these pathogens [1,3,4].

In this study, we reviewed the emergence events and recurring spillover events of bat-borne viral diseases in humans and livestock in the Asia–Pacific region and the Arabian Peninsula, i.e., two regions highly affected by El Niño/La Niña events [20]. Furthermore, we investigated the potential association between climate anomalies, El Niño/La Niña events, and these emergence and spillover events. First, we tested the potential association between the spillover events of HeV in Australia and NiV in Bangladesh, i.e., two bat-borne viruses characterized by a high number of recurring spillover events in these two regions, and climate factors (temperature, rainfall) and anomalies (ENSO and land surface temperature anomalies) using time-series analyses, logistic regression models, and structural equation modelling. Second, we assessed potential simultaneities between the emergence events of bat-borne viruses in human and livestock populations and El Niño/La Niña events using event coincidence analysis (ECA) [21]. We then discussed the potential ecological mechanisms that may explain the emergence and spillover events of bat-borne viruses in relation to climate factors and events of El Niño/La Niña.

2. Materials and Methods

2.1. Data

Bat-borne viral pathogens were defined as viruses with bats (Chiroptera) as their natural animal hosts, i.e., the long-term ecological niche of a viral population [22], or viruses whose closest viral relatives have bats as their natural hosts. The zoonotic sources of these viruses in human populations was either bats or another host species involved as an intermediate host in their emergence. Emergence events were defined as the first detection of a bat-borne viral pathogen in human or in livestock populations or the first detection of a bat-borne viral pathogen in a region significantly distant from any other regions where it was previously observed (e.g., Nipah virus emergence events in Malaysia, India, and the Philippines). Recurring spillover events correspond to subsequent pathogen detections after its first emergence in the same region. Data on the emergence of bat-borne viral diseases in human and livestock populations in the study area during the period 1990–2020 were gathered from original sources (Table 1) and from several databases (Emerging Infectious Diseases Repository (EIDR), https://eidr.ecohealthalliance.org/ (accessed on 17 May 2022); World Animal Health Information System (OIE-WAHIS), https://wahis.oie.int/#/home (accessed on 17 May 2022); PROMED). Data on recurring spillover events of the Nipah virus in Bangladesh and India (Table 2) were obtained from three studies [23–25], while data on recurring spillover events of the Hendra virus in Australia were obtained from the Queensland Government database (https://www.business.qld.gov.au/industries/service-industries-professionals/service-industries/veterinary-surgeons/guidelines-hendra/incident-summary (accessed on 17 May 2022)) (Table 3).

Table 1. Emergence of bat-borne viruses in the Asia–Pacific region and the Arabian Peninsula in relation to El Niño Southern Oscillation (ENSO)-driven climate anomalies (data on ENSO were retrieved from NOAA).

Emergence	Viral Family	Natural Reservoir	Intermediate host	Date, Location	ENSO Phase	References
Hendra virus	Paramyxoviridae	Pteropodid bats	Horse	Aug 1994, Australia	Warm Phase/El Niño	Giles et al., 2018 [18]
Australian bat lyssavirus	Rhabdoviridae	Pteropodid bats	None	Oct 1996, Australia	Neutral Phase	Field et al., 1999 [26]
Menangle virus	Paramyxoviridae	Pteropodid bats	Pig	Jun 1997, Australia	Warm Phase/El Niño	Chant et al., 1998 [27]
Nipah virus	Paramyxoviridae	Pteropodid bats	Pig	Sep 1998, Malaysia	Cool Phase/La Niña	Ang et al., 2018 [25]
Nipah virus	Paramyxoviridae	Pteropodid bats	None	Jan 2001, India	Cool Phase/La Niña	Ang et al., 2018 [25]

Table 1. *Cont.*

Emergence	Viral Family	Natural Reservoir	Intermediate host	Date, Location	ENSO Phase	References
SARS-CoV-1	Coronaviridae	Rhinolophid bats	Small carnivores	Nov 2002, China	Warm Phase/El Niño	Ge et al., 2013 [28]
Melaka virus	Reoviridae	Pteropodid bats	None	Mar 2006, Malaysia	Cool Phase/La Niña	Chua et al., 2008 [29]
Kampar virus	Reoviridae	Pteropodid bats	None	Aug 2006, Malaysia	Neutral Phase	Chua et al., 2008 [29]
MERS-CoV	Coronaviridae	Vespertilionid bats	Camel	Apr 2012, Middle East	Cool Phase/La Niña	Zaki et al., 2012 [30]
Nipah virus	Paramyxoviridae	Pteropodid bats	Horse	Mar 2014, The Philippines	Neutral phase	Ching et al., 2015 [31]
SADS-CoV	Coronaviridae	Rhinolophid bats	Pig (no human cases)	Oct 2016, China	Cool Phase/La Nina	Gong et al., 2017 [32]
SARS-CoV-2	Coronaviridae	Rhinolophid bats	?	December 2019, China	Warm Phase/El Niño	Zhu et al., 2020 [33]

Table 2. Recurring spillover events of the Nipah virus after its first emergence in South Asia (India and Bangladesh) in relation to El Niño Southern Oscillation (ENSO)-driven climate anomalies (data on ENSO were retrieved from NOAA). Data on outbreaks were retrieved from Rahman and Chakraborty (2012) [23], Ang et al. (2018) [25], and Rahman et al. (2021) [24].

Country	Date (Month/Year)	ENSO Phase
Bangladesh	April 2001	Neutral Phase
Bangladesh	January 2003	Warm Phase/El Niño
Bangladesh	January 2004	Neutral Phase
Bangladesh	April 2004	Neutral Phase
Bangladesh	January 2005	Warm Phase/El Niño
Bangladesh	January 2007	Warm Phase/El Niño
Bangladesh	March 2007	Neutral Phase
Bangladesh	April 2007	Neutral Phase
India	April 2007	Neutral Phase
Bangladesh	February 2008	Cool Phase/La Niña
Bangladesh	April 2008	Cool Phase/La Niña
Bangladesh	January 2009	Cool Phase/La Niña
Bangladesh	February 2010	Warm Phase/El Niño
Bangladesh	January 2011	Cool Phase/La Niña
Bangladesh	January 2012	Cool Phase/La Niña
Bangladesh	January 2013	Neutral Phase
Bangladesh	January 2014	Warm Phase/El Niño
Bangladesh	January 2015	Warm Phase/El Nino
Bangladesh	February 2015	Neutral Phase
Bangladesh	March 2015	Warm Phase/El Niño
Bangladesh	February 2017	Neutral Phase
Bangladesh	February 2018	Cool Phase/La Niña
Bangladesh	April 2018	Cool Phase/La Niña
India	May 2018	Neutral Phase
India	June 2019	Neutral Phase

Table 3. Recurring spillover events of the Hendra virus after its first emergence in Australia in relation to El Niño Southern Oscillation (ENSO)-driven climate anomalies (data on ENSO were retrieved from NOAA). Data on outbreaks were retrieved from the Queensland Government database (https://www.business.qld.gov.au/industries/service-industries-professionals/service-industries/veterinary-surgeons/guidelines-hendra/incident-summary, accessed on 17 May 2022).

Country	Date (Month/Year)	ENSO Phase
Australia	September 1994	Warm Phase/El Niño
Australia	January 1999	Cool Phase/La Niña
Australia	October 2004	Warm Phase/El Niño
Australia	December 2004	Warm Phase/El Niño
Australia	June 2006	Neutral Phase
Australia	October 2006	Warm Phase/El Niño
Australia	June 2007	Neutral Phase
Australia	July 2007	Cool Phase/La Niña
Australia	June 2008	Cool Phase/La Niña
Australia	July 2008,	Neutral Phase
Australia	July 2009	Neutral Phase
Australia	September 2009	Warm Phase/El Niño
Australia	May 2010	Neutral Phase
Australia	June 2011 (4 events)	Cool Phase/La Niña
Australia	July 2011 (8 events)	Neutral phase
Australia	August 2011 (5 events)	Cool Phase/La Niña
Australia	October 2011	Cool Phase/La Niña
Australia	January 2012	Cool Phase/La Niña
Australia	May 2012 (2 events)	Neutral Phase
Australia	June 2012	Neutral Phase
Australia	July 2012 (2 events)	Neutral Phase
Australia	September 2012	Neutral Phase
Australia	October 2012	Neutral Phase
Australia	January 2013	Neutral Phase
Australia	February 2013	Neutral Phase
Australia	June 2013 (2 events)	Neutral Phase
Australia	July 2013 (4 events)	Neutral Phase
Australia	March 2014	Neutral Phase
Australia	June 2014 (2 events)	Neutral Phase
Australia	July 2014	Neutral Phase
Australia	June 2015	Warm Phase/El Niño
Australia	July 2015	Warm Phase/El Niño
Australia	September 2015	Warm Phase/El Niño
Australia	December 2016	Cool Phase/La Niña
Australia	May 2017	Neutral Phase
Australia	July 2017	Neutral Phase
Australia	August 2017 (2 events)	Neutral Phase
Australia	September 2018	Neutral Phase
Australia	June 2019	Warm Phase/El Niño
Australia	June 2020	Neutral Phase

Data on ENSO values were retrieved from the National Oceanic and Atmospheric Administration (NOAA, https://www.noaa.gov, accessed on 17 May 2022). The 'NINO 3.4' index is the most commonly used index used to define El Niño and La Niña events and to study climate–rainfall or climate–disease connections [3]. The NINO 3.4 index is based on a 5-month running mean of the sea surface temperature (SST) in the region bounded by 5° N to 5° S, from 170° W to 120° W. El Niño (warm phase) and La Niña (cool phase) are defined when anomalies in the NINO 3.4 index exceeds +0.4 °C or −0.4 °C, respectively (NOAA, https://www.noaa.gov, accessed on 17 May 2022). The R package rsoi [34] was used to import the NINO 3.4 index values for the period 1990–2020 and the corresponding defined El Nino (warm phase) and La Niña (cool phase) phases from the NOAA website.

Data on the average monthly temperature and rainfall data in Australia and Bangladesh were also gathered from the World Bank database (https://climateknowledgeportal.worldbank.org, accessed on 17 May 2022) and data on the global land surface temperature anomalies were obtained from the NOAA (https://www.ncei.noaa.gov/access/monitoring/global-temperature-anomalies, accessed on 17 May 2022) to investigate the impact of other climate factors (rainfall, temperature) and climate variability (global land surface temperature anomalies), in addition to the ENSO anomalies, on the occurrence of recurring spillover events of HeV and NiV viruses.

2.2. Statistical Analyses

First, we investigated the potential association between the spillover events of HeV in Australia and NiV in Bangladesh and several climate factors. For this, we used: (1) time-series analyses to investigate the temporal association between HeV/NiV spillover events, and temperature, rainfall, ENSO, and land surface temperature anomalies and to estimate their time lag values; (2) logistic regression models to determine the significant factors (temperature, rainfall, ENSO, land surface temperature anomalies) explaining the spillover events using the time lag values computed from the results of the time-series analyses; and (3) structural equation modelling to test a causal chain of correlation that may explain the spillover events using the results of the logistic regression analyses. Second, we assessed potential simultaneities between the emergence events of all bat-borne viruses in human and livestock populations and El Niño/La Niña events using event coincidence analysis to test the hypothesis that HeV/NiV spillover events were statistically preceded by an event of El Niño/La Niña.

2.2.1. Time-Series Analysis

Time-series analyses were used to study the temporal patterns of ENSO anomalies (using the NINO 3.4 index), as well as the average monthly temperature and rainfall in Australia and Bangladesh using the ncf function implemented in R [35]. The time series included 330 months in total from January 1993, i.e., one year before the first spillover event recorded in our dataset, to June 2020. The residual autocorrelation function (ACF) was examined to determine the general form of the model to be fitted. A wavelet analysis was used to decompose a time series to reveal periodic signals at each time point in the series. The wavelet analysis coefficients show the correlation magnitudes of ENSO anomalies (NINO 3.4 index), temperature, or rainfall for each year and period length of the time series (i.e., 1993 to 2020), displayed using a power spectrum over the full time series using the biwavelet and WaveletComp packages [36,37] implemented in R [38]. The ccf function was then used to compute the cross-correlation or cross-covariance between univariate series, i.e., ENSO (NINO 3.4 index); the average monthly temperature; the average monthly rainfall; land surface temperature anomalies, and either HeV or NiV recurring spillover events in Australia and in Bangladesh, respectively.

2.2.2. Logistic Regression with Time Lag Analysis

Logistic regression modelling with a logit function and lag was used to test the significant effects of monthly rainfall; monthly temperature; anomalies in land surface temperature; and ENSO anomalies (NINO 3.4 index) on the recurring spillover events of HeV and NiV in Australia and Bangladesh, respectively. The most significant lag values computed by time-series cross-correlation analysis, as described above, and the glm function implemented in R with the family binomial [38] were used.

The initial general linear model with logit function was of the form:

Recurrent spillover of HeV/NiV ~ lag(NINO 3.4 index, lag value)
+ lag(average temperature Australia/Bangladesh, lag value)
+ lag(average rainfall Australia/Bangladesh, lag value)
+ lag(global land surface temperature anomalies, lag value)

Initial models included variables with significant lag values obtained from cross-correlation time-series analysis. Final models were selected using backward selection and AIC criterion using the stepAIC function of the MASS package [39] implemented in R.

2.2.3. Structural Equation Modelling

Structural equation modelling (SEM) was used to investigate the temporal relationships between recurring outbreaks of HeV and NiV, respectively, in Australia and Bangladesh, in relation to monthly rainfall, monthly temperature, anomalies in land surface temperature, and the NINO 3.4 index values. SEM combines measurement models (e.g., reliability) with structural models (e.g., regression), thus testing a chain of causality (path analysis) between outbreaks of HeV and NiV in Australia and Bangladesh and these climatic factors. SEM was performed using the 'piecewiseSEM' package [40]. The following structural equation model was tested for the period from January 1993 to June 2020 of the dataset:

F (outbreaks of HeV or NiV) = f1 (lag NINO 3.4 index) + f2 (lag temperature Australia or Bangladesh) + f2 (lag rainfall Australia or Bangladesh) + f2 (lag global land surface temperature anomalies) + b1

G (lag temperature Australia or Bangladesh) = g1 (lag NINO 3.4 index) + b2
H (lag rainfall Australia or Bangladesh) = h1 (lag NINO 3.4 index) + b3
I (lag anomalies of the land surface temperature) = i (lag NINO 3.4 index) + b4
with lag values computed by time-series cross-correlation analysis.

2.2.4. Event Coincidence Analysis

Finally, event coincidence analysis (ECA) was used to test if events of a given type are causally influenced by the timing of events of second type [41] and to investigate the statistical interdependence between emergence and spillover events of bat-borne viruses and El Niño/La Niña events. ECA was implemented in the CoinCalc R package [42] to test whether the observed coincidence rates are significantly different from two independent random events [41]. ECA defines the precursor coincidence rate (pcr) and the trigger coincidence rate (tcr). The pcr describes the fraction of first-type events, i.e., emergence/spillover events, preceded by at least one second-type event, i.e., El Niño/La Niña events. The tcr describes the fraction of second-type events, i.e., El Niño/La Niña events, followed by at least one first-type event, i.e., emergence/spillover events (see [41]). CoinCalc computed the probability of the precursor and trigger coincidence rates occurring by chance, with the null hypothesis that the observed precursor and trigger coincidence rates can be explained by two independent series of randomly distributed events [42]. The *p*-value of the corresponding analytical significance test corresponds to the probability that the two types of events are randomly distributed and independent of each other (following two independent Poisson processes) and sufficiently rare.

We tested the hypothesis that the pcr describing the emergence of bat-borne viruses or recurrent spillover events of NiV and HeV were statistically preceded by an El Niño/La Niña event, while the tcr did not depart from a random association. For the analysis of the recurrent spillover events of NiV in Bangladesh and HeV in Australia, the time lag values estimated in months using cross-correlation among the time series of outbreak events and NINO 3.4 index values were used. The time lag values were also moved around their estimates to explore the stability and constancy of the association given by ECA.

3. Results

A total of ten bat-borne viruses, belonging to the Coronaviridae ($n = 4$), Paramyxoviridae ($n = 3$), Reoviridae ($n = 2$), and Rhabdoviridae ($n = 1$) families, emerged in the Asia–Pacific region and the Arabian Peninsula in the period 1990–2020 (Table 1; Figure 1). Nine of these viruses emerged in humans, while the swine acute diarrhea syndrome coronavirus (SADS CoV) emerged in swine populations but was never detected in humans. Natural bat hosts of the Coronaviridae viruses were vespertilionid and rhinolophid bats, while their intermediate hosts included several mammal species. Pteropodid bats were the hosts of the emerging Paramyxoviridae, Reoviridae, and Rhabdoviridae, and livestock (horses and pigs) were involved as intermediate hosts in the emergence of HeV and NiV. Most of these viruses emerged in a single geographic area; only NiV emerged in several distant locations over a 16-year period. It first emerged in Malaysia in 1998, before then emerging in Bangladesh and India in 2001 and in the Philippines in 2014 (Table 1; Figure 1). After their first emergence, NiV and HeV then regularly spilled over in Bangladesh and Australia, respectively (Tables 2 and 3).

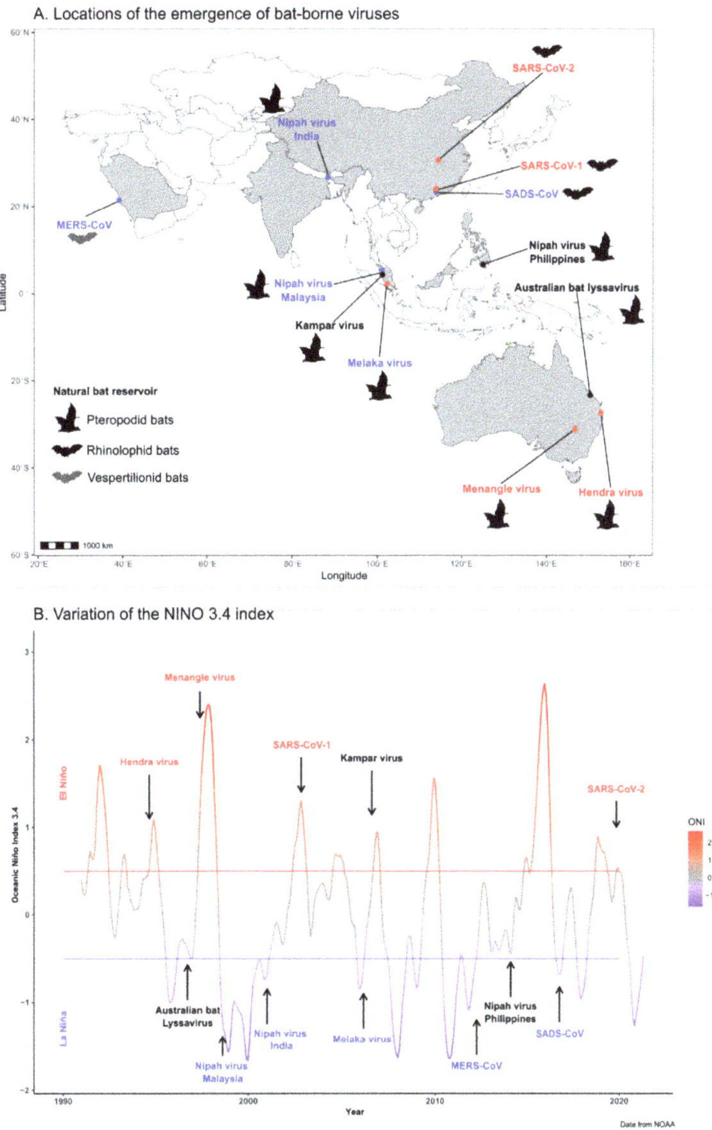

Figure 1. (**A**) Map showing the locations of emergence of bat-borne viruses in the Asia–Pacific region and the Arabian Peninsula (see Table 1) and the bat reservoir of each virus. Virus names are colored

according to the ENSO phase at the time of their emergence: neutral phase (black), cool-phase La Niña (blue), or warm-phase El Niño (red). (**B**) Variations of the NINO 3.4 index characterizing the El Niño Southern Oscillation (ENSO) retrieved from the National Oceanic and Atmospheric Administration (NOAA, https://www.noaa.gov, accessed on 17 May 2022) from 1990 to 2020. Red and blue threshold lines indicate warming El Niño or cooling La Niña climate anomalies, respectively. Arrows indicate the emergence time of new bat-borne viruses in the Asia–Pacific region and the Arabian Peninsula (see Table 1). Virus names are colored according to the ENSO phase at the time of their emergence: neutral phase (black), cool-phase La Niña (blue), or warm-phase El Niño (red).

Five emergence events, including NiV in Malaysia and India, Melaka virus, Middle East respiratory syndrome coronavirus (MERS-CoV), and SADS-CoV, occurred during a cool phase (La Niña event), while four of them, i.e., HeV, Menangle virus, and severe acute respiratory syndrome coronavirus 1 and 2 (SARS-CoV-1 and -2), occurred during a warm phase (El Niño event) (Figure 1; Table 1). The remaining three emergence events of bat-borne viruses, Kampar virus, Australian bat lyssavirus, and NiV in the Philippines occurred during a neutral phase. However, the spillover of the Australian bat lyssavirus to a human might not be considered as a natural emergence since the infection was acquired from a pet fruit bat living in captivity [26]. It should also be noted that the emergence of NiV in the Philippines in March 2014 followed the major volcanic activity of Mayon volcano that started in May 2013 [43].

3.1. Time-Series Analyses for HeV and NiV

There were strong and significant seasonal patterns of 12 months for temperature and rainfall, both in Bangladesh and Australia, as shown by the ACF and wavelet analysis (Figure 2A–D). Significant patterns over 24 and 36 months were observed for NINO 3.4 index (Figure 2E). Moreover, an increasing trend in the global land surface temperature anomalies was observed from 1990 to 2020 (Figure S1).

Cross-correlation analysis among pairs of temporal series of spillover events of NiV and HeV revealed several significant correlations with climate variables (Figure 3; Table 4). Significant correlations were observed for recurring spillover events of NiV with monthly rainfall (lag of 1 month), monthly temperature (lag of 1 month), and land surface temperature anomalies (lag of 10 months) (Figure 3A,B,D; Table 4). No significant correlation was observed for recurring spillover events of NiV with NINO 3.4 index values, although the best correlation was observed for no lag (Figure 3C; Table 4). Significant correlations were observed for recurring spillover events of HeV with NINO 3.4 index values (lag of 7 months), monthly rainfall (lag of 1 month), monthly temperature (no lag), and land surface temperature anomalies (lag of 3 months) (Figure 3E–H; Table 4). Cross-correlation analysis among pairs of temporal series of monthly rainfall, monthly temperature, and land surface temperature anomalies revealed few significant correlations with NINO 3.4 index values (Table 4). There was a significant correlation between land surface temperature anomalies and NINO 3.4 index values (lag of 3 months) and a significant correlation between monthly rainfall in Australia and NINO 3.4 index values (no lag) (Table 4). Non-significant correlations were observed for the monthly temperature in Australia (lag of 7 months), as well as for the monthly temperature and monthly rainfall in Bangladesh (with lags of 10 and 11 months, respectively) (Table 4).

3.2. Logistic Regression Analyses

The above results were used to build two initial logistic regression models based on the lag values obtained by the time-series cross-correlation analysis (see Figure 3 and Table 4). The selected model of the recurring spillover events of HeV in Australia show the significant effects of temperature (with no lag), the global land surface temperature anomalies (with a lag of 3 months), and NINO 3.4 index values (with a lag of 7 months). Rainfall was retained as a variable in the best explanatory model for HeV but had no significant effect(Table 5). The selected model of the recurrent spillover events of NiV in Bangladesh show the only significant effect of rainfall (with a lag of one month), but no effects of global land surface temperature anomalies, the NINO 3.4 index, and the mean temperature (Table 5).

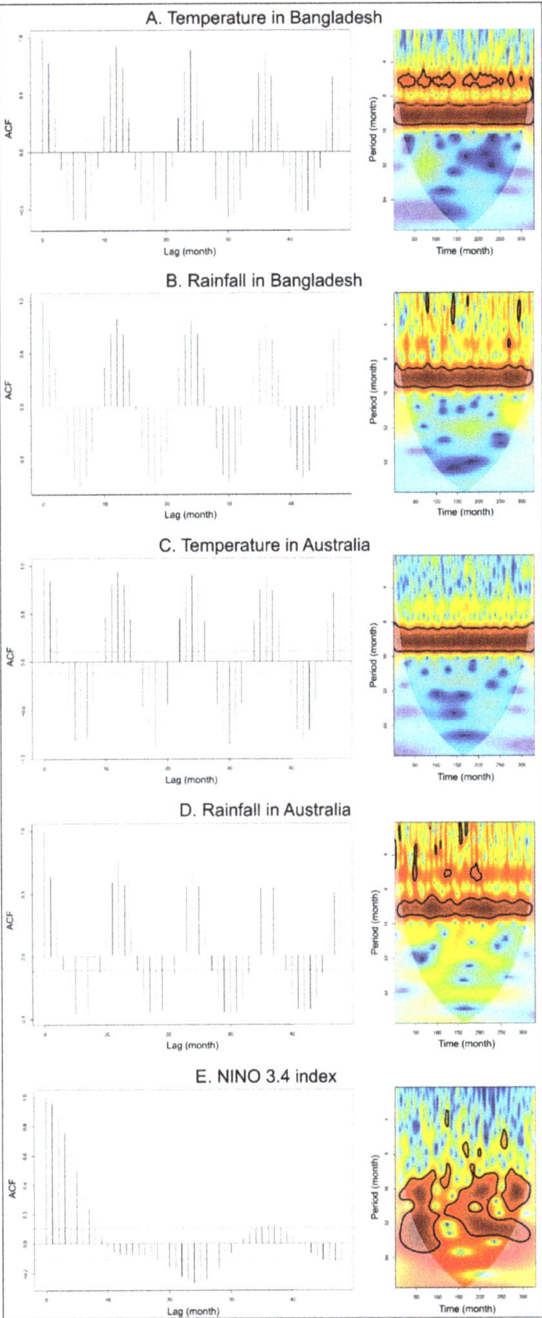

Figure 2. Time series and residual autocorrelation function (ACF) with significant auto-correlation values in dashed lines (left column) and wavelet power spectrum (right column) from January 1993 to

June 2020 (330 months) of (**A**) monthly temperature in Bangladesh, (**B**) monthly rainfall in Bangladesh, (**C**) monthly temperature in Australia, (**D**) monthly rainfall in Australia, and (**E**) NINO 3.4 index values decomposed in smooth trend and seasonal effect. Wavelet power values increased from blue to red, and black contour lines indicate a 5% significance level.

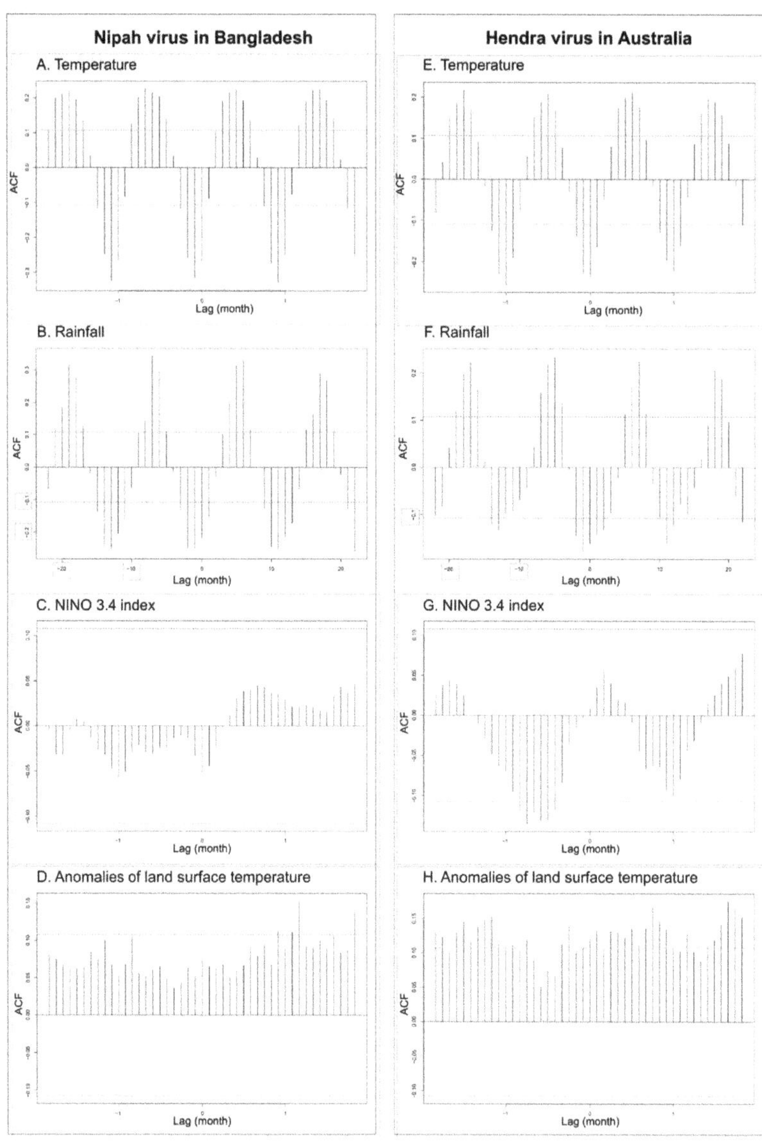

Figure 3. Temporal correlation from January 1993 to June 2020 (330 months) with significant autocorrelation values in dashed lines between spillover events of the Nipah virus in Bangladesh and the Hendra virus in Australia, as well as the monthly temperature (**A,E**), the monthly rainfall (**B,F**), the NINO 3.4 index (**C,G**), and anomalies of the land surface temperature (**D,H**).

Table 4. Results of temporal cross-association between spillover events of the Hendra virus (HeV), the Nipah virus (NiV), NINO 3.4 index, the average monthly temperature in Australia and Bangladesh, the average monthly rainfall in Australia and Bangladesh, and land surface temperature anomalies (monthly lag values were obtained from time-series cross-correlation analysis). Significant correlations are highlighted in bold.

First Time-Series	Second Time-Series	Lag	Correlation (p Value)
Spillover events of HeV (Australia)	NINO 3.4 index	7 months	**0.13 (0.018)**
	Rainfall (Australia)	1 month	**0.18 (0.002)**
	Temperature (Australia)	0 month	**0.24 (< 0.001)**
	Land surface temperature anomalies	3 months	**0.14 (0.012)**
Spillover events of NiV (Bangladesh)	NINO 3.4 index	0 month	0.05 (0.35
	Rainfall (Bangladesh)	1 month	**0.22 (0.008)**
	Temperature (Bangladesh)	1 month	**0.32 (0.008)**
	Land surface temperature anomalies	10 months	**0.10 (0.047)**
NINO 3.4 index	Rainfall (Australia)	0 month	**0.16 (0.007)**
	Temperature (Australia)	7 months	0.08 (0.35)
	Rainfall (Bangladesh)	10 months	0.03 (0.58)
	Temperature (Bangladesh)	11 months	0.06 (0.27)
	Land surface temperature anomalies	3 months	**0.40 (<0.0001)**

Table 5. Results of the logistic regression modelling with lags to explore the temporal association between spillover events of the Hendra virus (HeV) and the Nipah virus (NiV) from January 1993 to June 2020. The initial models included the following variables: the NINO 3.4 index, the average monthly temperature in Australia/Bangladesh, the average monthly rainfall in Australia/Bangladesh, and the land surface temperature anomalies with lag values obtained from time-series cross-correlation analysis (Table 4). The best explanatory models were selected using a backward procedure with AIC criterion. Significant p-values are highlighted in bold.

Response Variable	Predictor Variable	Estimate (Std Err)	Odds Ratio (2.5–97.5 %)	p	R2 (Global)
HeV spillover events (Australia)	NINO 3.4 index (lag = 7 months)	−0.72 (0.23)	0.49 (0.30–0.75)	**0.002**	
	Rainfall (lag = 1 month)	−0.01 (0.01)	0.99 (0.96–1.00	0.16	
	Temperature (lag = 0 month)	−0.11 (0.04)	0.90 (0.82–0.98)	**0.013**	
	Land surface temperature anomalies (lag = 3 months)	3.43 (1.03)	30.96 (4.32–254.30)	**0.001**	0.21
NiV spillover events (Bangladesh)	Rainfall (lag = 1 month)	−0.03 (0.01)	0.97 (0.95–0.99)	**0.008**	0.30

3.3. Structural Equation Modelling

SEM confirmed the above results for HeV. Significant correlations were observed between the series of recurring HeV spillovers and the mean monthly temperatures in Australia (with no lag, $p = 0.003$), the anomalies in the land surface temperature (with a lag of 3 months, $p < 0.001$), and NINO 3.4 index values (with a lag of 7 months, $p < 0.001$) (Figure 4A; Table 6). The mean monthly rainfall (with a lag of 1 month) and anomalies of the land surface temperature (with a lag of 3 months) were also significantly correlated with NINO 3.4 index values ($p = 0.003$ and $p < 0.001$, respectively) (Figure 4A; Table 6), with respective lags taking into account their estimated values given in Table 4.

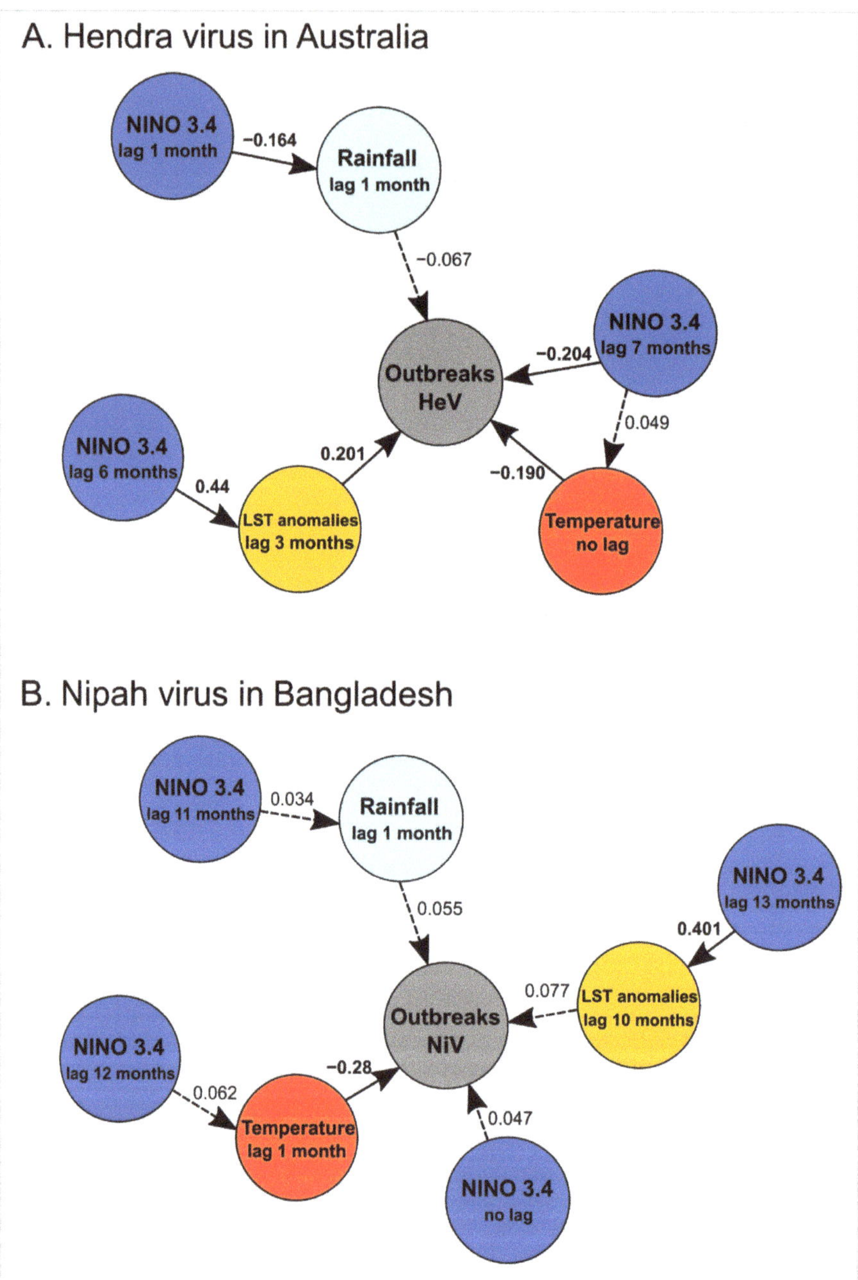

Figure 4. Results of structural equation modelling of (**A**) spillover events of the Hendra virus in Australia and (**B**) spillover events of the Nipah virus in Bangladesh on temporal trends from January 1993 to June 2020 (330 months) based on results obtained from temporal series analyses (Figure 2) and logistic regression analyses with lags (Table 4). Significant partial correlations are presented in continuous lines and non-significant partial correlations are presented in dashed lines with values of standardized estimates.

Table 6. Results of the structural equation modelling (SEM) to explore the temporal associations between spillover events of the Hendra virus (HeV) or the Nipah virus (NiV), the monthly temperature, the monthly rainfall, the NINO 3.4 index, and land surface temperature (LST) anomalies from January 1991 to June 2020. Lag values were obtained from the time-series analyses and the logistic regression analyses (Table 4). Significant *p*-values are highlighted in bold.

Model	Response Variable	Predictor Variable	Estimate (Std Err), df	Standardized Estimate	*p*	R2 (Individual)
HeV	Spillover events of HeV	'NINO 3.4' (lag = 7 months)	−0.077 (0.022), 313	−0.204	**<0.001**	
		Rainfall (lag = 1 month)	−0.001 (0.001), 313	−0.067	0.28	
		Temperature (lag = 0 month)	−0.078 (0.022), 313	−0.190	**0.003**	
		LST anom (lag = 3 months)	0.3123 (0.095), 313	0.201	**<0.001**	0.11
	Rainfall (lag = 1 month)	'NINO 3.4' (lag = 1 month)	−5.223 (1.7670), 316	−0.1638	**0.003**	0.027
	Temperature	'NINO 3.4' (lag = 7 months)	0.271 (0.313), 316	0.049	0.389	0.002
	LST anomalies (lag = 3 months)	'NINO 3.4' (lag = 6 months)	0.101 (0.012), 316	0.434	**<0.001**	0.19
NiV	Spillover events of NiV	'NINO 3.4' (lag = 0 month)	−0.014 (0.015), 312	−0.887	0.38	
		Rainfall (lag = 1 month)	−0.0001 (0.0001), 312	−0.718	0.47	
		Temperature (lag = 1 month)	−0.020 (0.005), 312	−3.689	**0.003**	
		LST anom (lag = 10 month)	0.098 (0.067), 312	1.443	0.15	0.12
	Rainfall (lag = 1 month)	NINO 3.4' (lag = 11 months)	6.730 (11.178), 315	0.602	0.55	0.001
	Temperature (lag = 1 month)	NINO 3.4' (lag = 12 months)	0.254 (0.231), 315	1.100	0.272	0.004
	LST anomalies (lag = 10 months)	'NINO 3.4' (lag = 13 months)	0.092 (0.012), 315	7.774	**<0.001**	0.16

We included the same variables in SEM for NiV and HeV using the lag with the best correlation, although some were non-significant (Table 4). SEM shows that the series of recurring NiV spillovers was only correlated with the mean monthly temperature in Bangladesh (with a lag of one month, *p* < 0.001) (Table 6; Figure 4B), while the monthly rainfall had no significant effect, contrary to what was suggested by the GLM (Table 5). The land surface temperature anomalies were evidently significantly correlated with NINO 3.4 index values (*p* < 0.001), with lag taking into account values given in Table 4.

3.4. Event Coincidence Analysis

Using the locations and dates of the emergence, as well as the series of spillover events of bat-borne viruses (Tables 1–3) and the corresponding ENSO phases (warm, neutral, cool) (Figure 1B; Table 1), we tested the hypothesis that the outbreaks of bat-borne viral diseases were directly preceded (no lag) by an ENSO-driven El Niño/La Niña climate event. The results of the ECA show a random association between an emergence event of bat-borne viral disease (Table 1) following an El Niño/La Niña event (*n* = 12) given by the non-significant value of the precursor coincidence rate (0.67, p= 0.066) and the non-significant value of the trigger coincidence rate (0.05, p = 0.28) (Figure 5A). However, a non-random association was observed with a significant value of precursor coincidence rate (0.80, p= 0.014) when the Australian bat lyssavirus (1996) and the Nipah virus in the Philippines (2014) were removed, as other factors may have impacted these emergence events, as explained above.

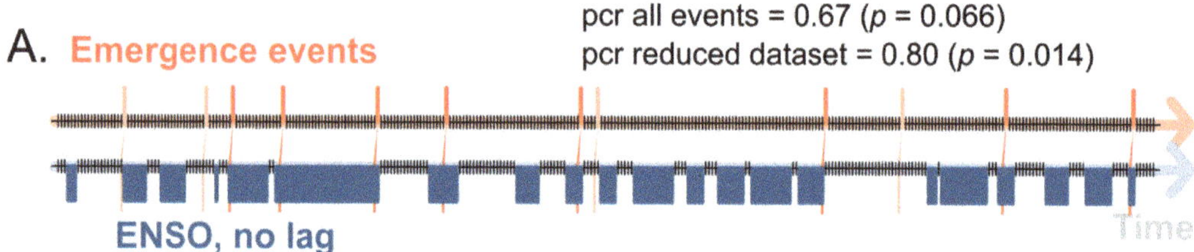

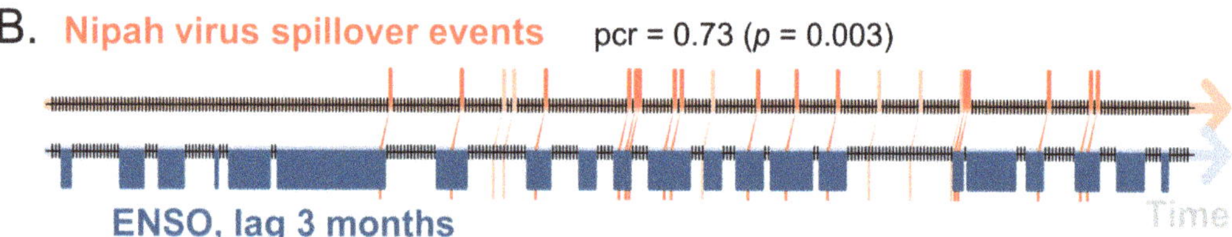

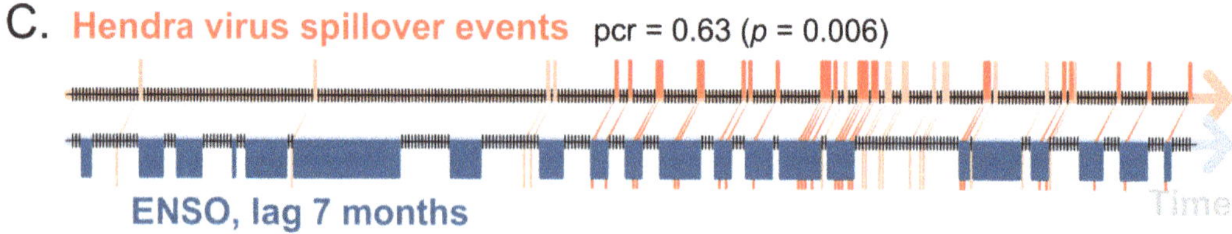

Figure 5. Event coincidence analyses of the association between (**A**) an emergence event of bat-borne virus in the Asia–Pacific region and the Arabian Peninsula, (**B**) a spillover event of the Nipah virus in Bangladesh, and (**C**) a spillover event of the Hendra virus in Australia with an ENSO event (El Niño or La Niña phases) with values of precursor coincidence rate (pcr) and its associated probability and lag value between paired events (lags values used for Nipah virus and Hendra virus analyses correspond to the results of cross-temporal series correlations, shown in Figure 3 and Table 4).

Even if there were no significant associations between NINO 3.4 index values and the NiV outbreak events using time-series analyses or SEM (see above), we explored a possible association using event coincidence analysis. We observed a random statistical relationship between an outbreak event of NiV in Bangladesh (Table 2) and events of El Niño/La Niña (n = 22) with no lag (best correlation observed in our cross-correlation analysis, as shown in Table 4). However, a non-random statistical and highly significant relationship between an outbreak event of NiV in Bangladesh and events of El Niño/La Niña using a lag of 3 months was observed, with a significant precursor coincidence rate (0.73, p= 0.003) and the non-significant value of a trigger coincidence rate (0.10, p – 0.09), suggesting a global lag effect of an ENSO event, whatever its phase (El Niño or La Niña) (Figure 5B). There was a non-random statistical relationship observed between an outbreak event of HeV in Australia (n = 40) following an event of El Niño/La Niña (a significant precursor coincidence rate = 0.63, *p*= 0.006; a trigger coincidence rate = 0.14, *p* = 0.15) (Figure 5C)), with a lag of 7 months estimated by cross-correlation time-series analysis (Figure 3G; Table 4).

4. Discussion

Numerous studies have investigated the origins and drivers of emergence of bat-borne viruses [44–48], with several pinpointing the importance of climate factors and their variability [49,50]. Abnormal rainfall, temperature, and vegetation development associated with ENSO climatic anomalies are known to create appropriate ecological conditions for pathogen emergence, transmission, and propagation [3]. Here, our main objective was to further investigate the potential statistical correlation between emergence and spillover events of several bat-borne viruses and climate factors, such as rainfall, temperature, global surface temperature anomalies, and ENSO events, using a single analytical framework. We used diverse methodologies, such as time-series analyses, logistic regression, SEM, and ECA, to better depict the complex relationships between these climate factors, variability, and anomalies, as well as the emergence and spillover events of bat-borne viruses in the Asia-Pacific region and the Arabian Peninsula.

While it has already been suggested that the emergences of HeV in Australia in 1994 and NiV in Malaysia in 1998 were associated with El Niño events [18,19,51], the long-term surveillance data of NiV in Bangladesh and HeV in Australia provided a better assessment of the influence of climate variability on the recurring spillovers of these bat-borne viruses. Our findings revealed that the spillover patterns of these two closely related paramyxoviruses, both belonging to the genus henipavirus and with closely related flying fox hosts (*Pteropus* spp.), are differently impacted by climate variability and with different time lags, according to our time-series cross-correlation analysis.

NiV outbreaks occurred almost annually in Bangladesh since 2001, following a seasonal pattern with most outbreaks occurring during the winter months [52]. Even if some livestock and domestic animals were found infected by NiV in Bangladesh [53], contact with sick livestock or domestic animals is not considered an important risk factor of spillover infections in Bangladesh [54]. The seasonal timing and spatial distribution of outbreaks coincide with patterns of raw date palm sap production and consumption [52], suggesting that human behavior and the consumption of date palm sap contaminated by *Pteropus* bats play an important role in these spillovers [54]. However, the fact that the number of NiV spillover events varies greatly from year to year suggests that additional factors influencing bat ecology and movement must be at play [55–57]. This was confirmed by a serological survey of *Pteropus medius* bats in Bangladesh which indicated that NiV viral shedding by bats can happen at any time of year and that viral dynamics are cyclical, but not annual or seasonal [58].

So far, a single climatic factor, winter temperature, was shown to be linked to NiV spillover, with colder winter temperatures being associated with more spillovers [56]. Our study confirms the influence of temperature, but our logistic regression models also suggest a correlation between NiV spillover and monthly rainfall, with a short time lag of one month, with lower rainfall being associated with more spillover events. Our cross-correlation analysis, logistic regression models, and SEM did not detect any significant direct correlation between the NINO 3.4 index and NiV spillover events. However, our ECA findings suggest that NiV spillovers in Bangladesh were significantly associated with ENSO events, either El Niño or La Niña phases, with a time lag of three months. This suggests that climate anomalies related to both warm and cool ENSO events may be linked to increased risk of NiV spillover from bats in Bangladesh, which may be explained by the fact that El Niño and La Niña events are characterized by similar rainfall and temperature anomalies in large regions of Bangladesh [59]. ENSO events are associated with the incidence of other diseases in Bangladesh, such as dengue and cholera [60,61]. Climate change will lead to warmer winter temperatures in Bangladesh over the next few decades [62], which may reduce the number of NiV spillover events in the country, as the negative correlation between temperature and NiV outbreaks was clearly demonstrated in this study and others [56]. However, droughts are expected to significantly increase in some regions of Bangladesh under global warming [62,63]. This could negatively impact bat food resources, induce increased bat movement, and potentially lead to more NiV spillover events, as shown by the negative correlation between rainfall and NiV spillover events observed in this study.

Our study revealed a stronger correlation between climate variability and the spillover pattern of HeV in eastern Australia than NiV in Bangladesh. HeV prevalence in flying foxes in Australia has shown multi-year inter-epidemic periods, suggesting that viral dynamics are not annual, but the ecological drivers and the climate influence behind this pattern remain unclear [64–66]. Numerous factors including food shortage, low concentration of nectar-based resources, extreme temperatures, dry conditions, phenology of eucalypt forests, physiological stress, flying fox foraging behavior, and use of wintering roosts in urban and agricultural areas were all suggested to be associated with increased HeV shedding in Australian flying foxes [18,66,67]. Our logistic regression models and SEM show that seasonal climate factors (monthly temperature), but also multi-annual climate variability (ENSO 3.4 index) and long-trend climate anomalies (land surface temperature anomalies), significantly influence the complex pattern of HeV spillover events in Australia. Our ECA also confirmed the hypothesis that HeV outbreaks were preceded by an ENSO-driven El Niño/La Niña climate event with a time lag of seven months. Interestingly, our results show no time lag between the mean monthly temperature recorded in Australia and the spillovers of HeV, suggesting a direct influence of climate seasonality. McMichael et al. [68] hypothesized that this correlation between lower winter temperature and increased HeV shedding in flying foxes could be mediated by the physiological cost of thermoregulation. The temporal lags observed between the ENSO 3.4 index (7 months) or the anomalies of the land surface temperature (3 months) suggest an indirect effect of the climate variability through ecological cascades that may affect food availability, bat migration patterns, and physiological stresses. A previous study [18] showed that the significant impact of ENSO on the flowering phenology of eucalypt, and consequently on bat foraging activities, was characterized by a time lag (3–8 months) similar to the one we observed between ENSO and HeV spillover events in this study (7 months). Lower eucalypt flowering and bat foraging activities induced by an El Niño event may lead to increased HeV prevalence a few months later [18].

Stress induced by climate variability can have a profound effect on disease dynamics in wild animal populations, mostly in relation to immune changes [69] or behavioral changes, such as climate-driven temporary migrations [18]. Bats undergo seasonal physiological changes, including immunological functions, which affect viral shedding [69]. Flying fox immunocompetence is challenged during food shortages driven by climatic anomalies, and HeV (sero) prevalence in Australian pteropid bats increased when their body condition decreased [67,70]. Immunological stress caused by physiological and behavioral changes during the breeding season has been suggested as a contributing factor in HeV shedding in some studies [67], but has been found to have no effect in others [70].

Beyond NiV and HeV, our findings suggest that the emergence of most viral diseases of bat origin was likely driven by ENSO climatic anomalies, as 9 out of 12 bat-borne viruses emerged in the Asia–Pacific region and the Arabian Peninsula after an ENSO event over the last three decades (Table 1; Figure 1). Removing the emergence of the Australian bat lyssavirus and NiV in the Philippines, given that other factors may have impacted these emergence events, gave a high prior probability for the emergence of bat virus after an event of El Niño/La Niña in our ECA (Figure 5). The recent emergence of SARS-CoV-2, responsible of the coronavirus disease COVID-19 in China in late 2019, also followed an important El Niño event, which had particularly affected China [71].

The emergence of Nipah in the Philippines in March 2014 was not linked to a warm or cool ENSO phase, but did occur following major volcanic activity at Mt Mayon that started one year prior to the emergence [43]. A study conducted after the more recent eruption of Mt. Mayon in 2018 showed large vegetation and environmental impacts of this eruption, which have affected the whole archipelago up to the northern part of Borneo [72]. Studies have also stressed the likely impacts of volcanic activities on disease outbreaks [73].

Retrospectively, our findings question the absence of emergence reports during major El Niño-La Niña events before 1994. This may be related to an important limitation of our study and the fact that several past spillover events of bat-borne viruses likely remained undetected. The successful detection of spillover events requires an efficient surveillance system adapted to wildlife, or bats in this case, and many countries are still lacking such a wildlife and human health surveillance system [74]. It is also important to note that climate is not the only factor influencing the emergence of bat-borne viruses. Several additional key drivers that promote cross-species transmission and emergence of zoonotic pathogens have been identified in Asia and include rapidly urbanizing populations, widespread wildlife trade and wildlife consumption, intensive livestock production, deforestation, habitat fragmentation, land-use change, and biodiversity loss [75,76]. Therefore, the absence of bat-borne virus emergence report before 1994 may also be explained by increased contact between bat and human/livestock populations over the last three decades due to land-use change, deforestation, intensification of farming practice, and the expansion of the distribution range of certain bat species linked to climate changes [49,75,76].

Climate modelling strongly suggests an intensification of extreme El Niño events in the future [77,78], which will potentially increase the occurrence and outbreaks of infectious diseases and the emergence of bat-borne viral diseases. Climate change will also continue to shift the global distribution of bats and drive changes in bat richness which will increase the risk of bat-borne coronaviruse emergence in the near future [49]. This study and its findings also stress the necessity of improving our knowledge of bat ecology. Close monitoring of bat populations will improve our understanding of viral spillover mechanisms, as demonstrated for HeV in eastern Australia and NiV in Bangladesh, and will contribute to better prediction and prevention strategies.

Supplementary Materials: The following supporting information can be downloaded at: https://www.mdpi.com/article/10.3390/v14051100/s1. Figure S1: Global land surface temperature anomalies 1993–2020. Table S1: Emergence and spillover data used in the study. R code S1: R code used to conduct the analyses described in the study. R code S2: R code used to obtain Figure 1B.

Author Contributions: Both authors contributed equally. All authors have read and agreed to the published version of the manuscript.

Funding: This work was part of the FutureHealthSEA project funded by the French ANR (ANR-17-CE35-0003-01).

Institutional Review Board Statement: Not applicable.

Informed Consent Statement: Not applicable.

Data Availability Statement: Data and codes used in the used in the study are available as Supplementary Materials (Table S1, R code S1, R code S2).

Conflicts of Interest: The authors declare no conflict of interest.

References

1. McIntyre, K.M.; Setzkorn, C.; Hepworth, P.J.; Morand, S.; Morse, A.P.; Baylis, M. Systematic Assessment of the Climate Sensitivity of Important Human and Domestic Animals Pathogens in Europe. *Sci. Rep.* **2017**, *7*, 7134. [CrossRef] [PubMed]
2. Anyamba, A.; Chretien, J.-P.; Small, J.; Tucker, C.J.; Linthicum, K.J. Developing global climate anomalies suggest potential disease risks for 2006–2007. *Int. J. Health Geogr.* **2006**, *5*, 60. [CrossRef] [PubMed]
3. Anyamba, A.; Chretien, J.-P.; Britch, S.C.; Soebiyanto, R.P.; Small, J.L.; Jepsen, R.; Forshey, B.M.; Sanchez, J.L.; Smith, R.D.; Harris, R.; et al. Global Disease Outbreaks Associated with the 2015–2016 El Niño Event. *Sci. Rep.* **2019**, *9*, 1930. [CrossRef] [PubMed]
4. Morand, S.; Owers, K.A.; Waret-Szkuta, A.; McIntyre, K.M.; Baylis, M. Climate variability and outbreaks of infectious diseases in Europe. *Sci. Rep.* **2013**, *3*, 1774. [CrossRef] [PubMed]

5. Hanley, D.E.; Bourassa, M.A.; O'Brien, J.J.; Smith, S.R.; Spade, E.R. A Quantitative Evaluation of ENSO Indices. *J. Clim.* **2003**, *16*, 1249–1258. [CrossRef]
6. Kogan, F.N. Satellite-Observed Sensitivity of World Land Ecosystems to El Niño/La Niña. *Remote Sens. Environ.* **2000**, *74*, 445–462. [CrossRef]
7. Rifai, S.W.; Girardin, C.A.J.; Berenguer, E.; del Aguila-Pasquel, J.; Dahlsjö, C.A.L.; Doughty, C.E.; Jeffery, K.J.; Moore, S.; Oliveras, I.; Riutta, T.; et al. ENSO Drives interannual variation of forest woody growth across the tropics. *Philos. Trans. R. Soc. B Biol. Sci.* **2018**, *373*, 20170410. [CrossRef]
8. McPhaden, M.J.; Santoso, A.; Cai, W. *El Niño Southern Oscillation in a Changing Climate*; John Wiley & Sons: Hoboken, NJ, USA, 2020.
9. Iizumi, T.; Luo, J.-J.; Challinor, A.J.; Sakurai, G.; Yokozawa, M.; Sakuma, H.; Brown, M.; Yamagata, T. Impacts of El Niño Southern Oscillation on the global yields of major crops. *Nat. Commun.* **2014**, *5*, 3712. [CrossRef]
10. Kovats, R.S.; Bouma, M.J.; Hajat, S.; Worrall, E.; Haines, A. El Niño and health. *Lancet* **2003**, *362*, 1481–1489. [CrossRef]
11. Rodó, X.; Pascual, M.; Fuchs, G.; Faruque, A. ENSO and cholera: A nonstationary link related to climate change? *Proc. Natl. Acad. Sci. USA* **2002**, *99*, 12901–12906. [CrossRef]
12. Nicholls, N. A method for predicting murray valley encephalitis in southeast australia using the southern oscillation. *Aust. J. Exp. Biol. Med Sci.* **1986**, *64*, 587–594. [CrossRef]
13. Chaves, L.F.; Calzada, J.E.; Valderrama, A.; Saldana, A. Cutaneous leishmaniasis and sand fly fluctuations are associated with El Niño in Panamá. *PLoS Negl. Trop. Dis.* **2014**, *8*, e3210. [CrossRef] [PubMed]
14. Vincenti-Gonzalez, M.F.; Tami, A.; Lizarazo, E.; Grillet, M.E. ENSO-driven climate variability promotes periodic major outbreaks of dengue in Venezuela. *Sci. Rep.* **2018**, *8*, 5727. [CrossRef] [PubMed]
15. Pramanik, M.; Singh, P.; Kumar, G.; Ojha, V.P.; Dhiman, R.C. El Niño Southern Oscillation as an early warning tool for dengue outbreak in India. *BMC Public Health* **2020**, *20*, 1498. [CrossRef] [PubMed]
16. Caminade, C.; Turner, J.; Metelmann, S.; Hesson, J.C.; Blagrove, M.S.C.; Solomon, T.; Morse, A.P.; Baylis, M. Global risk model for vector-borne transmission of Zika virus reveals the role of El Niño 2015. *Proc. Natl. Acad. Sci. USA* **2017**, *114*, 119–124. [CrossRef] [PubMed]
17. Dhiman, R.C.; Sarkar, S. El Niño Southern Oscillation as an early warning tool for malaria outbreaks in India. *Malar. J.* **2017**, *16*, 122. [CrossRef]
18. Giles, J.R.; Eby, P.; Parry, H.; Peel, A.J.; Plowright, R.K.; Westcott, D.A.; McCallum, H. Environmental drivers of spatiotemporal foraging intensity in fruit bats and implications for Hendra virus ecology. *Sci. Rep.* **2018**, *8*, 9555. [CrossRef]
19. Daszak, P.; Zambrana-Torrelio, C.; Bogich, T.L.; Fernandez, M.; Epstein, J.H.; Murray, K.A.; Hamilton, H. Interdisciplinary approaches to understanding disease emergence: The past, present, and future drivers of Nipah virus emergence. *Proc. Natl. Acad. Sci. USA* **2012**, *110* (Suppl. 1), 3681–3688. [CrossRef]
20. Rojas, O.; Li, Y.; Cumani, R. *Understanding the Drought Impact of El Niño on the Global Agricultural Areas: An Assessment Using FAO's Agricultural Stress Index (ASI)*; Food and Agriculture Organization of the United Nations (FAO): Rome, Italy, 2014.
21. Donges, J.F.; Schleussner, C.-F.; Siegmund, J.F.; Donner, R.V. Event coincidence analysis for quantifying statistical interrelationships between event time series. *Eur. Phys. J. Spéc. Top.* **2016**, *225*, 471–487. [CrossRef]
22. Corman, V.M.; Muth, D.; Niemeyer, D.; Drosten, C. Chapter Eight—Hosts and Sources of Endemic Human Coronaviruses. In *Advances in Virus Research 100*; Kielian, M., Mettenleiter, T.C., Roossinck, M.J., Eds.; Academic Press: Cambridge, MA, USA, 2018; pp. 163–188.
23. Rahman, M.; Chakraborty, A. Nipah virus outbreaks in Bangladesh: A deadly infectious disease. *WHO South-East Asia J. Public Health* **2012**, *1*, 208–212. [CrossRef]
24. Rahman, M.Z.; Islam, M.M.; Hossain, M.E.; Rahman, M.M.; Islam, A.; Siddika, A.; Sultana, S.; Klena, J.; Flora, M.; Daszak, P.; et al. Genetic diversity of Nipah virus in Bangladesh. *Int. J. Infect. Dis.* **2021**, *102*, 144–151. [CrossRef] [PubMed]
25. Ang, B.S.P.; Lim, T.C.C.; Wang, L.; Kraft, C.S. Nipah Virus Infection. *J. Clin. Microbiol.* **2018**, *56*, e01875-17. [CrossRef] [PubMed]
26. Field, H.; McCall, B.; Barrett, J. Australian Bat Lyssavirus Infection in a Captive Juvenile Black Flying Fox. *Emerg. Infect. Dis.* **1999**, *5*, 438–440. [CrossRef] [PubMed]
27. Chant, K.; Chan, R.; Smith, M.; Dwyer, D.E.; Kirkland, P. Probable human infection with a newly described virus in the family Paramyxoviridae. *Emerg. Infect. Dis.* **1998**, *4*, 273–275. [CrossRef]
28. Ge, X.-Y.; Li, J.-L.; Yang, X.-L.; Chmura, A.A.; Zhu, G.; Epstein, J.H.; Mazet, J.K.; Hu, B.; Zhang, W.; Peng, C.; et al. Isolation and characterization of a bat SARS-like coronavirus that uses the ACE2 receptor. *Nature* **2013**, *503*, 535–538. [CrossRef]
29. Chua, K.B.; Voon, K.; Crameri, G.; Tan, H.S.; Rosli, J.; McEachern, J.A.; Suluraju, S.; Yu, M.; Wang, L.-F. Identification and Characterization of a New Orthoreovirus from Patients with Acute Respiratory Infections. *PLoS ONE* **2008**, *3*, e3803. [CrossRef]
30. Zaki, A.M.; Van Boheemen, S.; Bestebroer, T.M.; Osterhaus, A.D.M.E.; Fouchier, R.A.M. Isolation of a Novel Coronavirus from a Man with Pneumonia in Saudi Arabia. *N. Engl. J. Med.* **2012**, *367*, 1814–1820. [CrossRef]
31. Ching, P.K.G.; de los Reyes, V.C.; Sucaldito, M.N.; Tayag, E.; Columna-Vingno, A.B.; Malbas, F.F.; Bolo, G.C.; Sejvar, J.J.; Eagles, D.; Playford, G.; et al. Outbreak of Henipavirus Infection, Philippines, 2014. *Emerg. Infect. Dis.* **2015**, *21*, 328–331. [CrossRef]
32. Gong, L.; Li, J.; Zhou, Q.; Xu, Z.; Chen, L.; Zhang, Y.; Xue, C.; Wen, Z.; Cao, Y. A New Bat-HKU2-like Coronavirus in Swine, China, 2017. *Emerg. Infect. Dis.* **2017**, *3*, 1607–1609. [CrossRef]

33. Zhu, N.; Zhang, D.; Wang, W.; Li, X.; Yang, B.; Song, J.; Zhao, X.; Huang, B.; Shi, W.; Lu, R.; et al. A Novel Coronavirus from Patients with Pneumonia in China, 2019. *N. Engl. J. Med.* **2020**, *382*, 727–733. [CrossRef]
34. Albers, S.; Campitelli, E. Rsoi: Import Various Northern and Southern Hemisphere Climate Indices. R Package Version 05. Available online: https://cran.r-project.org/web/packages/rsoi (accessed on 17 May 2022).
35. Bjornstad, O.N.; Cai, J. ncf: Spatial Covariance Functions. R Package Version 12-8. Available online: https://cran.r-project.org/web/packages/ncf (accessed on 17 May 2022).
36. Rösch, A.; Schmidbauer, H. WaveletComp 1.1: A Guided Tour through the R Package. Available online: http://www.hs-stat.com/projects/WaveletComp/WaveletComp_guided_tour.pdf (accessed on 17 May 2022).
37. Gouhier, T.; Grinsted, A.; Simko, V. biwavelet: Conduct Univariate and Bivariate Wavelet Analyses. R Package Version 02017. Available online: https://cran.r-project.org/web/packages/biwavelet/ (accessed on 17 May 2022).
38. R Core Team. *R: A language and Environment for Statistical Computing*; R Foundation for Statistical Computing: Vienna, Austria, 2020.
39. Venables, W.; Ripley, B.D. *Statistics Complements to Modern Applied Statistics with S*, 4th ed.; Springer: Berlin/Heidelberg, Germany, 2002.
40. Lefcheck, J.S. piecewiseSEM: Piecewise structural equation modelling in r for ecology, evolution, and systematics. *Methods Ecol. Evol.* **2016**, *7*, 573–579. [CrossRef]
41. Siegmund, J.F.; Sanders, T.G.; Heinrich, I.; Van der Maaten, E.; Simard, S.; Helle, G.; Donner, R. Meteorological Drivers of Extremes in Daily Stem Radius Variations of Beech, Oak, and Pine in Northeastern Germany: An Event Coincidence Analysis. *Front. Plant Sci.* **2016**, *7*, 733. [CrossRef] [PubMed]
42. Siegmund, J.F.; Siegmund, N.; Donner, R.V. CoinCalc—A new R package for quantifying simultaneities of event series. *Comput. Geosci.* **2017**, *98*, 64–72. [CrossRef]
43. Global Volcanism Program. Report on Mayon (Philippines). In *Bulletin of the Global Volcanism Network*; Wunderman, R., Ed.; Smithsonian Institution: Washington, DC, USA, 2013; Volume 38. [CrossRef]
44. Brierley, L.; Vonhof, M.J.; Olival, K.J.; Daszak, P.; Jones, K.E. Quantifying Global Drivers of Zoonotic Bat Viruses: A Process-Based Perspective. *Am. Nat.* **2016**, *187*, E53–E64. [CrossRef] [PubMed]
45. Olival, K.J.; Hosseini, P.R.; Zambrana-Torrelio, C.; Ross, N.; Bogich, T.L.; Daszak, P. Host and viral traits predict zoonotic spillover from mammals. *Nature* **2017**, *546*, 646–650. [CrossRef]
46. Letko, M.; Seifert, S.N.; Olival, K.J.; Plowright, R.K.; Munster, V.J. Bat-borne virus diversity, spillover and emergence. *Nat. Rev. Genet.* **2020**, *18*, 461–471. [CrossRef]
47. Latinne, A.; Hu, B.; Olival, K.J.; Zhu, G.; Zhang, L.; Li, H.; Chmura, A.A.; Field, H.E.; Zambrana-Torrelio, C.; Epstein, J.H.; et al. Origin and cross-species transmission of bat coronaviruses in China. *Nat. Commun.* **2020**, *11*, 4235. [CrossRef]
48. Wells, K.; Morand, S.; Wardeh, M.; Baylis, M. Distinct spread of DNA and RNA viruses among mammals amid prominent role of domestic species. *Glob. Ecol. Biogeogr.* **2020**, *29*, 470–481. [CrossRef]
49. Beyer, R.M.; Manica, A.; Mora, C. Shifts in global bat diversity suggest a possible role of climate change in the emergence of SARS-CoV-1 and SARS-CoV-2. *Sci. Total Environ.* **2021**, *767*, 145413. [CrossRef]
50. Buceta, J.; Johnson, K. Modeling the Ebola zoonotic dynamics: Interplay between enviroclimatic factors and bat ecology. *PLoS ONE* **2017**, *12*, e0179559. [CrossRef]
51. McFarlane, R.; Becker, N.; Field, H. Investigation of the Climatic and Environmental Context of Hendra Virus Spillover Events 1994–2010. *PLoS ONE* **2011**, *6*, e28374. [CrossRef] [PubMed]
52. Gurley, E.S.; Hegde, S.T.; Hossain, K.; Sazzad, H.M.S.; Hossain, M.J.; Rahman, M.; Sharker, M.A.Y.; Salje, H.; Islam, M.S.; Epstein, J.H.; et al. Convergence of Humans, Bats, Trees, and Culture in Nipah Virus Transmission, Bangladesh. *Emerg. Infect. Dis.* **2017**, *23*, 1446–1453. [CrossRef] [PubMed]
53. Chowdhury, S.; Khan, S.U.; Crameri, G.; Epstein, J.H.; Broder, C.C.; Islam, A.; Peel, A.J.; Barr, J.; Daszak, P.; Wang, L. F.; et al. Serological Evidence of Henipavirus Exposure in Cattle, Goats and Pigs in Bangladesh. *PLOS Neglected Trop. Dis.* **2014**, *8*, e3302. [CrossRef] [PubMed]
54. Hegde, S.T.; Sazzad, H.M.S.; Hossain, M.J.; Alam, M.-U.; Kenah, E.; Daszak, P.; Rollin, P.; Rahman, M.; Luby, S.; Gurley, E.S. Investigating Rare Risk Factors for Nipah Virus in Bangladesh: 2001–2012. *EcoHealth* **2016**, *13*, 720–728. [CrossRef]
55. Cortes, M.C.; Cauchemez, S.; Lefrancq, N.; Luby, S.P.; Hossain, M.J.; Sazzad, H.; Rahman, M.; Daszak, P.; Salje, H.; Gurley, E.S. Characterization of the Spatial and Temporal Distribution of Nipah Virus Spillover Events in Bangladesh, 2007–2013. *J. Infect. Dis.* **2018**, *217*, 1390–1394. [CrossRef]
56. McKee, C.D.; Islam, A.; Luby, S.P.; Salje, H.; Hudson, P.J.; Plowright, R.K.; Gurley, E. The Ecology of Nipah Virus in Bangladesh: A Nexus of Land-Use Change and Opportunistic Feeding Behavior in Bats. *Viruses* **2021**, *13*, 169. [CrossRef]
57. Olival, K.J.; Latinne, A.; Islam, A.; Epstein, J.H.; Hersch, R.; Engstrand, R.C.; Gurley, E.S.; Amato, G.; Luby, S.P.; Daszak, P. Population genetics of fruit bat reservoir informs the dynamics, distribution and diversity of Nipah virus. *Mol. Ecol.* **2020**, *29*, 970–985. [CrossRef]
58. Epstein, J.H.; Anthony, S.J.; Islam, A.; Kilpatrick, A.M.; Khan, S.A.; Balkey, M.D.; Ross, N.; Smith, I.; Zambrana-Torrelio, C.; Tao, Y.; et al. Nipah virus dynamics in bats and implications for spillover to humans. *Proc. Natl. Acad. Sci. USA* **2020**, *117*, 29190–29201. [CrossRef]
59. Wahiduzzaman, M.; Luo, J.-J. A statistical analysis on the contribution of El Niño–Southern Oscillation to the rainfall and temperature over Bangladesh. *Meteorol. Atmos. Phys.* **2021**, *133*, 55–68. [CrossRef]

60. Cash, B.A.; Rodó, X.; Kinter, J.L.; Yunus, M. Disentangling the Impact of ENSO and Indian Ocean Variability on the Regional Climate of Bangladesh: Implications for Cholera Risk. *J. Clim.* **2010**, *23*, 2817–2831. [CrossRef]
61. Sharmin, S.; Glass, K.; Viennet, E.; Harley, D. Interaction of Mean Temperature and Daily Fluctuation Influences Dengue Incidence in Dhaka, Bangladesh. *PLOS Neglected Trop. Dis.* **2015**, *9*, e0003901. [CrossRef] [PubMed]
62. Chowdhury, M.R.; Ndiaye, O. Climate change and variability impacts on the forests of Bangladesh—A diagnostic discussion based on CMIP5 GCMs and ENSO. *Int. J. Clim.* **2017**, *37*, 4768–4782. [CrossRef]
63. Islam, A.R.M.T.; Salam, R.; Yeasmin, N.; Kamruzzaman, M.; Shahid, S.; Fattah, M.A.; Uddin, A.S.; Shahariar, M.H.; Mondol, A.H.; Jhajharia, D.; et al. Spatiotemporal distribution of drought and its possible associations with ENSO indices in Bangladesh. *Arab. J. Geosci.* **2021**, *14*, 2681. [CrossRef]
64. Plowright, R.K.; Peel, A.J.; Streicker, D.G.; Gilbert, A.T.; McCallum, H.; Wood, J.; Baker, M.; Restif, O. Transmission or Within-Host Dynamics Driving Pulses of Zoonotic Viruses in Reservoir–Host Populations. *PLOS Neglected Trop. Dis.* **2016**, *10*, e0004796. [CrossRef] [PubMed]
65. Páez, D.J.; Giles, J.; Mccallum, H.; Field, H.; Jordan, D.; Peel, A.J.; Plowright, R.K. Conditions affecting the timing and magnitude of Hendra virus shedding across pteropodid bat populations in Australia. *Epidemiol. Infect.* **2017**, *145*, 3143–3153. [CrossRef]
66. Becker, D.; Eby, P.; Madden, W.; Peel, A.; Plowright, R. Ecological conditions experienced by bat reservoir hosts predict the intensity of Hendra virus excretion over space and time. *bioRxiv* **2021**. [CrossRef]
67. Plowright, R.K.; Field, H.E.; Smith, C.; Divljan, A.; Palmer, C.; Tabor, G.; Daszak, P.; Foley, J.E. Reproduction and nutritional stress are risk factors for Hendra virus infection in little red flying foxes (*Pteropus scapulatus*). *Proc. R. Soc. B Boil. Sci.* **2008**, *275*, 861–869. [CrossRef]
68. McMichael, L.; Edson, D.; Smith, C.; Mayer, D.; Smith, I.; Kopp, S.; Meers, J.; Field, H. Physiological stress and Hendra virus in flying-foxes (*Pteropus* spp.), Australia. *PLoS ONE* **2017**, *12*, e0182171. [CrossRef]
69. Banerjee, A.; Baker, M.L.; Kulcsar, K.; Misra, V.; Plowright, R.; Mossman, K. Novel Insights into Immune Systems of Bats. *Front. Immunol.* **2020**, *11*, 26. [CrossRef]
70. Edson, D.; Peel, A.J.; Huth, L.; Mayer, D.G.; Vidgen, M.E.; McMichael, L.; Broos, A.; Melville, D.; Kristoffersen, J.; de Jong, C.; et al. Time of year, age class and body condition predict Hendra virus infection in Australian black flying foxes (*Pteropus alecto*). *Epidemiol. Infect.* **2019**, *147*, e240. [CrossRef]
71. Liqiang, H. El Nino makes trouble for broad areas of China, may hurt rice farmers. *China Daily.* 2019. Available online: http://www.chinadaily.com.cn/a/201903/01/WS5c7888eba3106c65c34ec1b7.html (accessed on 17 May 2022).
72. Tiwari, A.; Singh, S.; Soni, V.; Kumar, R. Environmental Impact of Recent Volcanic Eruption from Mt. Mayon Over 1 South-East Asia. *J. Geogr. Nat. Disast.* **2021**, *10*, 547.
73. Batumbo Boloweti, D.; Giraudoux, P.; Deniel, C.; Garnier, E.; Mauny, F.; Kasereka, C.M.; Kizungu, R.; Muyembe, J.J.; Bompangue, D.; Bornette, G. Volcanic activity controls cholera outbreaks in the East African Rift. *PLoS Negl. Trop. Dis.* **2020**, *14*, e0008406. [CrossRef] [PubMed]
74. Miranda, A.V.; Wiyono, L.; Rocha, I.C.N.; Cedeño, T.D.D.; Lucero-Prisno, D.E.I. Strengthening Virology Research in the Association of Southeast Asian Nations: Preparing for Future Pandemics. *Am. J. Trop. Med. Hyg.* **2021**, *105*, 1141–1143. [CrossRef] [PubMed]
75. Coker, R.J.; Hunter, B.M.; Rudge, J.W.; Liverani, M.; Hanvoravongchai, P. Emerging infectious diseases in southeast Asia: Regional challenges to control. *Lancet* **2011**, *377*, 599–609. [CrossRef]
76. Keesing, F.; Ostfeld, R.S. Impacts of biodiversity and biodiversity loss on zoonotic diseases. *Proc. Natl. Acad. Sci. USA* **2021**, *118*, e2023540118. [CrossRef]
77. Cai, W.; Ng, B.; Wang, G.; Santoso, A.; Wu, L.; Yang, K. Increased ENSO sea surface temperature variability under four IPCC emission scenarios. *Nat. Clim. Chang.* **2022**, *12*, 228–231. [CrossRef]
78. Hu, K.; Huang, G.; Huang, P.; Kosaka, Y.; Xie, S.-P. Intensification of El Niño-induced atmospheric anomalies under greenhouse warming. *Nat. Geosci.* **2021**, *14*, 377–382. [CrossRef]

Article

Risk of Viral Infectious Diseases from Live Bats, Primates, Rodents and Carnivores for Sale in Indonesian Wildlife Markets

Thais Q. Morcatty [1,*], Paula E. R. Pereyra [2], Ahmad Ardiansyah [1,3,4], Muhammad Ali Imron [5], Katherine Hedger [3], Marco Campera [3,6], K. Anne-Isola Nekaris [1,3] and Vincent Nijman [1,*]

[1] Oxford Wildlife Trade Research Group, School of Social Sciences, Oxford Brookes University, Oxford OX3 0BP, UK
[2] Department of Ecology, Institute of Biosciences, Federal University of Rio Grande do Sul, 9500 Ave., Bento Gonçalves 90010-150, Brazil
[3] Little Fireface Project, Cipaganti 44163, Indonesia
[4] Forest and Nature Conservation Policy Group, Wageningen University, 6708 PD Wageningen, The Netherlands
[5] Faculty of Forestry, Universitas Gajah Madah, Yogyakarta 55281, Indonesia
[6] Faculty of Life Sciences, Oxford Brookes University, Oxford OX3 0FL, UK
* Correspondence: tqueiroz-morcatty@brookes.ac.uk (T.Q.M.); vnijman@brookes.ac.uk (V.N.)

Abstract: Southeast Asia is considered a global hotspot of emerging zoonotic diseases. There, wildlife is commonly traded under poor sanitary conditions in open markets; these markets have been considered 'the perfect storm' for zoonotic disease transmission. We assessed the potential of wildlife trade in spreading viral diseases by quantifying the number of wild animals of four mammalian orders (Rodentia, Chiroptera, Carnivora and Primates) on sale in 14 Indonesian wildlife markets and identifying zoonotic viruses potentially hosted by these animals. We constructed a network analysis to visualize the animals that are traded alongside each other that may carry similar viruses. We recorded 6725 wild animals of at least 15 species on sale. Cities and markets with larger human population and number of stalls, respectively, offered more individuals for sale. Eight out of 15 animal taxa recorded are hosts of 17 zoonotic virus species, nine of which can infect more than one species as a host. The network analysis showed that long-tailed macaque has the greatest potential for spreading viral diseases, since it is simultaneously the most traded species, sold in 13/14 markets, and a potential host for nine viruses. It is traded alongside pig-tailed macaques in three markets, with which it shares six viruses in common (Cowpox, Dengue, Hepatitis E, Herpes B, Simian foamy, and Simian retrovirus type D). Short-nosed fruit bats and large flying foxes are potential hosts of Nipah virus and are also sold in large quantities in 10/14 markets. This study highlights the need for better surveillance and sanitary conditions to avoid the negative health impacts of unregulated wildlife markets.

Keywords: zoonosis; Nipah; One Health; pandemic; COVID-19; wildlife trade; wet market; mammals

Citation: Morcatty, T.Q.; Pereyra, P.E.R.; Ardiansyah, A.; Imron, M.A.; Hedger, K.; Campera, M.; Nekaris, K.A.-I.; Nijman, V. Risk of Viral Infectious Diseases from Live Bats, Primates, Rodents and Carnivores for Sale in Indonesian Wildlife Markets. *Viruses* **2022**, *14*, 2756. https://doi.org/10.3390/v14122756

Academic Editors: Myriam Ermonval and Serge Morand

Received: 3 November 2022
Accepted: 7 December 2022
Published: 10 December 2022

Publisher's Note: MDPI stays neutral with regard to jurisdictional claims in published maps and institutional affiliations.

1. Introduction

Infectious diseases are responsible for more than 7 million deaths annually, causing negative impacts on global health and substantial economic losses [1,2]. Seventy-five percent of all emerging infectious diseases are zoonotic, i.e., diseases that have originated from an animal and crossed the species barrier to infect humans, of which many have their origins in wildlife [3,4]. Most of those emerging zoonotic diseases are caused by viruses. All recent pandemic diseases allegedly originated from wildlife, such as HIV, SARS, and COVID-19, are caused by viruses with a long history of adaptation to their natural hosts, suggesting that investigations about activities that bring wild animals and humans in close contact are urgently needed [5].

Global trade and commerce, including wildlife trade, are recognized as key factors to the increase in emerging viral infectious diseases [6–8]. Specifically for wildlife, trade usually involves close contact between humans and animals (or their products) during the harvest, processing and exchange, raising the risk of a zoonotic pathogen crossing species lines [9,10]. The potential for viral infections in wildlife markets is enhanced because animals are slaughtered on the spot to be either legally or illegally traded as medicines, meat and pets [11]. These wild animals are often originated from areas many hundreds of miles far from the market [12,13], are offered for sale alongside domestic animals and kept in cramped conditions with little regard for hygiene or welfare. In addition, during transportation or sale, species that would not naturally have contact with each other are often kept close together in the facilities [14,15]. Those contacts break existent geographical, ecological or behavioural separations of humans and domestic animals with wildlife, increasing the likelihood of cross-species pathogen transmission [16].

In recent decades, increased global human population and recently established domestic and international travel networks have escalated the commercialization of wildlife, creating a situation in which the extent and velocity of zoonotic pathogen movement are historically unmatched [17,18]. Much of the wild species are traded illegally, but even for most of the legally sold there is no mandatory testing for pathogens [19]; this means that once a pathogen has crossed the species boundary, the risk of the infection spreading to susceptible populations is elevated [20].

For instance, the outbreak of SARS-CoV, linked to civets in China's wildlife markets, spread to 37 countries, affected 8,096 people (774 died) [21], and is estimated to have led to the loss of $40 billion to the global economy [22]. Moreover, in China, natural infections by SARS-CoV were detected in wild-caught masked palm civets (*Paguma larvata*) for sale in wildlife markets but were not present in farmed civets [23]. Examples of zoonotic diseases related to wildlife markets from the past few decades also include the Ebola virus in primates, monkeypox in African rodents and possibly HIV in chimpanzees (*Pan troglodytes*) [24,25].

China and South and Southeast Asia are considered global hotspots of emerging zoonotic diseases [26]. In those regions, wildlife is commonly traded in open markets; these markets have been considered 'the perfect storm' for zoonotic disease transmission. Events of pathogen transfer to humans could be avoided or greatly reduced if transmission was better understood and practices adjusted to mitigate risk, but the composition of species sold in markets and the potential cross-transmission of pathogens among them and to humans is still poorly investigated. In this study, we surveyed 14 wildlife markets in 10 cities of Indonesia to estimate volumes and composition of live wild species on sale and assess the potential of wildlife trade in disseminating zoonotic diseases and facilitating a spill over of viruses to humans and across wild species.

2. Materials and Methods

We collected data on wildlife trade in markets located on the islands of Java and Bali, Indonesia. There are at least 53 animal markets on those islands, viz. nine large (50 to over 200 stalls or shops), 22 medium (20–49 stalls or shops) and 22 small (less than 20 stall or shops) [27]. For this study, we surveyed 14 markets of 10 cities ranging from 15 to 100 stalls that were visited for a total of 179 times, an average of 14.1 ± SE 3.5 times each over the period of February 2016–February 2020 (Table 1). Here, we focused on four mammalian Orders (primates, bats, rodents and carnivores) due to their phylogenetic relatedness with humans, high prevalence in the markets, and for being among the main mammalian orders hosting viruses, meaning that the susceptibility of a pathogen cross-transmission among them and with humans is more likely [28]. Each market was visited by one or two of the authors. By slowly walking through the market all live animals on sale (excluding domesticated ones) were identified and counted. Species identification was normally done in situ, mostly at the species level or, less frequently, at the genus level. This information was logged into a mobile phone in the market or recorded in a notebook after leaving the market. We also counted the number of stalls selling wild mammals as a measure of the size of the market. Trade was open and there was no need to resort to undercover techniques; no animals were purchased.

Table 1. Details on location (city) with population size, names of the markets sampled, the sample size in number of visits to the market, number of stalls, and richness of mammalian taxa sold in each of the 14 wildlife markets surveyed in Indonesia (Java and Bali). Cities are listed from west to east.

City	Population Size (Million)	Market (Number of Stalls)	Visits (N)	Richness of Taxa (N)
Jakarta	10.562	Pramuka (100)	16	10
		Jatinegara (55)	28	15
		Barito (20)	28	14
Bogor	1.127	Tj Empang (15)	5	6
Bandung	2.510	Sukahaji (40)	40	13
Garut	0.065	Kerkhof (17)	35	11
Cirebon	0.322	Plered (40)	8	10
Semarang	1.654	Karimata (35)	10	9
Yogyakarta	0.436	Pasty (60)	7	10
Surakarta	0.522	Depok (70)	7	4
Surabaya	2.874	Bratang (75)	2	7
		Kupang (25)	2	8
Denpasar	0.963	Satria (25)	6	8
TOTAL		(647)	179	15

Information on viruses potentially infecting the different species on sale was obtained from Johnson et al. [16], who catalogued the presence (or lack thereof) of 139 zoonotic viruses in 5,335 wild terrestrial animal species. In cases when the animals were identified at the genus level, we considered those species with known distribution in Java and Bali and summed up the number of viruses species for the genus.

We summed the number of taxa on sale as a measure of species richness, and for each taxon we calculated the mean number and standard deviation of animals detected across all markets where they were present. We used Generalized Linear Models (GLM) with Gamma family of distribution to assess the relationship between the population size (in ln scale) of the city surveyed and both mean number of animals and richness of taxa on sale in the markets, using individual markets in each city as replicates. Similarly, we used GLMs to assess the relationship between the number of stalls selling wildlife in each market (in ln scale) and both the mean number of individuals and richness of taxa on sale. GLMs were performed using the "gamlss" package. We obtained the human population size in 2020 of each surveyed city from Statistics Indonesia [29]. To build the interaction network representing the interactions between markets and animals and the main associated diseases, a weighted matrix was constructed with the number of individuals that was recorded in each market. In order to visualize the animals that are traded in the markets and the potential diseases that animals carry and share, we built a diagram using the Sankey Network function of the "networkD3" package [30], where the animal taxa sold, market and virus species are the nodes, and the links are the number of animals offered for sale. All statistical analyses were performed in R statistical software v4.1.0 (R Foundation for Statistical Computing, Vienna, Austria) [31].

3. Results

3.1. Animals Sold in Markets

We recorded 6,725 wild animals of at least 15 species within the order Rodentia, Chiroptera, Carnivora and Primates for sale in the 14 wildlife markets (Figure 1; Table 2). Five taxa accounted for almost 90% of traded animals, namely long-tailed macaque (*Macaca fascicularis*) (24%), Asian palm civet (*Paradoxurus hermaphroditus*) (22%), and plantain squirrel (*Callosciurus notatus*) (20%), followed by the large flying fox (*Pteropus vampytus*) (13%) and Indonesian short-nosed fruit bat [*Cynopterus titthaecheilus*; possibly in western Java also greater short-nosed fruit bat (*Cynopterus sphinx*) (10%)]. Informal conversations with sellers indicate that most of the animals must have been collected within Indonesia, mostly on Java, Bali and Sumatra, but also Borneo and possibly Sulawesi. None, or very few may have come from abroad.

Figure 1. Trade in wild mammals in Java, Indonesia. (**A**) Indonesian short-nosed fruit bat (*Cynopterus titthaecheilus*); (**B**) plantain squirrel (*Callosciurus notatus*), (**C**) long-tailed macaque (*Macaca fascicularis*) and Asian palm civet (*Paradoxurus hermaphroditus*); (**D**) long-tailed macaque and giant fruit bat (*Pteropus vampyrus*); (**E**) long-tailed macaque; (**F**) Javan mongoose (*Urva javanica*).

Table 2. Average number of individuals sold in the 14 wildlife markets surveyed in Indonesia, and number of zoonotic viruses that are able to infect these taxa as hosts.

Taxon	Number of Individuals (Mean When Present ± SD)	Number of Markets with Presence (% of Total)	Number of Zoonotic Viruses (% of Total)
Plantain squirrel *Callosciurus notatus*	1313 (14.4 ± 11.0)	14 (100)	0 (0)
Prevost's squirrel *Callosciurus prevostii*	115 (4.5 ± 1.6)	6 (43)	0 (0)
Masked palm civet *Paguma larvata*	46 (1.5 ± 0.6)	7 (50)	1 (6)
Javan leopard cat *Prionailurus bengalensis*	111 (1.8 ± 0.5)	6 (43)	1 (6)
Small Indian civet *Viverricula indica*	63 (1.7 ± 1.4)	11 (79)	0 (0)
Javan mongoose *Herpestes javanicus*	57 (1.2 ± 0.4)	9 (64)	2 (12)
Asian palm civet *Paradoxurus hermaphroditus*	1501 (7.5 ± 7.9)	14 (100)	0 (0)
Asiatic small-clawed otter *Aonyx cinereus*	45 (1.5 ± 0.8)	7 (50)	0 (0)

Table 2. *Cont.*

Taxon	Number of Individuals (Mean When Present ± SD)	Number of Markets with Presence (% of Total)	Number of Zoonotic Viruses (% of Total)
Javan ferret badger *Melogale orientalis*	40 (1.8 ± 0.7)	9 (64)	0 (0)
Indonesian short-nosed fruit bat *Cynopterus titthaecheilus*	656 (19.9 ± 17.9)	10 (71)	5 (29)
Large flying fox *Pteropus vampyrus*	907 (9.5 ± 9.0)	13 (93)	1 (6)
Long-tailed macaque *Macaca fascicularis*	1620 (10.3 ± 15.6)	13 (93)	9 (53)
Southern pig-tailed macaque *Macaca nemestrina*	70 (2.8 ± 1.9)	3 (21)	6 (35)
Slow loris *Nycticebus* spp.	141 (2.9 ± 2.8)	5 (36)	0 (0)
Langur *Trachypithecus* spp.	40 (2.2 ± 1.3)	5 (36)	1 (6)
Total	6725	14 (100)	17 (100)

The mean number of individuals offered for sale showed a positive trend related with the human population in the respective city (Estimate = 14.91 ± 4.10, t-value = 3.64, *p*-value = 0.004) and the number of market stalls selling wildlife (Estimate = 39.54 ± 14.02, t-value = 2.82, *p*-value = 0.02) (Figure 2). Conversely, no clear trend was found between human population (Estimate = 0.53 ± 0.44, t-value = 1.21, *p*-value = 0.25) or number of stalls in the market (Estimate = 0.05 ± 1.45, t-value = 0.035, *p*-value = 0.97) with taxa richness on sale.

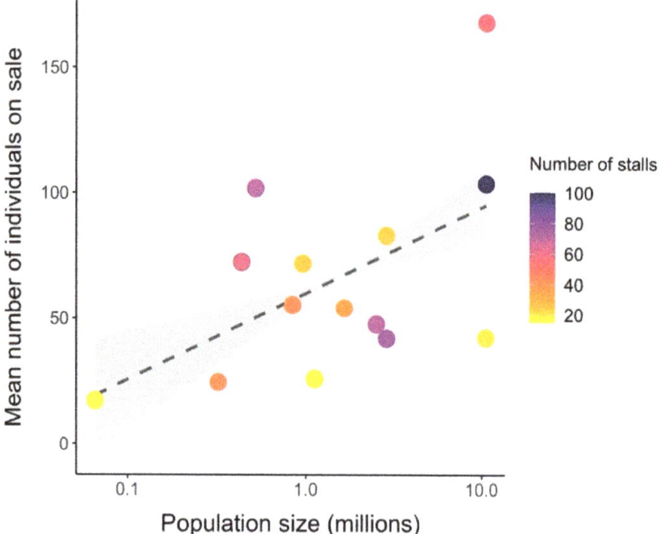

Figure 2. Relationship between human population size (in ln scale) of the surveyed city and mean number of primates, rodents, bats and carnivores on sale in wildlife markets in Indonesia. Each point is a sampled market, and the colour gradient refers to the estimated number of stalls selling wildlife in each market.

3.2. Virus Species and Network of Markets and Hosts

At least eight (53%) out of 15 animal taxa recorded for sale are hosts of virus species listed in the database consulted for known zoonotic diseases (Table 2). We recorded 17 different zoonotic viruses that can affect these taxa. Of those, nine viruses can infect more than one recorded species as a host, including Influenza A, Nipah, Cowpox, Herpes B; these viruses are all easily transmitted to humans with no need of vectors.

The taxa with the highest number of zoonotic viruses recorded are both macaques; the long-tailed macaque, a potential host of 9 viruses (53% of the total number recorded), including Cowpox, Dengue, Hepatitis E, Herpes B, Monkeypox, Reston, Ebola, Simian Foamy, Simian retrovirus type D and Vesicular stomatitis viruses; and southern pig-tailed macaque, with 6 (35%) virus species, including Cowpox, Dengue, Herpes B, Simian Foamy, Simian retrovirus type D and St. Louis encephalitis viruses. Short-nosed fruit bats also stand out by being potentially infected by 5 (29%) of the viruses recorded, but different species from those infecting macaques, which include Influenza A, Issyk-Kul, Japanese encephalitis, Kyasanur forest disease, and Nipah viruses. Of the remaining mammal taxa sold, masked palm civet is a host of SARS-CoV (and SARS-CoV related) virus, Javan leopard cat is a host of Influenza A virus, Javan mongoose is a host of Hepatitis E virus and Rabies, large flying fox is a host of Nipah virus, and *Trachypithecus* langurs are a host of Dengue virus. No virus was reported for the two recorded squirrels (Prevost's squirrel and plantain squirrel) and two civets (masked palm civet and small Indian civet) on trade, neither for Asiatic small-clawed otter (*Aonyx cinereus*), Javan ferret badger (*Melogale orientalis*) and slow lorises.

The Sankey network diagram shows that long-tailed macaque is the species with the greatest potential for spreading diseases (Figure 3), since it is both a host for a large number of viruses and the most traded species, being sold in large quantities in 13 (93%) out of 14 surveyed markets.

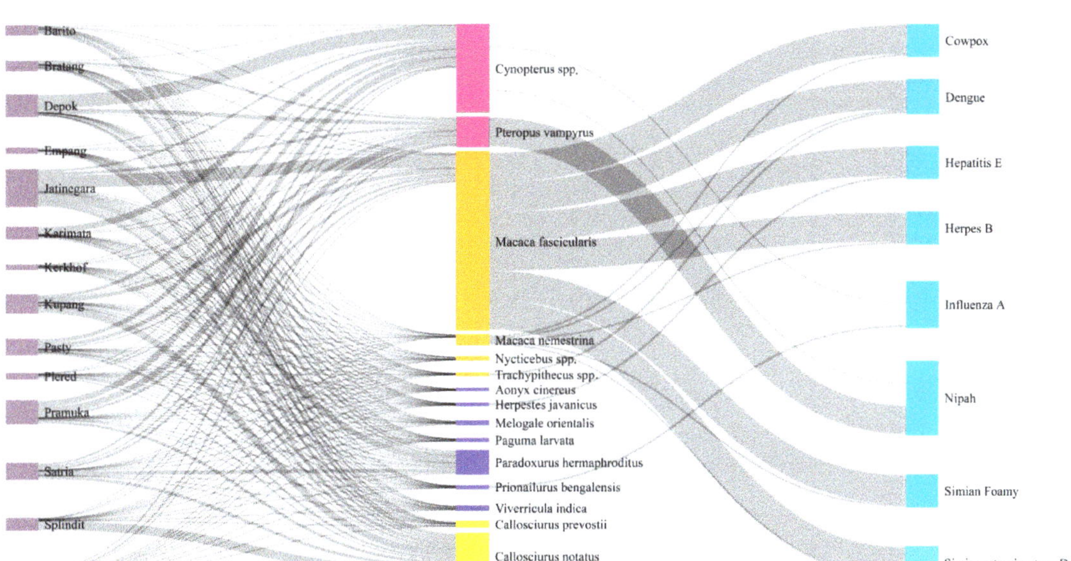

Figure 3. Sankey network diagram illustrating the 14 surveyed wildlife markets (purple), the mammal taxa (pink = bats, orange = primates, navy blue = carnivores, yellow = rodents), and the viruses (blue) these animals have the potential to host and share by co-occurring in the same markets.

In addition, long-tailed macaques are traded simultaneously with pig-tailed macaques in three markets, with the potential to spread six diseases in common (Cowpox, Dengue, Hepatitis E, Herpes B, Simian foamy, and Simian retrovirus type D). Indonesian short-nosed fruit bats and the large flying fox are also sold in large quantities and traded together in 10 markets, increasing the potential for the spread of Nipah virus. Furthermore, short-nosed fruit bats and the Javan leopard cat are traded in four markets in common and are potential hosts of Influenza A.

4. Discussion

4.1. Links between Wildlife Trade and Transmission of Viruses

Using the widespread and open trade in live wild mammals in markets on the two main islands of Indonesia as our case study, we show that live wildlife markets may provide optimal conditions for the spill over and spread of viral diseases. More populated cities and larger markets had larger quantities of animals being sold, and those same markets had on sale several mammal species that potentially share viruses in common. In addition, some of the most recorded species, such as long-tailed macaque and short-nosed fruit bats, are hosts of several virus species and are traded in large quantities across almost all markets, posing a risk of disease outbreaks to millions of people inhabiting those highly populated islands. These findings are of great concern in terms of public health, because there is a higher likelihood for an infected animal to end up in the market of a city where (i) there is a wider range of species, increasing the chance of a spill over, and (ii) that is more populated by humans, increasing the chance of fast pathogen spread and consequently for outbreaks and epidemics to occur.

The number of zoonotic viruses potentially hosted by the traded species (17) in this study is very similar to the number (16) identified in wildlife traded as wild meat in Malaysia [32]; of these, nine are common between studies. This means that the risk of these viruses in infecting humans by contact in wildlife markets is not exclusive of Indonesia, but potentially widespread across Southeast Asia. For most zoonotic viruses reported here, the type of contact that sellers, buyers and even visitors have with the animals is enough to enable transmission. SARS-CoV, Influenza, Hepatitis E, Issyk-Kul, Nipah viruses can be transmitted through infected respiratory secretions and/or exposition to contaminated faeces and urine [33,34], which is very likely to occur in a crowded market, often with low levels of hygiene [35]. Others, such as Rabies, Cowpox, Herpes B Cercopithecine herpesvirus 1, Simian Foamy and Simian retrovirus type D can be transmitted transcutaneous through animal bites and scratches, which are also possible to happen when handling the animals or by humans having their mucous membranes or damaged skin exposed to animal body fluids [36]. Only 4/17 (23.5%) of the viruses identified cause vector-borne diseases, requiring the presence of a vector to be transmitted from infected animals to humans. These include Dengue, Japanese encephalitis and St. Louis encephalitis, which have mosquitoes as a vector, and Kyasanur forest disease virus that needs ticks as vectors to be transmitted [37,38]. Although it is not impossible for vectors to be present at the markets or, later, at the place where the bought animal is kept (especially ticks carried by the animals), it is reasonable to assume that the need of a vector decreases substantially the chance of those diseases to occur due to wildlife trade.

It is of great concern that the recorded virus species are potentially circulating in wildlife markets in Southeast Asia, since among them are some viruses responsible for causing serious and deadly diseases to humans. Nipah virus, in particular, is a bat-borne virus that has caused severe disease outbreaks in Asia, with mortality rates reaching over 90% in some cases, consisting of one of the deadliest viruses affecting humans [39]. This virus causes acute respiratory infection and fatal encephalitis, and yet there is no treatment for infected individuals or vaccine available [40]. The virus can spill over to domestic animals, such as pigs, horses, cats and dogs. In the surveyed markets in Indonesia, domestic animals, mostly dogs and cats, were seen being sold as pets in large quantities; free roaming cats and dogs, in addition to rats, are frequently encountered in the markets. These animals may be infected especially by contact with traded bats' urine. Outbreaks of Nipah have happened in Malaysia, Bangladesh and India due to contact of humans with infected domestic animals and contaminated food [40,41]. Therefore, Indonesia and other countries presenting fruit bats of the Pteropodidae family may be at a very high risk, especially with the facilitation of spill over through trade of these species. Herpes B Cercopithecine herpesvirus 1 is also a pathogen worth of concern, since apparently healthy macaques can host it without any overt signs of disease [36], increasing the chance of infected animals to end up in the market and to be sold. Stress or immunosuppression, both common health issues due to poor enclosure conditions, was observed to increase the chance of macaques in shedding the virus [42]. Conversely, Herpes B in humans usually results in fatal encephalomyelitis or severe neurologic impairment, with a death rate of >70% when there is limited availability of antiviral therapy [36,42].

4.2. Sanitary and Health Implications

A number of species we encountered in the markets are included on Indonesia's list of protected species and no wild-caught individuals are allowed to enter the trade. These include the Javan leopard cat, slow lorises and *Trachypithecus* langurs. In Indonesia trade in species that are not legally protected is regulated through a harvest quota system [43]. For mammals these are mostly set at zero (i.e., no wild harvest is allowed) or only small numbers are allowed to be harvested and traded for specific purposes. The numbers we observed in the markets greatly exceeded these harvest quotas, as for instance for the year 2020 a total of one small Indian civet was allowed to be traded for pets for all of Indonesia, in addition to five Prevost squirrels, five large flying foxes, six masked palm civets, 29 palm civets and 135 plantain squirrels. Single visits to the animal markets on Java and Bali often recorded these species in numbers far exceeding this. Hence, most of the trade in wild mammals in the markets in Java and Bali is illegal and in violation of Indonesia's domestic legislation and regulations. The maintenance of an illegal trade not only increases the number of species and individuals on sale, creating more situations of contact for potential sharing and spreading of pathogens, but also hamper proper control and establishment of sanitary and hygienic measures to avoid viral transmissions and infections.

While the illegal trade certainly aggravates the potential of wildlife markets in spreading diseases, the legal trade, if not properly monitored, also poses a similar threat to humans. Reducing this threat is not a simple task, once ownership, consumption and trade of wild species are usually part of the local culture, play important roles in local economy and occur more frequently in developing, but megadiverse, countries where surveillance is often insufficient [44,45]. It is, therefore, essential to consider this issue in the light of One Health approach and reinforce the appreciation that human, animal and ecosystem health are interdependent [46]. The implementation of measures to prevent new viral emergences and protect human health requires a holistic approach that embraces all three components. Wild population declines due to overexploitation and reduction in wildlife habitat quality were strongly related to a higher risk of disease transmission of animal viruses to humans [16]. Accordingly, any proposals should to integrate these multiple dimensions and take into account environmental, social and economic issues.

Sanitary measures should focus on the most affected species and largest markets and cities. For instance, the most commercialized species in our study, the long-tailed macaque, a potential host of nine virus species, has been recently classified by IUCN as Endangered [47], especially due to high utilization by humans as a meat source for local populations. During the COVID-19 pandemic, the demand for the species increased [47,48], consequently raising the risk of disease transmission. In addition, the species is being threatened due to its widespread use for biomedical and toxicological research [48]. Hence, most efforts should be done for controlling its trade and prevent viral diseases to spread from its use. However, for a more comprehensive analysis of risks, more information should be available on the pathogen loads in traded animals, transmission risks at different contact points, and potentials for animals to be taken to different parts of the country. One suggestion is an implementation of zoonotic vigilance, where species in markets can be periodically monitored for early detection of occurrence of main pathogens, allowing the prevention of outbreaks in both humans and animals [49,50]. A functional early alert system may attenuate the health, social and economic impacts of epidemics and pandemics [51].

In 2021, the World Health Organisation (WHO), the World Organisation for Animal Health (OIE) and the United Nations Environment Programme (UNEP) have issued guidance to help reduce public health risks associated with the sale of live wild mammals [52], prompted by the COVID-19 pandemic and the allegedly role of a wildlife market in Wuhan, China, in its origin. Although this attribution has been questioned, between 2017 and 2019 around 47 thousand individuals from 38 species were kept under poor welfare and hygiene conditions and sold in Wuhan's markets prior to the COVID-19 pandemic [53]. This fact shows the potential of these markets in breaking the barriers of contact between humans and wild animals, even in large urban areas of the world. The guidance from WHO is focused on the sale of live species in food markets, but some may apply to the sale of live species for broader purposes, such as pet ownership. Among the measures are the obvious improvement of standards of hygiene and sanitation in these markets, which may include avoidance of keeping animals in overpopulated cages and regular cleaning and disinfection of animal enclosures, pest control and waste management with special attention to animal urine, faeces and other secretions. Traceability of farmed wild animals, where this is an option, can also contribute to curb the trade of animals illegally sourced from the wild that are more likely to be shedding a pathogen [52]. The document also recommends the development and implementation of campaigns for market traders, stallholders, consumers and the wide general public that can bring information about the risks of transmission of zoonotic pathogens at the human–animal interface, safety practices in handling and keeping live wild animals and what to consider when selling or buying an animal in order to reduce the likelihood of spreading zoonotic diseases.

In addition to wildlife hunting and trade, the modification of environments by deforestation, intensification of agricultural production and urbanization increases the possibility of interspecies transmission due to higher probability of contact between humans and animals [54,55]. In fact, forestation has been advocated as one potential ecological measure to prevent virus outbreaks [56]. Climate change may also expand the habitat range of some of the recorded taxa, such as bats, resulting in modifications in interactions among species and facilitating cross-species spillover [57]. Our study demonstrated the potential of applying interaction networks to better visualize the wildlife trade in Indonesia, as well as the potential diseases that may emerge through interactions between organisms. In the face of manyfold environmental changes happening in the world, new studies could use this approach to advance the understanding of the use and trade of animals in other regions of Asia, as well as other tropical areas in Africa and Latin America that are considered infectious disease hotspots [3,58].

In this study we highlight the need for better surveillance and sanitary conditions to avoid the negative health impacts of unregulated wildlife markets. More than half of the species traded in wildlife markets in Indonesia are in fact hosts of zoonotic virus species. This study could be used in the development of public health strategies in Southeast Asia, such as implementing sanitary measures and standards in wildlife markets. This information is also useful to develop awareness campaigns to educate people about the numerous health risks from trading or buying wildlife highlighted, encouraging them to buy wildlife that are legally sourced and surveilled for pathogens, or the consumption of alternative foods whenever possible. Such initiatives could have additional benefits for the conservation of threatened species by helping reduce the illegal domestic and international trade of species in and from Southeast Asia [59].

Author Contributions: Conceptualization, T.Q.M., V.N. and K.A.-I.N.; methodology, T.Q.M., V.N., M.C. and K.A.-I.N.; formal analysis, T.Q.M., P.E.R.P. and V.N.; investigation, A.A., K.H. and V.N.; data curation, T.Q.M., P.E.R.P., K.H., K.A.-I.N. and V.N.; writing—original draft preparation, T.Q.M. and V.N.; writing—review and editing, P.E.R.P., A.A., M.A.I., K.H., M.C., K.A.-I.N. and V.N.; project administration, M.A.I., K.A.-I.N. and V.N.; funding acquisition, T.Q.M., V.N. and K.A.-I.N. All authors have read and agreed to the published version of the manuscript.

Funding: Our research in Indonesia was funded by Cleveland Zoo and Zoo Society, Columbus Zoo and Aquarium, Disney Worldwide Conservation Fund, Global Challenges Fund, Henry Doorly Zoo, Lee Richardson Zoo, Little Fireface Project, Mohamed bin al Zayed Species Conservation Fund (152511813), Moody Gardens Zoo, Naturzoo Rhein, Paradise Wildlife Park, People's Trust for Endangered Species, Sacramento Zoo, Shaldon Wildlife Trust, and ZGAP. TQM and VN are funded by an Oxford Brookes University Research Excellence Award. PERP is funded by Conselho Nacional de Desenvolvimento Científico e Tecnológico (CNPq) (151005/2021-4).

Institutional Review Board Statement: This study did not involve any experiment on animals or research involving human participants. Part of the research was underpinned by a Memorandum of Understanding between Universitas Gadjah Mada and Oxford Brookes University. In our United Kingdom and Indonesian Institutes, we did not require institutional permission for observational research in markets; they were, however, added to Oxford Brookes University's Register of Activities Involving Animals (2016–2021). Informal discussions with traders followed the ethical guidelines proposed by the Association of Social Anthropologists of the UK and Commonwealth.

Data Availability Statement: The data presented in this study that are not yet included in the paper are available on request from the corresponding authors.

Conflicts of Interest: The authors declare no conflict of interests. The funders had no role in the design of the study; in the collection, analyses, or interpretation of data; in the writing of the manuscript, or in the decision to publish the results.

References

1. Morens, D.; Folkers, G.; Fauci, A. The challenge of emerging and re-emerging infectious diseases. *Nature* **2004**, *430*, 242–249. [CrossRef] [PubMed]

2. Fonkwo, P.N. Pricing infectious disease: The economic and health implications of infectious diseases. *EMBO Rep.* **2008**, *9*, S13–S17. [CrossRef] [PubMed]

3. Jones, K.E.; Patel, N.G.; Levy, M.A.; Storeygard, A.; Balk, D.; Gittleman, J.L.; Daszak, P. Global trends in emerging infectious diseases. *Nature* **2008**, *451*, 990–993. [CrossRef]

4. Taylor, L.H.; Latham, S.M.; Woolhouse, M.E.J. Risk Factors for Human Disease Emergence. *Philos. Trans. R. Soc. B Biol. Sci.* **2001**, *356*, 983–989. [CrossRef] [PubMed]

5. Wolfe, N.D.; Dunavan, C.P.; Diamond, J. Origins of major human infectious diseases. *Nature* **2007**, *447*, 279–283. [CrossRef]

6. Hui, E.K.W. Reasons for the increase in emerging and re-emerging viral infectious diseases. *Microbes Infect.* **2006**, *8*, 905–916. [CrossRef]

7. Bengis, R.G.; Leighton, F.A.; Fischer, J.R.; Artois, M.; Mörner, T.; Tate, C.M. The Role of Wildlife in Emerging and Re-Emerging Zoonoses. *OIE Rev. Sci. Tech.* **2004**, *23*, 497–511.

8. Piret, J.; Boivin, G. Pandemics Throughout History. *Front. Microbiol.* **2021**, *11*, 631736. [CrossRef]

9. Smith, K.M.; Anthony, S.J.; Switzer, W.M.; Epstein, J.H.; Seimon, T.; Jia, H.; Sanchez, M.D.; Huynh, T.T.; Galland, G.G.; Shapiro, S.E.; et al. Zoonotic viruses associated with illegally imported wildlife products. *PLoS ONE* **2012**, *7*, e29505. [CrossRef]

10. Can, Ö.E.; D'Cruze, N.; Macdonald, D.W. Dealing in deadly pathogens: Taking stock of the legal trade in live wildlife and potential risks to human health. *Glob. Ecol. Conserv.* **2019**, *17*, e00515. [CrossRef]

11. Webster, R.G. Wet markets—A continuing source of severe acute respiratory syndrome and influenza? *Lancet* **2004**, *363*, 234–236. [CrossRef] [PubMed]

12. Bell, D.; Roberton, S.; Hunter, P.R. Animal origins of SARS coronavirus: Possible links with the international trade in small carnivores. *Philos. Trans. R. Soc. Lond. Ser. B Biol. Sci.* **2004**, *359*, 1107–1114. [CrossRef] [PubMed]

13. Bush, E.R.; Baker, S.E.; Macdonald, D.W. Global trade in exotic pets 2006–2012. *Conserv. Biol.* **2014**, *28*, 663–676. [CrossRef] [PubMed]

14. Woo, P.C.; Lau, S.K.; Yuen, K.Y. Infectious diseases emerging from Chinese wet-markets: Zoonotic origins of severe respiratory viral infections. *Curr. Opin. Infect. Dis.* **2006**, *19*, 401. [CrossRef]

15. Reed, K.D.; Melski, J.W.; Graham, M.B.; Regnery, R.L.; Sotir, M.J.; Wegner, M.V.; Kazmierczak, J.J.; Stratman, E.J.; Li, Y.; Fairley, J.A.; et al. The detection of monkeypox in humans in the Western Hemisphere. *N. Engl. J. Med.* **2004**, *350*, 342–350. [CrossRef]

16. Johnson, C.K.; Hitchens, P.L.; Pandit, P.S.; Rushmore, J.; Evans, T.S.; Young, C.C.; Doyle, M.M. Global shifts in mammalian population trends reveal key predictors of virus spillover risk. *Proc. Royal. Soc. B* **2020**, *287*, 20192736. [CrossRef]

17. Karesh, W.B.; Cook, R.A.; Bennett, E.L.; Newcomb, J. Wildlife trade and global disease emergence. *Emerg. Infect. Dis.* **2005**, *11*, 1000–1002. [CrossRef]

18. Cleaveland, S.; Haydon, D.; Taylor, L. Overviews of pathogen emergence: Which pathogens emerge, when and why. *Curr. Topics Microbiol. Immunol.* **2007**, *315*, 85–111.

19. Smith, F.; Behrens, M.; Schloegel, L.M.; Marano, N.; Burgiel, S.; Daszak, P. Reducing the risks of the wildlife trade. *Science* **2009**, *324*, 594–595. [CrossRef]

20. Perrings, C.; Levin, S.; Daszak, P. The economics of infectious disease, trade and pandemic risk. *EcoHealth* **2018**, *15*, 241–243. [CrossRef]

21. Rabaan, A.A.; Al-Ahmed, S.H.; Haque, S.; Sah, R.; Tiwari, R.; Malik, Y.S.; Dhama, K.; Yatoo, M.I.; Bonilla-Aldana, D.K.; Rodriguez-Morales, A.J. SARS-CoV-2, SARS-CoV, and MERS-CoV: A comparative overview. *Infez. Med.* **2020**, *28*, 174–184. [PubMed]

22. Fehr, A.R.; Perlman, S. Coronaviruses: An overview of their replication and pathogenesis. In *Coronaviruses*; Humana Press: New York, NY, USA, 2015; pp. 1–23.

23. Guan, Y.; Zheng, B.J.; He, Y.Q.; Liu, X.L.; Zhuang, Z.X.; Cheung, C.L.; Luo, S.W.; Li, P.H.; Zhang, L.J.; Guan, Y.J.; et al. Isolation and characterization of viruses related to the sars coronavirus from animals in Southern China. *Science* **2003**, *302*, 276–278. [CrossRef] [PubMed]

24. Guarner, J.; Johnson, B.J.; Paddock, C.D.; Shieh, W.J.; Goldsmith, C.S.; Reynolds, M.G.; Damon, I.K.; Regnery, R.L.; Zaki, S.R. Veterinary Monkeypox Virus Working Group. Monkeypox transmission and pathogenesis in prairie dogs. *Emerg. Infect. Dis.* **2004**, *10*, 426–431. [CrossRef] [PubMed]

25. Keele, B.F.; Jones, J.H.; Terio, K.A.; Estes, J.D.; Rudicell, R.S.; Wilson, M.L.; Li, Y.; Learn, G.H.; Beasley, T.M.; Schumacher-Stankey, J.; et al. Increased mortality and AIDS-like immunopathology in wild chimpanzees infected with SIVcpz. *Nature* **2009**, *460*, 515–519. [CrossRef]

26. Allen, T.; Murray, K.A.; Zambrana-Torrelio, C.; Morse, S.S.; Rondinini, C.; Di Marco, M.; Breit, N.; Olival, K.J.; Daszak, P. Global hotspots and correlates of emerging zoonotic diseases. *Nat. Commun.* **2017**, *8*, 1124. [CrossRef]

27. Nijman, V.; Langgeng, A.; Birot, H.; Imron, M.A.; Nekaris, K.A.I. Wildlife trade, captive breeding and the imminent extinction of a songbird. *Glob. Ecol. Conserv.* **2018**, *15*, e00425. [CrossRef]

28. Shivaprakash, K.N.; Sen, S.; Paul, S.; Kiesecker, J.M.; Bawa, K.S. Mammals, wildlife trade, and the next global pandemic. *Curr. Biol.* **2021**, *31*, 3671–3677. [CrossRef]

29. Statistics Indonesia. Indonesia Population Census 2020 (Hasil Sensus Penduduk 2020). 2021. Available online: https://www.bps.go.id/publication/2021/01/21/213995c881428fef20a18226/potret-sensus-penduduk-2020-menuju-satu-data-kependudukan-indonesia.html (accessed on 1 September 2022).

30. Allaire, J.; Gandrud, C.; Kenton, R.; Yetman, C. networkD3: D3 JavaScript Network Graphs from R. 2017. Available online: https://cran.r-project.org/package=networkD3 (accessed on 15 August 2022).

31. R Core Team. *R: A Language and Environment for Statistical Computing*; R Foundation for Statistical Computing: Vienna, Austria, 2021; Available online: https://www.R-project.org/ (accessed on 1 October 2022).

32. Cantlay, J.C.; Ingram, D.J.; Meredith, A.L. A review of zoonotic infection risks associated with the wild meat trade in Malaysia. *EcoHealth* **2017**, *14*, 361–388. [CrossRef]

33. Kenney, S.P. The current host range of Hepatitis E viruses. *Viruses* **2019**, *11*, 452. [CrossRef]

34. Desvaux, S.; Marx, N.; Ong, S.; Gaidet, N.; Hunt, M.; Manuguerra, J.C.; Sorn, S.; Peiris, M.; Van der Werf, S.; Reynes, J.M. Highly pathogenic avian influenza virus (H5N1) outbreak in captive wild birds and cats, Cambodia. *Emerg. Infect. Dis.* **2009**, *15*, 475–478. [CrossRef]

35. Gao, X.L.; Xao, M.F.; Luo, Y.; Dong, Y.F.; Ouyang, F.; Dong, W.Y.; Li, J. Airborne bacterial contaminations in typical Chinese wet market with live poultry trade. *Sci. Total Environ.* **2016**, *572*, 681–687. [CrossRef] [PubMed]

36. Huff, J.L.; Barry, P.A. B-Virus (Cercopithecine herpesvirus 1) infection in humans and macaques: Potential for zoonotic disease. *Emerg. Infect. Dis.* **2003**, *9*, 246–250. [CrossRef]

37. Shah, S.Z.; Jabbar, B.; Ahmed, N.; Rehman, A.; Nasir, H.; Nadeem, S.; Jabbar, I.; Rahman, Z.; Azam, S. Epidemiology, pathogenesis, and control of a tick-borne disease-Kyasanur forest disease: Current status and future directions. *Front. Cell. Infect. Microbiol.* **2018**, *8*, 149. [CrossRef] [PubMed]

38. Mulvey, P.; Duong, V.; Boyer, S.; Burgess, G.; Williams, D.T.; Dussart, P.; Horwood, P.F. The Ecology and evolution of Japanese encephalitis virus. *Pathogens* **2021**, *10*, 1534. [CrossRef] [PubMed]

39. Sharma, V.; Kaushik, S.; Kumar, R.; Yadav, J.P.; Kaushik, S. Emerging trends of Nipah virus: A review. *Rev. Med. Virol.* **2019**, *29*, e2010. [CrossRef]

40. Skowron, K.; Bauza-Kaszewska, J.; Grudlewska-Buda, K.; Wiktorczyk-Kapischke, N.; Zacharski, M.; Bernaciak, Z.; Gospodarek-Komkowska, E. Nipah virus–Another threat from the world of zoonotic viruses. *Front. Microbiol.* **2022**, *12*, 811157. [CrossRef]

41. Uwishema, O.; Wellington, J.; Berjaoui, C.; Muoka, K.O.; Onyeaka, C.V.P.; Onyeaka, H. A short communication of Nipah virus outbreak in India: An urgent rising concern. *Ann. Med. Surg.* **2022**, *82*, 104599. [CrossRef]

42. Weigler, B.J.; Hird, D.W.; Hilliard, J.K.; Lerche, N.W.; Roberts, J.A.; Scott, L.M. Epidemiology of cercopithecine herpesvirus 1 (B virus) infection and shedding in a large breeding cohort of rhesus macaques. *J. Infect. Dis.* **1993**, *167*, 257–263. [CrossRef]

43. Nijman, V.; Morcatty, T.Q.; Feddema, K.; Campera, M.; Nekaris, K.A.I. Disentangling the Legal and Illegal Wildlife Trade–Insights from Indonesian Wildlife Market Surveys. *Animals* **2022**, *12*, 628. [CrossRef]

44. Macedo Couto, R.; Ranzani, O.T.; Waldman, E.A. Zoonotic tuberculosis in humans: Control, surveillance, and the one health approach. *Epidemiol. Rev.* **2019**, *41*, 130–144. [CrossRef]

45. Coker, R.; Rushton, J.; Mounier-Jack, S.; Karimuribo, E.; Lutumba, P.; Kambarage, D.; Pfeiffer, D.U.; Stärk, K.; Rweyemamu, M. Towards a conceptual framework to support one-health research for policy on emerging zoonoses. *Lancet Infect. Dis.* **2011**, *11*, 326–331. [CrossRef] [PubMed]

46. Cunningham, A.A.; Daszak, P.; Wood, J.L. One Health, emerging infectious diseases and wildlife: Two decades of progress? *Philos. Trans. R. Soc. Lond. B Biol. Sci.* **2017**, *372*, 20160167. [CrossRef] [PubMed]

47. Hansen, M.F.; Ang, A.; Trinh, T.; Sy, E.; Paramasiwam, S.; Ahmed, T.; Dimalibot, J.; Jones-Engel, L.; Ruppert, N.; Griffioen, C.; et al. *Macaca Fascicularis*, Long-Tailed Macaque. *IUCN Red List Threat. Species* **2022**, *8235*, 6–12.

48. Hansen, M.F.; Gill, M.; Nawangsari, V.A.; Sanchez, K.L.; Cheyne, S.M.; Nijman, V.; Fuentes, A. Conservation of Long-Tailed Macaques: Implications of the Updated IUCN Status and the COVID-19 Pandemic. *Primate Conserv.* **2021**, *35*, 1–11.

49. Córdoba-Aguilar, A.; Ibarra-Cerdeña, C.N.; Castro-Arellano, I.; Suzan, G. Tackling zoonoses in a crowded world: Lessons to be learned from the COVID-19 pandemic. *Acta Trop.* **2021**, *214*, 105780. [CrossRef]

50. Naguib, M.M.; Li, R.; Ling, J.; Grace, D.; Nguyen-Viet, H.; Lindahl, J.F. Live and wet markets: Food access versus the risk of disease emergence. *Trends Microbiol.* **2021**, *29*, 573–581. [CrossRef]

51. Charlier, J.; Barkema, H.W.; Becher, P.; De Benedictis, P.; Hansson, I.; Hennig-Pauka, I.; La Ragione, R.; Larsen, L.E.; Madoroba, E.; Maes, D.; et al. Disease control tools to secure animal and public health in a densely populated world. *Lancet Planet. Health* **2022**, *6*, 812–824. [CrossRef] [PubMed]

52. WHO-World Health Organization. *Reducing Public Health Risks Associated with the Sale of Live Wild Animals of Mammalian Species in Traditional Food Markets*; World Health Organization, WHO Headquarters (HQ): Geneva, Switzerland, 2021; pp. 1–8.

53. Xiao, X.; Newman, C.; Buesching, C.D.; Macdonald, D.W.; Zhou, Z.M. Animal sales from Wuhan wet markets immediately prior to the COVID-19 pandemic. *Sci. Rep.* **2021**, *11*, 11898. [CrossRef]

54. Lambin, E.F.; Tran, A.; Vanwambeke, S.O.; Linard, C.; Soti, V. Pathogenic Landscapes: Interactions between Land, People, Disease Vectors, and Their Animal Hosts. *Int. J. Health Geogr.* **2010**, *9*, 54. [CrossRef]

55. Rohr, J.R.; Barrett, C.B.; Civitello, D.J.; Craft, M.E.; Delius, B.; DeLeo, G.A.; Hudson, P.J.; Jouanard, N.; Nguyen, K.H.; Ostfeld, R.S.; et al. Emerging Human Infectious Diseases and the Links to Global Food Production. *Nat. Sustain.* **2019**, *2*, 445–456. [CrossRef]

56. Nabi, G.; Siddique, R.; Ali, A.; Khan, S. Preventing bat-born viral outbreaks in future using ecological interventions. *Environ. Res.* **2020**, *185*, 109460. [CrossRef] [PubMed]
57. Martin, G.; Yanez-Arenas, C.; Chen, C.; Plowright, R.K.; Webb, R.J.; Skerratt, L.F. Climate Change Could Increase the Geographic Extent of Hendra Virus Spillover Risk. *EcoHealth* **2018**, *15*, 509–525. [CrossRef] [PubMed]
58. Albery, G.F.; Eskew, E.A.; Ross, N.; Olival, K.J. Predicting the Global Mammalian Viral Sharing Network Using Phylogeography. *Nat. Commun.* **2020**, *11*, 2260. [CrossRef] [PubMed]
59. Nijman, V. An overview of international wildlife trade from Southeast Asia. *Biodivers. Conserv.* **2010**, *19*, 1101–1114. [CrossRef]

Article

Prevalence, Seroprevalence and Risk Factors of Avian Influenza in Wild Bird Populations in Korea: A Systematic Review and Meta-Analysis

Eurade Ntakiyisumba [1,†], Simin Lee [1,†], Byung-Yong Park [2], Hyun-Jin Tae [2] and Gayeon Won [1,*]

[1] College of Veterinary Medicine, Jeonbuk National University, Iksan Campus, Gobong-ro 79, Iksan 54596, Republic of Korea
[2] Department of Veterinary Medicine and Bio-Safety Research Institute, Jeonbuk National University, Iksan 54596, Republic of Korea
* Correspondence: gywon@jbnu.ac.kr; Tel.: +82-63-850-0944
† These authors contributed equally to this work.

Abstract: Since the first recorded outbreak of the highly pathogenic avian influenza (HPAI) virus (H5N1) in South Korea in 2003, numerous sporadic outbreaks have occurred in South Korean duck and chicken farms, all of which have been attributed to avian influenza transmission from migratory wild birds. A thorough investigation of the prevalence and seroprevalence of avian influenza viruses (AIVs) in wild birds is critical for assessing the exposure risk and for directing strong and effective regulatory measures to counteract the spread of AIVs among wild birds, poultry, and humans. In this study, we performed a systematic review and meta-analysis, following the PRISMA guidelines, to generate a quantitative estimate of the prevalence and seroprevalence of AIVs in wild birds in South Korea. An extensive search of eligible studies was performed through electronic databases and 853 records were identified, of which, 49 fulfilled the inclusion criteria. The pooled prevalence and seroprevalence were estimated to be 1.57% (95% CI: 0.98, 2.51) and 15.91% (95% CI: 5.89, 36.38), respectively. The highest prevalence and seroprevalence rates were detected in the Anseriformes species, highlighting the critical role of this bird species in the dissemination of AIVs in South Korea. Furthermore, the results of the subgroup analysis also revealed that the AIV seroprevalence in wild birds varies depending on the detection rate, sample size, and sampling season. The findings of this study demonstrate the necessity of strengthening the surveillance for AIV in wild birds and implementing strong measures to curb the spread of AIV from wild birds to the poultry population.

Keywords: avian influenza virus; wild birds; prevalence; seroprevalence; systematic review; meta-analysis; South Korea

Citation: Ntakiyisumba, E.; Lee, S.; Park, B.-Y.; Tae, H.-J.; Won, G. Prevalence, Seroprevalence and Risk Factors of Avian Influenza in Wild Bird Populations in Korea: A Systematic Review and Meta-Analysis. *Viruses* **2023**, *15*, 472. https://doi.org/10.3390/v15020472

Academic Editors: Myriam Ermonval and Serge Morand

Received: 10 January 2023
Revised: 3 February 2023
Accepted: 7 February 2023
Published: 8 February 2023

1. Introduction

Avian influenza (AI), also known as the "bird flu," a disease caused by influenza type A viruses, affects a wide variety of domestic and wild birds. Based on their pathogenicity in birds, influenza A viruses are classified as either highly pathogenic or low pathogenic avian influenza viruses, known as HPAI and LPAI viruses, respectively [1,2]. Wild birds, particularly migratory aquatic birds of the order Anseriformes (ducks, geese, and swans) and Charadriiformes (shorebirds and gulls) are natural reservoirs of LPAI viruses [3–5]. As LPAI viruses primarily replicate in duck intestinal tracts, their transmission among wild birds occurs primarily through the fecal-oral route [1]. LPAI viruses are excreted in feces and have been demonstrated to survive in water for an extended period of time [6]. Thus, waterborne transmission could play a significant role in the spread of LPAI viruses among migratory waterbirds.

Generally, AI viruses do not cause disease in wild birds, although subtypes of HPAI viruses can invade and replicate in different organs and may cause severe infections [4,7].

HPAI viruses evolve by mutation when the virus, carried in its mild form by a wild bird, is introduced into poultry [5,8]. These viruses are capable of infecting a wide range of animal species, such as swine, birds, companion animals, marine animals, and humans [7,9]. The transmission of avian influenza viruses (AIVs) from infected wild birds to domestic birds is perceived to occur through the sharing of water sources or the contamination of feed [10,11]. In humans, zoonotic subtypes of AIVs are transmitted mainly through direct contact with infected domestic poultry [11,12]. To date, eight AIVs have been reported to infect humans, of which, the H5N1 and H7N9 subtypes are associated with high morbidity and mortality in a large number of humans [11,13,14]. In South Korea, the HPAI subtype, H5N1, was first detected in duck meat imported from mainland China in 2000, which resulted in the loss of 4588 tons of meat [15]. From 2003 to 2004, an HPAI outbreak affected 392 chicken and duck farms in South Korea, causing a total discard of 5,285,000 birds, which was equivalent to $458 million [15,16].

In wild birds, the first cases of the H5N1 HPAI virus infection were primarily observed in Hong Kong in late 2002 [17,18]. Since then, multiple AI outbreaks associated with the H5N1 subtype have been reported in Asia, Africa, and Europe, all of which have been ascribed to wild migratory birds [19,20]. These documented cases imply that wild aquatic birds may play a major role in carrying AIVs over long distances via migration. Waterfowl are the most observed migratory birds, and winter birds are predominantly associated with the occurrence of AI in South Korea [2,16]. Although many countries have been able to halt the spread of H5N1 in animal and human populations by conducting regular surveillance and enforcing strict animal health regulations, the virus remains endemic to poultry populations, primarily in low-income countries with inadequate animal health and surveillance facilities. Owing to the rapid evolution of HPAI viruses, their devastating impact on the global poultry industry, and the threat they pose to public health, it is critical to understand the prevalence of AIVs in wild birds for risk assessment and preparedness against future outbreaks.

The prevalence and seroprevalence of AIVs in the wild bird populations of South Korea have been reported in various individual studies; however, no attempt has been made to consolidate these studies to derive a robust prevalence estimate of AIVs using a meta-analytical approach. The crucial benefit of meta-analysis is that it combines evidence to achieve a more robust point estimate with a higher statistical power as compared with that obtained from a single study from where the data originated [21,22]. Currently, systematic reviews and meta-analyses are perceived as the best available knowledge sources to make decisions regarding treatment choices [23], and meta-analyses are broadly used to calculate precise estimates of disease frequency, such as disease incidence rates and prevalence proportions [21,24]. In various studies, meta-analysis and regression analysis techniques have been used to generate overall prevalence estimates of infectious agents in animal populations and provide empirical evidence on associated risk factors [11,25,26]. In this study, we performed a systematic review and meta-analysis to estimate the overall prevalence and seroprevalence of AI in wild birds, using data from available studies conducted in South Korea. We hypothesized that the detection rate of AI in wild birds would depend on the sampling period, detection method, sample size, and sample type. Thus, subgroup analysis was adapted to investigate the sources of heterogeneity between the reported prevalence from individual studies using the above-mentioned variables.

2. Materials and Methods

2.1. Study Design and Systematic Review Protocol

A systematic review and meta-analysis were performed in accordance with the Preferred Reporting Items for Systematic Review and Meta-Analysis Protocols (PRISMA-P) guidelines [27] to determine the prevalence and seroprevalence of AI in wild birds in South Korea (Table S1). The review question was structured in accordance with the "population, exposure, comparator, and outcome" (PECO) format. In this systematic review, the "population of interest" refers to the wild birds, and "exposure" refers to the AIVs. As this study is a systematic review and meta-analysis of prevalence, the category of "comparator" was not relevant to this study. The "outcomes of interest" included the detected prevalence and seroprevalence of AIVs in wild birds in South Korea.

2.2. Literature Search Strategy

An extensive literature search was conducted with no language restriction using MEDLINE (via PubMed), Scopus, Web of Science, and South Korean databases, such as RISS and KISS, to identify studies published between 1980 and 2021. The last literature search was conducted on 23 December 2021. The following keywords: (wild bird* OR migratory bird* OR waterfowl OR Galliformes OR Charadriiformes OR Anseriformes) AND (avian influenza* OR AI OR bird flu OR avian flu OR influenza A virus OR AIV) AND (Korea OR South Korea) AND (prevalence OR inciden* OR proportion OR cases OR surveillance OR seroprevalence) were used to find eligible studies on the prevalence and seroprevalence of AIVs in wild birds in South Korea. An asterisk was used to extend a search term to related words with the same meaning (e.g., inciden* for incidence and incident).

2.3. Eligibility and Exclusion Criteria

In this systematic review and meta-analysis, the inclusion criteria were as follows: cross-sectional studies, primary studies conducted in South Korea, studies that assessed the prevalence and/or seroprevalence of AI in wild birds, studies that reported the sample size and the number of positive samples or the prevalence/seroprevalence rate, and studies with virus-isolation data. Studies were excluded if they were not conducted in South Korea, if samples were collected from animals other than wild birds, and if they did not report the total number of samples alongside the number of positive samples detected or the exact calculated prevalence rate. The titles and abstracts were screened for suitability using predetermined criteria. The full texts of potentially relevant articles were obtained and evaluated.

2.4. Data Extraction

Data on the prevalence and seroprevalence of AI in wild birds in South Korea were extracted by two independent reviewers, and any disagreements were resolved through discussion and consensus. From all eligible studies, information regarding the first author, year of publication, publication status (i.e., published or non-published), sample type (i.e., feces, cloacal swabs, carcass, or blood), detection method, sampling season, sampling location, bird species, detected AI subtype, sample size, and the number of positive samples was extracted. Data were extracted and organized into a pre-developed Microsoft Excel spreadsheet.

2.5. Risk of Bias Assessment

The eligible studies were assessed for internal and external validity by two independent reviewers using the Joanna Briggs Institute (JBI) critical appraisal tools for prevalence studies [28,29]. Each study was classified as having a low, high, or unclear risk of bias. The checklist contained nine questions, but only eight were evaluated because one question (regarding the response rate) was irrelevant to this study.

2.6. Data Synthesis

Data analysis was conducted using R version 4.1.2 (R Studio version 1.4) software [30,31]. The meta-analysis was performed and the forest plots were generated using the "meta" and "metafor" packages [32–34]. The total number of samples collected and the number of positive samples detected in each study were used to calculate overall prevalence estimates.

To fulfill the assumption of a normal distribution, the logit transformation method was applied to the data [24,26,35] using the following formula:

$$logit\ p = \ln\left(\frac{p}{1-p}\right)$$
$$\text{with variance}:\ var(logit\ p) = \frac{1}{np} + \frac{1}{n(1-p)}$$

where "n" is the total sample size and "p" is the prevalence of the pathogen under study. A generalized linear mixed model, together with a logit transformation, demonstrates better performance; different studies recommend the use of this approach, which was adapted in this study to pool the data [35,36]. A random effects model was used to generate the pooled prevalence and seroprevalence of AIV in wild birds in South Korea. To combine the study estimates, the between-study variance (τ^2) was estimated using the maximum likelihood method. The overall effect size of the logit model and its corresponding 95% confidence interval (CI) were calculated and back-transformed to prevalence rates for ease of interpretation. The between-study heterogeneity was assessed using the Q test and I^2 statistic, which accounts for the amount of the observed variance that reflects the variance in true effects rather than sampling error [37]. The heterogeneity between studies was considered substantially high if the Q test yielded a statistically significant p-value ($p < 0.05$) and I^2 was greater than 50%.

To investigate the reason for heterogeneity, a subgroup analysis was undertaken using four pre-specified variables, including sampling season (i.e., fall/winter and spring/summer), sample size (i.e., more than 1000 or less than 1000), sample type (i.e., feces, cloacal swabs, carcass, and blood), detection method (i.e., ELISA, reverse transcription-polymerase chain reaction (RT-PCR), rRT-PCR, hemagglutination (HA) test, virus isolation, hemagglutinin inhibition (HI) test, and agar gel precipitation test (AGPT)) that could potentially affect the reported prevalence in the literature. Publication bias was assessed through visual inspection of the symmetry of the contour-enhanced funnel plots, and a quantitative estimate of publication bias was performed using Egger's regression test [38,39]. After confirming publication bias, the Duval and Tweedie trim-and-fill method was used to estimate an unbiased effect by imputing missing studies in the funnel plot [40].

3. Results

3.1. Search Results

Initially, 853 records were obtained by conducting an electronic database search. After duplicates were removed, 434 studies remained, and their titles and abstracts were reviewed for eligibility. After the title and abstract screening, 337 of the 434 records were removed. The remaining 97 studies were subjected to full-text screening, of which 48 were deemed irrelevant to this study, and the remaining 49 were finally included in the quantitative synthesis (meta-analysis). The study selection process is illustrated in Figure 1.

3.2. Study Characteristics

Among the 49 studies eligible for the meta-analysis, 39 assessed the prevalence and 10 assessed the seroprevalence of AIVs in wild birds in South Korea. Of the prevalence studies, 24 studies were published in peer-reviewed journals and 15 were non-published records (e.g., government reports, research institute reports, and student dissertations). Regarding the sampling season, 16 studies collected samples in fall and winter, whereas the other 23 studies did not report the sampling season. The sample types included feces (30 trials), carcasses (10 trials), cloacal swabs (10 trials), and combinations of samples (2 trials); two trials did not specify the type of samples used. Samples were collected from the Anseriformes (10 trials), Charadriiformes (5 trials), other species (9 trials), and non-reported bird species (36 trials). Regarding seroprevalence studies, three studies were published in South Korean or international academic journals, and the other seven were non-published records. Regarding the sampling season, three studies collected samples in the fall and winter, whereas the other seven studies did not report the sampling season. Blood samples were collected from the Anseriformes species (8 trials), Charadriiformes (5 trials), other species (8 trials), and non-reported species (3 trials). The characteristics of studies included in this systematic review and meta-analysis are summarized in Tables S2 and S3.

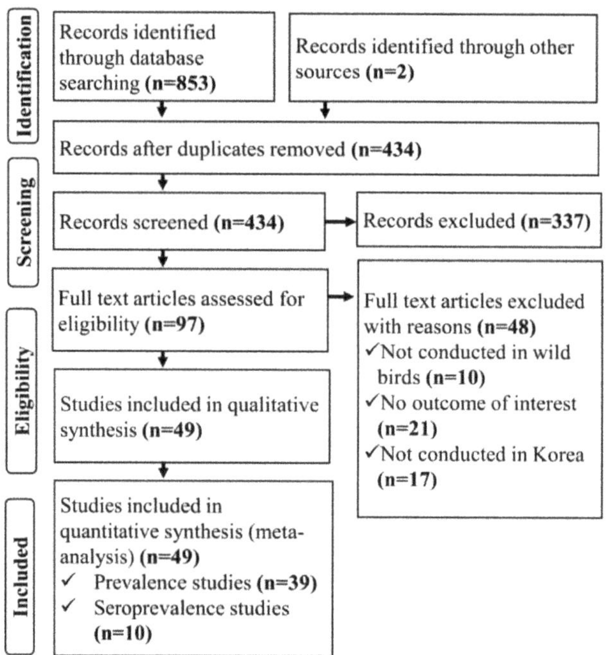

Figure 1. PRISMA flow diagram: Selection of studies on the prevalence and seroprevalence of avian influenza virus in wild bird populations in South Korea for use in a systematic review and meta-analysis.

3.3. Risk of Bias Assessment

The results of the quality assessment of relevant studies that reported the prevalence and seroprevalence of AIVs in wild birds are shown in Figure 2. Studies that reported both prevalence and seroprevalence had a risk of bias, assessed by sorting each result. Prevalence studies that used samples from the entire nation [8,17,41–56] or major migratory bird habitats [2,57–63] determined that the sampling frame was properly chosen, as samples were taken from within the pertinent regions to calculate the prevalence therein. The sampling frame made for the studies whose primary goal was to identify the characteristics of isolated AIVs [64–67] was judged to be at a high risk of bias because the isolation rate reported in the studies was constrained to a particular region. It was challenging to assume that the sample frame of the studies reporting the prevalence of carcasses referred for diagnosis was representative of the general wild bird population [68,69]. However, these prevalence studies used census, a suitable sampling method that examined all samples within a predetermined sampling frame. Seven primary studies that evaluated the prevalence of AIVs were judged to have a low bias in the sampling method, owing to proper capture methods from randomized wild birds [2,41–43,48,50,57]. All studies, except for two [59,66] that did not describe the isolation method in detail, recorded the condition of the samples by examining them with the proper techniques, such as RT-PCR [2,8,17,41–44,48–52,54–56,58,60–62,65,67–74], rRT-PCR [53,63,64,75,76], or HA tests [41,42,46,47,57,77]. All investigations, except for 12 studies [8,46,55,57–59,65–68,75,77], had read the experimental results with distinguishing criteria. Prevalence studies with small [56–59,64–69,71,73,74] and large sample sizes [2,8,17,41–55,60–63,70,72,75,76] were differentiated according to adequate sample size (1000 samples).

All studies unambiguously stated the number of examined samples and the number of positive or virus-detected samples. Studies that assessed prevalence in subgroups subdivided into sampling month [46,60,64,68,70,72], sampling year [2,8,48,49,51,54,55,72], province [17,41–43,51,58–60,68,75], and bird species [42,43,46,48–52,57–59,68,75] allowed for comparisons between the study sample and the population of interest. A low coverage bias was determined in prevalence studies [2,41–43,48,50,52,60,72,75] that used a similar sample number for each distinct subgroup. Otherwise, coverage bias was assessed as high [8,17,46,49,51,54,55,57–59,64,68,70].

Seroprevalence studies, in which samples were collected from the entire country [41–43,46,48,50,51] and major migratory bird habitats [2,49,78], were judged to set a suitable sample frame. Seroprevalence studies [2,41–43] that sampled random subjects using proper capture methods were judged to have low sampling bias. All experiments detected antibodies in serum using appropriate methods, including the HI assay [46,48,49] and ELISA [2,41–43,50,51,78]. All investigations, except for one [46], read the experimental results with distinguishing criteria. Seroprevalence studies with small [2,41–43,78] and large sample numbers [46,48–51] were differentiated according to adequate sample size (1000 samples). All the studies unambiguously stated the number of examined and positive samples. Studies that assessed prevalence into subgroups subdivided into sampling month [2], sampling year [2,48,51,78], province [43,49], and bird species [41,42,46,48–51,78] allowed for comparisons between the study sample and the population of interest. Low coverage bias was observed in prevalence studies [2,48,50] that used a similar sample size for each distinct subgroup. Otherwise, coverage bias was assessed to be high [41–43,46,49,51,78].

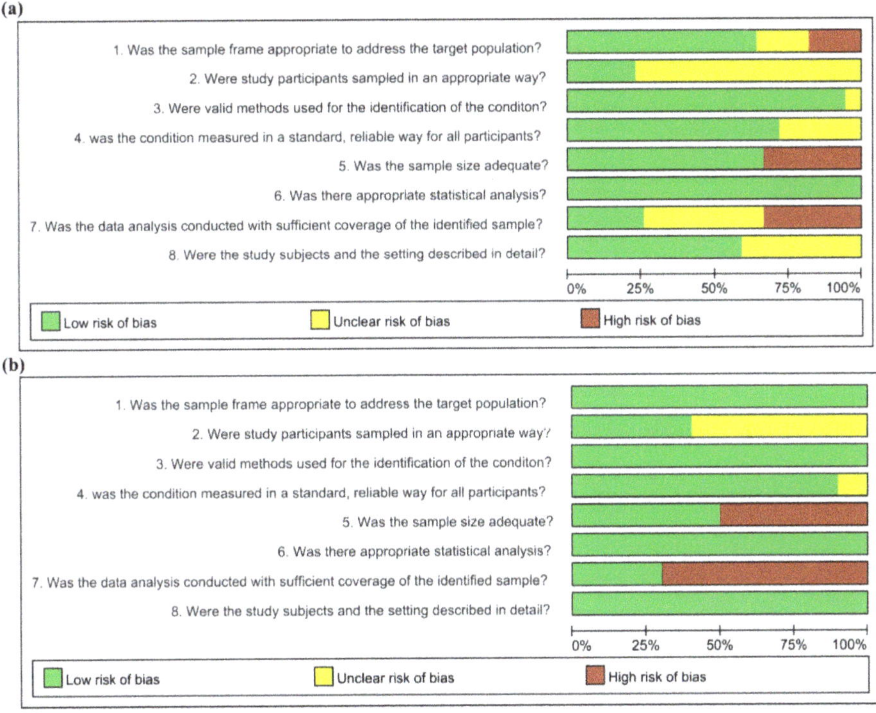

Figure 2. Risk-of-bias assessment of eligible studies on prevalence (**a**) and seroprevalence (**b**) of avian influenza viruses in wild birds of South Korea using the Joanna Briggs Institute (JBI) critical appraisal tools for prevalence studies.

3.4. Meta-Analysis Results

3.4.1. Prevalence Estimates

Thirty-nine studies investigated the prevalence of AIVs in wild birds in South Korea (Figure 3). Overall, the pooled prevalence was estimated to be 1.57% (95% CI: 0.98, 2.51) with high between-study heterogeneity ($I^2 = 100\%$). Subgroup analyses were performed to investigate the source of heterogeneity using different variables that could potentially affect the prevalence rates among individual studies. Regarding bird species, the highest prevalence was detected in the Anseriformes species (4.34% (95% CI: 1.44, 12.30)), followed by a group of non-reported species (1.20% (95% CI: 0.74, 1.94)). The lowest prevalence was detected among Charadriiformes and other species (rather than Anseriformes and Charadriiformes), with a prevalence of 0.19% (95% CI: 0.03, 1.33) and 0.22% (95% CI: 0.04, 1.20), respectively. However, the heterogeneity was still high within the Anseriformes ($I^2 = 98\%$) and non-reported ($I^2 = 100$) subgroups, whereas no heterogeneity was observed within the Charadriiformes and other species subgroups ($I^2 = 0$ for each). Based on the sample type, the highest prevalence of 4.59% (95% CI: 0.76, 23.07) was detected in carcasses, followed by 1.58% (95% CI: 1.06, 2.36) and 1% (95% CI: 0.36, 2.75) in feces and cloacal swabs, respectively. The lowest prevalence, 0.63% (95% CI: 0.56, 0.71) and 0.97% (95% CI: 0.73, 1.28), was reported in mixed samples and non-reported sample types, respectively. The variables of sample size, sampling season, detection method, and publication status, had no significant influence on the prevalence rates ($p > 0.05$). All the variables assessed are presented in Table 1.

Table 1. Results of subgroup analysis (prevalence rates) based on six potential effect modifiers.

Variables	Prevalence Estimates (95%)	I^2 (%)	τ^2	$P_{subgroup}$
1. Bird species				<0.01 [a]
Anseriformes	4.34 [1.44; 12.30]	98	3.2233	
Charadriiformes	0.19 [0.03; 1.33]	0	0	
Other species	0.22 [0.04; 1.20]	0	1.9568	
Not reported	1.20 [0.74; 1.94]	100	2.1236	
Overall prevalence	1.14 [0.72; 1.82]	99	2.8912	
2. Sample type				<0.01 [a]
Feces	1.58 [1.06; 2.36]	99	1.2236	
Carcass	4.59 [0.76; 23.07]	95	7.1496	
Cloacal swabs	1.00 [0.36; 2.75]	100	2.6837	
Mixed samples	0.63 [0.56; 0.71]	0	0	
Not reported	0.97 [0.73; 1.28]	96	0.039	
Overall prevalence	1.7 [1.10; 2.64]	100	2.6417	
3. Detection method				=0.25 [b]
RT-PCR	1.98 [1.11; 3.53]	100	2.3799	
HA-test	1.06 [0.59; 1.90]	95	0.5665	
rRT-PCR	0.75 [0.32; 1.75]	85	0.8644	
Virus isolation	1.25 [0.14; 10.25]	88	2.0493	
Overall prevalence	1.54 [0.99; 2.38]	100	2.0266	
4. Sample size				=0.05 [b]
Less than 1000	3.28 [1.20; 8.68]	97	3.3332	
More than 1000	1.10 [0.74; 1.65]	100	1.1152	
Overall prevalence	1.57 [0.99; 2.47]	100	2.1073	

Table 1. *Cont.*

Variables	Prevalence Estimates (95%)	I^2 (%)	τ^2	$P_{subgroup}$
5. Sampling season				=0.48 [b]
Fall to winter	1.94 [0.80; 4.61]	98	3.1168	
Not reported	1.35 [0.82; 2.20]	100	1.4628	
Overall prevalence	1.57 [0.99; 2.47]	100	2.1073	
6. Publication status				=0.82 [b]
Published	1.63 [0.80; 3.27]	100	3.07	
Non-published	1.47 [0.95; 2.28]	100	0.7476	
Overall prevalence	1.57 [0.99; 2.47]	100	2.1073	

[a] The difference in prevalence estimates between subgroups was statistically significant. [b] There was no significant difference in prevalence estimates between subgroups.

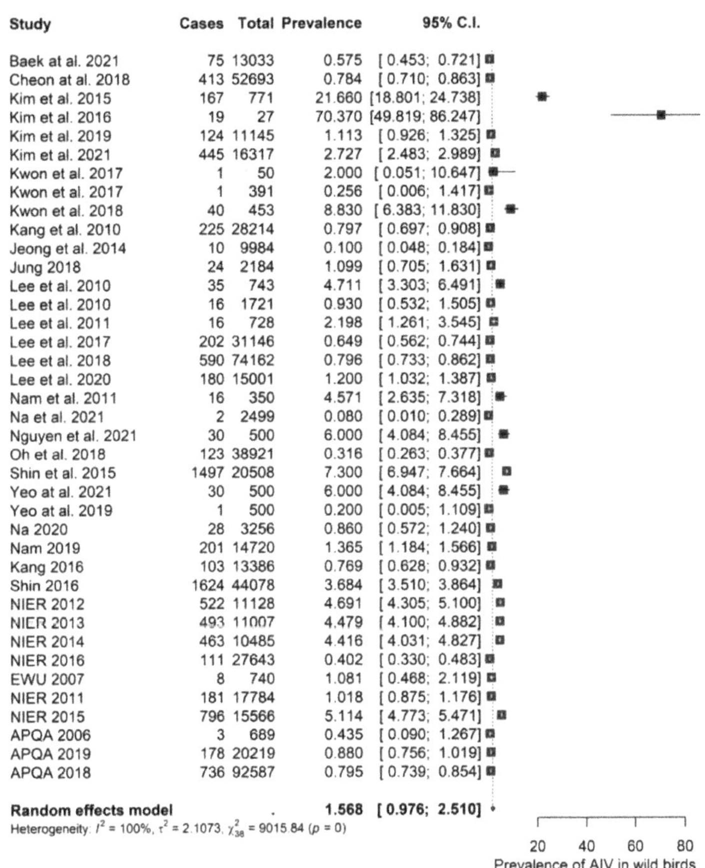

Figure 3. Forest plot of 39 studies assessing the prevalence of avian influenza virus in the wild bird populations of South Korea [2,8,17,41–76].

3.4.2. Seroprevalence Estimates

Ten studies assessed the seroprevalence of AIVs in wild birds in South Korea. The pooled seroprevalence estimate was 15.91% with a 95% CI of 5.89–36.38 (Figure 4). Between-study heterogeneity was significantly high (I^2 = 100%). To identify the reasons for heterogeneity, we conducted subgroup analysis using bird species, detection method, sample size, publication status, and sampling season as potential effect modifiers. All variables had a significant influence on seroprevalence rates, except for publication status (Table 2). Regarding bird species, the highest seroprevalence was detected in the Anseriformes species (30.45% (95% CI:18.97, 45.03)), followed by the Charadriiformes and non-reported species with seroprevalence estimates of 2.95% (95% CI: 0.24, 27.43)) and 2.85% (95% CI: 1.17, 6.76), respectively. The lowest seroprevalence was detected among other bird species (rather than Anseriformes and Charadriiformes), with an estimate of 2.83% (95% CI: 0.40, 17.26). Heterogeneity was still high within all subgroups ($I^2 > 90\%$), except for the Charadriiformes species ($I^2 = 49\%$). Based on the detection method, the highest seroprevalence, 31.47% (95% CI: 20.47, 45.02), was detected by ELISA, whereas the lowest seroprevalence of 2.46% (95% CI: 1.12, 5.31) was indicated by the HI test. The sample size also demonstrated a significant association with the seroprevalence rate, with the highest seroprevalence (30.93% (18.48, 46.93)) observed in studies with less than 1000 samples compared with 5.03% (1.25, 18.20) observed in those with more than 1000 samples ($p < 0.01$). Subgroup analyses also revealed that the highest seroprevalence was detected among studies that collected samples from fall to winter (36.48% (24.05, 51.01)) than in studies that did not report the sampling season (10.48% (3.46, 27.66)) ($p < 0.02$). Regarding the publication status, no significant difference in seroprevalence was observed between the published (9.07% (1.91, 33.78)) and non-published studies (19.85% (7.52, 43.01)) ($p < 0.37$). The results of the subgroup analysis are presented in Table 2.

Table 2. Results of subgroup analysis (seroprevalence rates) based on five potential effect modifiers.

Variables	Seroprevalence Estimates (95%)	I^2 (%)	τ^2	$P_{subgroup}$
1. Bird species				<0.01 [a]
Anseriformes	30.45 [18.97; 45.03]	100	0.7793	
Charadriiformes	2.95 [0.24; 27.43]	49	5.2003	
Other species	2.83 [0.40; 17.26]	94	7.0157	
Not reported	2.85 [1.17; 6.76]	99	0.6122	
Overall prevalence	7.71 [3.33; 16.86]	99	4.3612	
2. Detection method				<0.01 [a]
ELISA	31.47 [20.47; 45.02]	97	0.5904	
HI test	2.46 [1.12; 5.31]	99	0.475	
Overall prevalence	15.90 [6.76; 33.01]	100	2.3668	
3. Sample size				<0.01 [a]
Less than 1000	30.93 [18.48; 46.93]	98	0.7001	
More than 1000	5.03 [1.25; 18.20]	100	2.1246	
Overall prevalence	15.90 [6.76; 33.01]	100	2.3668	
4. Sampling season				=0.02 [a]
Fall to winter	36.48 [24.05; 51.01]	98	0.2553	
Not reported	10.48 [3.46; 27.66]	100	2.53	
Overall prevalence	15.90 [6.76; 33.01]	100	2.3668	
5. Publication status				=0.37 [b]
Published	9.07 [1.91; 33.78]	99	2.0361	
Non-published	19.85 [7.52; 43.01]	100	2.2482	
Overall prevalence	15.9 [6.76; 33.01]	100	2.2482	

[a] The difference in seroprevalence estimates between subgroups was statistically significant. [b] There was no significant difference in seroprevalence estimates between subgroups.

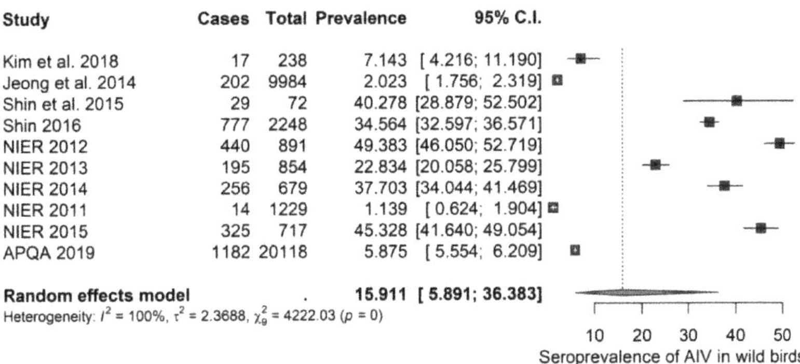

Figure 4. Forest plot of 10 studies assessing the seroprevalence of avian influenza virus in the wild bird populations of South Korea [2,41–43,46,48–51,78].

3.5. Publication Bias

Publication bias occurs when the likelihood of a study being published is influenced by its findings. In contrast to smaller studies with low effects, larger studies with relatively high effects are more likely to be published because they are statistically significant. This results in publication bias. To assess the presence of publication bias, contour-enhanced funnel plots were generated with the effect sizes on the x-axis and their standard errors on the y-axis (Figure 5). On visual inspection, the studies were symmetrically distributed on both sides of the mean effect and demonstrated significant results ($p < 0.05$). This symmetrical pattern suggests that a publication bias is unlikely. To avoid subjective inferences from funnel plot visualizations, Egger's regression test was applied to quantify the presence of funnel plot asymmetry. Egger's regression test yielded p-values of 0.094 and 0.506 for prevalence and seroprevalence outcomes, respectively, indicating no funnel plot asymmetry; hence, publication bias was not confirmed.

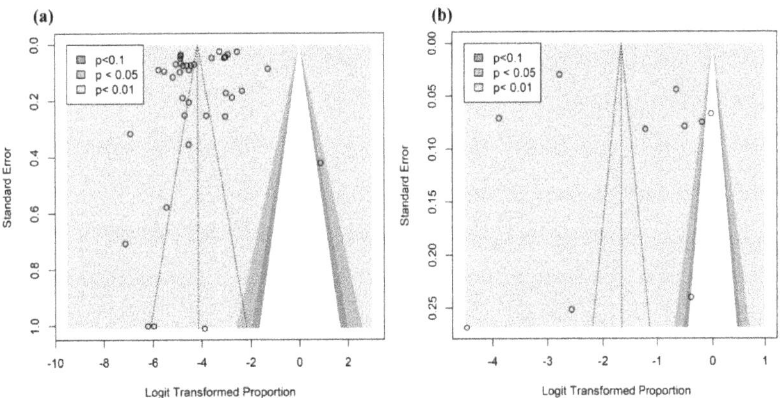

Figure 5. Contour-enhanced funnel plots for publication-bias assessment of the prevalence (**a**) and seroprevalence (**b**) of avian influenza virus in wild birds in South Korea.

4. Discussion

AIVs in wild birds pose a pandemic threat to humans and the poultry industry worldwide. Previous studies have confirmed the relationship between the wild bird migratory route and AIV prevalence in South Korea by evaluating the geographical distributions of HPAI outbreaks and cases of mortality in wild birds [2,79]. Therefore, it is of critical importance to understand the current status of AI prevalence and seroprevalence in wild birds for use as an early warning system. In this study, we performed a systematic review and meta-analysis to consolidate the data from individual primary studies that evaluated the prevalence and seroprevalence of AIVs in wild birds in South Korea. The overall prevalence was estimated to be 1.568% (0.976; 2.510), indicating that approximately 2% of the wild bird population in South Korea are carriers of AIVs.

According to the census of winter migratory birds conducted in South Korea, approximately 1.63 million winter birds visited South Korea in 2020 [80]. Of these, 850,000 birds belonged to the order Anseriformes and accounted for 52% of the total. Based on the results of this meta-analysis, it can be estimated that approximately 32,600 migratory birds in South Korea carry AIVs. Chen et al. (2019) discovered that the prevalence and seroprevalence of AI were 2.5% and 26.5%, respectively, in wild birds in China [81]. The relatively low prevalence of AIVs in wild birds in South Korea is consistent with the knowledge that South Korea is not a breeding site, but rather a wintering area for adult wild birds, particularly waterfowl, such as ducks and geese [17]. On the other hand, the seroprevalence estimate was 15.911% (5.891; 36.383), suggesting that approximately 16% of the wild bird population in South Korea has been exposed to AIVs. As the antibody-positive cases included individuals that had recovered from AIV, the seroprevalence would tend to be relatively high compared with the prevalence. Another possible explanation for the high seroprevalence of AIVs in wild birds is that during the migration route, the migratory birds aggregate at nesting and feeding sites, which results in high rates of contact between birds, facilitating AIV transmission and a high prevalence of antibodies in the bird population [82,83]. It is, therefore, likely that wild birds arriving in South Korea will have had repeated exposure to AIVs, which leads to the persistence of anti-AIV antibodies over long periods in their bodies. The antibodies detected in wild birds could only be the result of seroconversion induced by a natural viral infection, as they are not immunized against AIV. Thus, they could play a critical role in spreading the virus to the surrounding environment, livestock, and humans.

Of the six variables used in the subgroup analysis, two (bird species and sample type) showed a significant influence on the prevalence rate of AIVs in wild birds. In contrast, four variables (bird species, detection method, sample size, and sampling season) showed a significant relationship with seroprevalence rates. Small studies (less than 1000 samples) demonstrated higher prevalence and seroprevalence rates than large studies (more than 1000 samples). This could be related to the fact that studies with small sample sizes are associated with higher effect sizes than bigger studies. Another possible reason is that the larger studies included in the analysis are mostly the non-published government reports that collected samples from different provinces of the country as part of a normal AIV surveillance routine, thus reducing the chance of getting positive samples compared with small studies that mainly collected samples from specific locations during or after an HPAI outbreak, thus increasing the probability of getting more positive samples. Regarding the species of wild birds, the highest AIV prevalence and seroprevalence rates were detected in the Anseriformes species compared to others. These results are in line with previous reports that waterfowl are the predominant migratory birds and are primarily associated with AI occurrence in South Korea [2,16]. Similar findings were also reported in China, where the highest AIV prevalence (6.8%) and seroprevalence (41.8%) were observed in the Anseriformes species compared with that in non-Anseriformes species [81]. Based on the sample type, the highest prevalence rate was revealed in carcasses compared to other sample types ($p < 0.01$). One possible reason for these results is that carcass samples were collected during or shortly after the 2014 HPAI outbreak (H5N8) in South Korean duck and chicken farms and wild birds found in the Donglim reservoir, Jeonbuk province [2,51,68,69]; most of these carcasses were confirmed to have died from the HPAI virus (H5N8) clade 2.3.4.6 [69,84].

Considering the detection method, the highest seroprevalence was detected by ELISA rather than the HI test ($p < 0.01$). This difference in performance could be due to the low sensitivity of the HI test for detecting AIV antibodies, particularly the H5N1 and

H3N2 serotypes [85,86]. Furthermore, the highest seroprevalence rate detected during the fall-to-winter season is consistent with the National Institute of Environmental Research report "Surveillance and monitoring of wildlife diseases in Korea, 2012," which states that the prevalence of AIVs increases from October to December (stage 1), when waterfowl migrate from the north, and in April (stage 4), when passing migratory birds are moving to the north [2]. Surprisingly, the results of the subgroup analysis confirmed no significant influence of the sampling season on the prevalence estimates. However, this should be interpreted with caution, as many studies included in the assessment of the prevalence did not clearly report the sampling season, and many studies fell into a subgroup of "not reported." Consequently, this could have limited the power of the statistical tests to detect the significance while it was present. In addition to the above-mentioned moderators, prevalence rates in birds are likely to vary depending upon the surveillance period, the sampling region, and whether surveillance was performed in response to an outbreak or conducted as routine surveillance [11]. As most of the studies included in this meta-analysis did not provide clear information about these variables, we could not evaluate their contribution to the observed prevalence and seroprevalence rates. Although the prevalence and seroprevalence estimates between subgroups were significantly different, the within-subgroup heterogeneity was substantially high, indicating that none of the variables could entirely explain the reasons for between-study heterogeneity.

This study had a few limitations. First, there was substantial variation in the prevalence rates among individual studies. Although we used a couple of moderators to investigate the source of heterogeneity, only a few studies clearly reported on these variables, and a large number fell into the "not reported" subgroup. Furthermore, insufficient information was available to adequately categorize studies based on the reason for surveillance (in response to an outbreak or as routine surveillance), surveillance period, and sampling region, which are also relevant covariates that could possibly demonstrate significant relationships with the observed pooled prevalence estimates. Despite these limitations, the findings of this meta-analysis provide a more robust estimate of AIV prevalence and seroprevalence in wild birds in South Korea than that obtained from a single study.

5. Conclusions

In conclusion, this study provides solid evidence for the current prevalence of AIVs among the South Korean wild bird population. These findings demonstrated that a large number of wild birds in South Korea, particularly those of the order Anseriformes, are carriers of AIVs, and others have already been exposed to AI because of the high detection rate of anti-AIV antibodies. This poses a threat to the poultry industry and, potentially, to humans in South Korea, due to the critical role of wild birds in the spread of AIV. Furthermore, migratory wild birds have different flyways, which affect the distribution of AIVs in different countries. A multi-country surveillance system would provide detailed information on the prevalence and distribution of AIVs in this region. The evidence from this study highlights the need to strengthen existing preventive measures and increase surveillance activities to impede the risk of AIV transmission from wild birds to domestic poultry and human beings.

Supplementary Materials: The following supporting information can be downloaded at: https://www.mdpi.com/article/10.3390/v15020472/s1, Table S1. PRISMA 2020 check list; Table S2. Characteristics of 39 studies included in the meta-analysis of prevalence of avian influenza viruses in wild birds in South Korea; Table S3. Characteristics of 10 studies included in the meta-analysis of seroprevalence of avian influenza viruses in wild birds in South Korea.

Author Contributions: Conceptualization, E.N. and G.W.; methodology, E.N., G.W. and S.L.; software, E.N. and S.L.; validation, G.W., B.-Y.P. and H.-J.T.; formal analysis, E.N., S.L. and B.-Y.P.; resources, B.-Y.P., H.-J.T. and G.W.; data curation, E.N. and S.L.; writing—original draft preparation, E.N.; writing—review and editing, G.W., S.L. and E.N.; supervision, G.W.; project administration, E.N., B.-Y.P. and H.-J.T.; funding acquisition, G.W. All authors have read and agreed to the published version of the manuscript.

Funding: This subject was supported by the National Institute of Wildlife Disease Control and Prevention as a Specialized Graduate School Support Project for Wildlife Disease Specialists. This research was supported by Basic Science Research Program through the National Research Foundation (NRF), funded by the Ministry of Education (2019R1A6A1A03033084).

Institutional Review Board Statement: Not applicable.

Informed Consent Statement: Not applicable.

Data Availability Statement: All data used or analyzed during this study are included in this article and its Supplementary Information Files.

Acknowledgments: The authors sincerely thank Min Kang, Junghwan Suh, and Suyeon Lee for their significant contributions to the data provision and extraction.

Conflicts of Interest: The authors declare no conflict of interest.

References

1. Bodewes, R.; Kuiken, T. Changing role of wild birds in the epidemiology of avian influenza A viruses. In *Advances in Virus Research*; Elsevier: Amsterdam, The Netherlands, 2018; pp. 279–307.
2. Shin, J.H.; Woo, C.; Wang, S.J.; Jeong, J.; An, I.J.; Hwang, J.K.; Jo, S.D.; Yu, S.D.; Choi, K.; Chung, H.M.; et al. Prevalence of avian influenza virus in wild birds before and after the HPAI H5N8 outbreak in 2014 in South Korea. *J. Microbiol.* **2015**, *53*, 475–480. [CrossRef] [PubMed]
3. Webster, R.G.; Bean, W.J.; Gorman, O.T.; Chambers, T.M.; Kawaoka, Y. Evolution and ecology of influenza A viruses. *Microbiol. Rev.* **1992**, *56*, 152–179. [CrossRef]
4. Vandegrift, K.J.; Sokolow, S.; Daszak, P.; Kilpatrick, A.M. Ecology of avian influenza viruses in a changing world. *Ann. N. Y. Acad. Sci.* **2010**, *1195*, 113–128. [CrossRef] [PubMed]
5. World Health Organization. *Avian Influenza: Assessing the Pandemic Threat*; World Health Organization: Geneva, Switzerland, 2005.
6. Stallknecht, D.E.; Brown, J.D. Wild bird infections and the ecology of avian influenza viruses. In *Animal Influenza*; Wiley Online Library: Hoboken, NJ, USA, 2017; pp. 153–176.
7. MCEIRS. *Avian Influenza: Detection in Wild Birds-Field Collection of Samples*; MCEIRS: Minneapolis, MN, USA, 2017.
8. Lee, Y.-N.; Cheon, S.-H.; Lee, E.-K.; Heo, G.-B.; Bae, Y.-C.; Joh, S.-J.; Lee, M.-H.; Lee, Y.-J. Pathogenesis and genetic characteristics of novel reassortant low-pathogenic avian influenza H7 viruses isolated from migratory birds in the Republic of Korea in the winter of 2016–2017. *Emerg. Microbes Infect.* **2018**, *7*, 1–13. [CrossRef] [PubMed]
9. Kim, H.K.; Jeong, D.G.; Yoon, S.W. Recent outbreaks of highly pathogenic avian influenza viruses in South Korea. *Clin. Exp. Vaccine Res.* **2017**, *6*, 95–103. [CrossRef]
10. Alexander, D.J. An overview of the epidemiology of avian influenza. *Vaccine* **2007**, *25*, 5637–5644. [CrossRef] [PubMed]
11. Bui, C.; Rahman, B.; Heywood, A.E.; MacIntyre, C.R. A Meta-Analysis of the Prevalence of Influenza A H5N1 and H7N9 Infection in Birds. *Transbound Emerg. Dis.* **2017**, *64*, 967–977. [CrossRef]
12. Rabinowitz, P.; Perdue, M.; Mumford, E. Contact variables for exposure to avian influenza H5N1 virus at the human-animal interface. *Zoonoses Public Health* **2010**, *57*, 227–238. [CrossRef]
13. Lu, L.; Lycett, S.J.; Leigh Brown, A.J. Reassortment patterns of avian influenza virus internal segments among different subtypes. *BMC Evol. Biol.* **2014**, *14*, 16. [CrossRef]
14. To, K.K.; Chan, J.F.; Chen, H.; Li, L.; Yuen, K.-Y. The emergence of influenza A H7N9 in human beings 16 years after influenza A H5N1: A tale of two cities. *Lancet Infect. Dis.* **2013**, *13*, 809–821. [CrossRef]
15. Mo, I.-P.; Bae, Y.-J.; Lee, S.-B.; Mo, J.-S.; Oh, K.-H.; Shin, J.-H.; Kang, H.-M.; Lee, Y.-J. Review of Avian Influenza Outbreaks in South Korea from 1996 to 2014. *Avian Dis.* **2016**, *60* (Suppl. S1), 172–177. [CrossRef] [PubMed]
16. Lee, Y.-J.; Choi, Y.-K.; Kim, Y.-J.; Song, M.-S.; Jeong, O.-M.; Lee, E.-K.; Jeon, W.-J.; Jeong, W.; Joh, S.-J.; Choi, K.-S.; et al. Highly pathogenic avian influenza virus (H5N1) in domestic poultry and relationship with migratory birds, South Korea. *Emerg. Infect. Dis.* **2008**, *14*, 487–490. [CrossRef] [PubMed]
17. Kang, H.M.; Jeong, O.M.; Kim, M.C.; Kwon, J.S.; Paek, M.R.; Choi, J.G.; Lee, E.K.; Kim, Y.J.; Lee, Y.J. Surveillance of avian influenza virus in wild bird fecal samples from South Korea, 2003–2008. *J. Wildl. Dis.* **2010**, *46*, 878–888. [CrossRef] [PubMed]
18. Ellis, T.M.; Bousfield, R.B.; Bissett, L.A.; Dyrting, K.C.; Luk, G.S.M.; Tsim, S.T.; Sturm-Ramirez, K.; Webster, R.G.; Guan, Y.; Peiris, J.S.M. Investigation of outbreaks of highly pathogenic H5N1 avian influenza in waterfowl and wild birds in Hong Kong in late 2002. *Avian Pathol.* **2004**, *33*, 492–505. [CrossRef] [PubMed]
19. Chen, H.; Li, Y.; Li, Z.; Shi, J.; Shinya, K.; Deng, G.; Qi, Q.; Tian, G.; Fan, S.; Zhao, H.; et al. Properties and dissemination of H5N1 viruses isolated during an influenza outbreak in migratory waterfowl in western China. *J. Virol.* **2006**, *80*, 5976–5983. [CrossRef] [PubMed]
20. Chen, H.; Smith, G.J.; Zhang, S.Y.; Qin, K.; Wang, J.; Li, K.S.; Webster, R.G.; Peiris, J.S.; Guan, Y. Avian flu: H5N1 virus outbreak in migratory waterfowl. *Nature* **2005**, *436*, 191–192. [CrossRef]

21. Paudyal, N.; Pan, H.; Liao, X.; Zhang, X.; Li, X.; Fang, W.; Yue, M. A Meta-Analysis of Major Foodborne Pathogens in Chinese Food Commodities Between 2006 and 2016. *Foodborne Pathog. Dis.* **2018**, *15*, 187–197. [CrossRef]
22. Field, A.P.; Gillett, R. How to do a meta-analysis. *Br. J. Math Stat. Psychol.* **2010**, *63*, 665–694. [CrossRef]
23. Haidich, A.B. Meta-analysis in medical research. *Hippokratia* **2010**, *14*, 29–37.
24. Barendregt, J.J.; Doi, S.A.; Lee, Y.Y.; Norman, R.E.; Vos, T. Meta-analysis of prevalence. *J. Epidemiol. Community Health* **2013**, *67*, 974–978. [CrossRef]
25. Islam, M.Z.; Musekiwa, A.; Islam, K.; Ahmed, S.; Chowdhury, S.; Ahad, A.; Biswas, P.K. Regional variation in the prevalence of E. coli O157 in cattle: A meta-analysis and meta-regression. *PLoS ONE* **2014**, *9*, e93299. [CrossRef]
26. Ferreira, G.C.M.; Canozzi, M.E.A.; Peripolli, V.; Moura, G.d.P.; Sánchez, J.; Martins, C.E.N. Prevalence of bovine Babesia spp., Anaplasma marginale, and their co-infections in Latin America: Systematic review-meta-analysis. *Ticks Tick Borne Dis.* **2022**, *13*, 101967. [CrossRef]
27. Page, M.J.; McKenzie, J.E.; Bossuyt, P.M.; Boutron, I.; Hoffmann, T.C.; Mulrow, C.D.; Shamseer, L.; Tetzlaff, J.M.; Akl, E.A.; Brennan, S.E.; et al. The PRISMA 2020 statement: An updated guideline for reporting systematic reviews. *Syst. Rev.* **2021**, *10*, 89. [CrossRef]
28. Munn, Z.; Moola, S.; Lisy, K.; Riitano, D.; Tufanaru, C. Methodological guidance for systematic reviews of observational epidemiological studies reporting prevalence and cumulative incidence data. *Int. J. Evid. Based Health C* **2015**, *13*, 147–153. [CrossRef]
29. Porritt, K.; Gomersall, J.; Lockwood, C. JBI's Systematic Reviews: Study selection and critical appraisal. *Am. J. Nurs.* **2014**, *114*, 47–52. [CrossRef] [PubMed]
30. Team, R.C. *R: A Language and Environment for Statistical Computing*; R Foundation for Statistical Computing: Vienna, Austria, 2013; Available online: http://www.R-project.org/ (accessed on 24 December 2021).
31. Balduzzi, S.; Rucker, G.; Schwarzer, G. How to perform a meta-analysis with R: A practical tutorial. *Evid. Based Ment. Health* **2019**, *22*, 153–160. [CrossRef] [PubMed]
32. Alboukadel, K. ggcorrplot: Visualization of a Correlation Matrix using 'ggplot2'. *CRAN* **2019**.
33. Harrer, M.; Cuijpers, P.; Furukawa, T.A.; Ebert, D.D. *Doing Meta-Analysis with R: A Hands-On Guide*; Chapman and Hall/CRC: London, UK, 2021.
34. Viechtbauer, W. Conducting Meta-Analyses in R with the metafor Package. *J. Stat. Softw.* **2010**, *36*, 1–48. [CrossRef]
35. Golden, C.E.; Mishra, A. Prevalence of *Salmonella* and *Campylobacter* spp. in Alternative and Conventionally Produced Chicken in the United States: A Systematic Review and Meta-Analysis. *J. Food Prot.* **2020**, *83*, 1181–1197. [CrossRef]
36. Lin, L.F.; Chu, H.T. Meta-analysis of Proportions Using Generalized Linear Mixed Models. *Epidemiology* **2020**, *31*, 713–717. [CrossRef]
37. Higgins, J.P.T.; Thompson, S.G.; Deeks, J.J.; Altman, D.G. Measuring inconsistency in meta-analyses. *BMJ* **2003**, *327*, 557–560. [CrossRef] [PubMed]
38. Sabarimurugan, S.; Kumarasamy, C.; Baxi, S.; Devi, A.; Jayaraj, R. Systematic review and meta-analysis of prognostic microRNA biomarkers for survival outcome in nasopharyngeal carcinoma. *PLoS ONE* **2019**, *14*, e0209760. [CrossRef] [PubMed]
39. Egger, M.; Smith, G.D.; Schneider, M.; Minder, C. Bias in meta-analysis detected by a simple, graphical test. *BMJ* **1997**, *315*, 629–634. [CrossRef] [PubMed]
40. Duval, S.; Tweedie, R. Trim and fill: A simple funnel-plot-based method of testing and adjusting for publication bias in meta-analysis. *Biometrics* **2000**, *56*, 455–463. [CrossRef]
41. National Institute of Environmental Research. *Surveillance and Mornitoring of Wildlife Diseases in Korea—Avian Influenza Surveillance in Wild Animals*; National Institute of Environmental Research: Incheon, Republic of Korea, 2012.
42. National Institute of Environmental Research. *Avian Influenza Surveillance in Wild Animals*; National Institute of Environmental Research: Incheon, Republic of Korea, 2013.
43. National Institute of Environmental Research. *Avian Influenza Surveillance and Molecular-Biological Characterization of the Spatiotemporal Distribution of Avian Influenza Viruses in 2014*; National Institute of Environmental Research: Incheon, Republic of Korea, 2014.
44. National Institute of Environmental Research. *Research on Early Detection and Characterization of Zoonotic Diseases in Wild Birds (I)*; National Institute of Environmental Research: Incheon, Republic of Korea, 2016.
45. Animal and Plant Quarantine Agency. *A Study on the Epidemiology of Highly Pathogenic Avian Influenza*; Animal and Plant Quarantine Agency: Gimcheon, Republic of Korea, 2018.
46. Animal and Plant Quarantine Agency. *A Study on the Epidemiology of Highly Pathogenic Avian Influenza*; Animal and Plant Quarantine Agency: Gimcheon, Republic of Korea, 2019.
47. Cheon, S.-H.; Lee, Y.-N.; Kang, S.-I.; Kye, S.-J.; Lee, E.-K.; Heo, G.-B.; Lee, M.-H.; Kim, J.-W.; Lee, K.-N.; Son, H.-M.; et al. Genetic evidence for the intercontinental movement of avian influenza viruses possessing North American-origin nonstructural gene allele B into South Korea. *Infect. Genet. Evol.* **2018**, *66*, 18–25. [CrossRef]
48. National Institute of Environmental Research. Monitoring the Infection and Transmission Status of Avian Influenza in the Migratory and Resident Wild Birds. In *Monitoring the Infection and Transmission Status of Avian Influenza in the Migratory and Resident Wild Birds*; National Institute of Environmental Research: Incheon, Republic of Korea, 2011.

49. Jeong, J.; Kang, H.-M.; Lee, E.-K.; Song, B.-M.; Kwon, Y.-K.; Kim, H.-R.; Choi, K.-S.; Kim, J.-Y.; Lee, H.-J.; Moon, O.-K.; et al. Highly pathogenic avian influenza virus (H5N8) in domestic poultry and its relationship with migratory birds in South Korea during 2014. *Vet. Microbiol.* **2014**, *173*, 249–257. [CrossRef]
50. National Institute of Environmental Research. *Surveillance on Avian Influenza of Wild Birds and Molecular-Biological Characterization of the Spatiotemporal Distribution of Avian Influenza Viruses in 2015*; National Institute of Environmental Research: Incheon, Republic of Korea, 2015.
51. Shin, J.-H. *Spatiotemporal Distribution of Avian Influenza Viruses in Wild Birds and Tracking Their Migration Route*; Chungbuk National University: Chungju, Republic of Korea, 2016.
52. Lee, E.-K.; Kang, H.-M.; Song, B.-M.; Lee, Y.-N.; Heo, G.-B.; Lee, H.-S.; Lee, Y.-J.; Kim, J.-H. Surveillance of avian influenza viruses in South Korea between 2012 and 2014. *Virol. J.* **2017**, *14*, 54. [CrossRef]
53. Kim, G.-S.; Kim, T.-S.; Son, J.-S.; Lai, V.D.; Park, J.-E.; Wang, S.-J.; Jheong, W.-H.; Mo, I.-P. The difference of detection rate of avian influenza virus in the wild bird surveillance using various methods. *J. Vet. Sci.* **2019**, *20*, 1134050. [CrossRef]
54. Nam, J.-H. *Surveillance and Genetic Characterization of Avian and Canine Influenza Viruses Isolated from South Korea*; Korea University: Seoul, Republic of Korea, 2019.
55. Lee, Y.-N.; Lee, D.-H.; Cheon, S.-H.; Park, Y.-R.; Baek, Y.-G.; Si, Y.-J.; Kye, S.-J.; Lee, E.-K.; Heo, G.-B.; Bae, Y.-C. Genetic characteristics and pathogenesis of H5 low pathogenic avian influenza viruses from wild birds and domestic ducks in South Korea. *Sci. Rep.* **2020**, *10*, 12151. [CrossRef]
56. Yeo, S.-J.; Hoang, V.T.; Duong, T.B.; Nguyen, N.M.; Tuong, H.T.; Azam, M.; Sung, H.W.; Park, H. Emergence of a Novel Reassortant H5N3 Avian Influenza Virus in Korean Mallard Ducks in 2018. *Intervirology* **2021**, *65*, 1–16. [CrossRef]
57. Animal and Plant Quarantine Agency. *Investigation and Epidemiological Analysis of Avian Influenza Infection in Ornamental Birds and Birds Inhabiting Dadohaehaesang National Park*; Animal and Plant Quarantine Agency: Gimcheon, Republic of Korea, 2006.
58. *Studies on Findings of Mechanism and Separation of Avian Influenza Virus*; Ministry of Agriculture and Forestry: Gwacheon, Republic of Korea, 2007.
59. Lee, D.-H.; Lee, H.-J.; Kang, H.-M.; Jeong, O.-M.; Kim, M.-C.; Kwon, J.-S.; Kwon, J.-H.; Kim, C.-B.; Lee, J.-B.; Park, S.-Y.; et al. DNA Barcoding Techniques for Avian Influenza Virus Surveillance in Migratory Bird Habitats. *J. Wildl. Dis.* **2010**, *46*, 649–654. [CrossRef] [PubMed]
60. Kang, M. *Characterization of Influenza A Virus Derived from Animals and Risk Assessment for Interspecies Transmissibility*; Jeonbuk National Uuniversity: Iksan, Republic of Korea, 2016.
61. Jung, S.-H. *Surveillance and Genetic Characterization of Avian Influenza Viruses in South Korea During the Winter of 2016–2017*; Korea University: Seoul, Republic of Korea, 2018.
62. Kim, E.-H.; Kim, Y.-L.; Kim, S.M.; Yu, K.-M.; Casel, M.A.B.; Jang, S.-G.; Pascua, P.N.Q.; Webby, R.J.; Choi, Y.K. Pathogenic assessment of avian influenza viruses in migratory birds. *Emerg. Microbes Infect.* **2021**, *10*, 565–577. [CrossRef] [PubMed]
63. Na, E.-J.; Kim, Y.-S.; Kim, Y.-J.; Park, J.-S.; Oem, J.-K. Genetic Characterization and Pathogenicity of H7N7 and H7N9 Avian Influenza Viruses Isolated from South Korea. *Viruses* **2021**, *13*, 2057. [CrossRef] [PubMed]
64. Lee, D.-H.; Park, J.-K.; Youn, H.-N.; Lee, Y.-N.; Lim, T.-H.; Kim, M.-S.; Lee, J.-B.; Park, S.-Y.; Choi, I.-S.; Song, C.-S. Surveillance and Isolation of HPAI H5N1 from Wild Mandarin Ducks (*Aix galericulata*). *J. Wildl. Dis.* **2011**, *47*, 994–998. [CrossRef] [PubMed]
65. Nam, J.-H.; Kim, E.-H.; Song, D.; Choi, Y.K.; Kim, J.-K.; Poo, H. Emergence of Mammalian Species-Infectious and -Pathogenic Avian Influenza H6N5 Virus with No Evidence of Adaptation. *J. Virol.* **2011**, *85*, 13271–13277. [CrossRef]
66. Kwon, J.-H.; Lee, D.-H.; Swayne, D.; Noh, J.-Y.; Yuk, S.-S.; Erdene-Ochir, T.-O.; Hong, W.-T.; Jeong, J.-H.; Jeong, S.; Gwon, G.-B.; et al. Reassortant Clade 2.3.4.4 Avian Influenza A(H5N6) Virus in a Wild Mandarin Duck, South Korea. *Emerg. Infect. Dis.* **2017**, *23*, 5.
67. Kwon, J.-H.; Jeong, S.; Lee, D.-H.; Swayne, D.E.; Kim, Y.; Lee, S.; Noh, J.-Y.; Erdene-Ochir, T.-O.; Jeong, J.-H.; Song, C.-S. New Reassortant Clade 2.3.4.4b Avian Influenza A(H5N6) Virus in Wild Birds, South Korea, 2017–2018. *Infect. Dis. J. CDC* **2018**, *24*, 10.
68. Kim, H.-R.; Kwon, Y.-K.; Jang, I.; Lee, Y.-J.; Kang, H.-M.; Lee, E.-K.; Song, B.-M.; Lee, H.-S.; Joo, Y.-S.; Lee, K.-H.; et al. Pathologic Changes in Wild Birds Infected with Highly Pathogenic Avian Influenza A(H5N8) Viruses, South Korea, 2014. *Emerg. Infect. Dis.* **2015**, *21*, 5. [CrossRef]
69. Kim, S.-H.; Hur, M.; Suh, J.-H.; Woo, C.; Wang, S.-J.; Park, E.-R.; Hwang, J.; An, I.-J.; Jo, S.-D.; Shin, J.-H.; et al. Molecular characterization of highly pathogenic avian influenza H5N8 viruses isolated from Baikal teals found dead during a 2014 outbreak in Korea. *J. Vet. Sci.* **2016**, *17*, 299–306. [CrossRef]
70. Baek, Y.-G.; Lee, Y.-N.; Lee, D.-H.; Shin, J.-I.; Lee, J.-H.; Chung, D.; Lee, E.-K.; Heo, G.-B.; Sagong, M.; Kye, S.-J.; et al. Multiple Reassortants of H5N8 Clade 2.3.4.4b Highly Pathogenic Avian Influenza Viruses Detected in South Korea during the Winter of 2020–2021. *Viruses* **2021**, *13*, 490. [CrossRef]
71. Kwon, J.-H.; Lee, D.-H.; Jeong, J.-H.; Yuk, S.-S.; Erdene-Ochir, T.-O.; Noh, J.-Y.; Hong, W.-T.; Jeong, S.; Gwon, G.-B.; Lee, S.-W.; et al. Isolation of an H5N8 Highly Pathogenic Avian Influenza Virus Strain from Wild Birds in Seoul, a Highly Urbanized Area in South Korea. *J. Wildl. Dis.* **2017**, *53*, 630–635. [CrossRef]
72. Oh, K.-H.; Mo, J.-S.; Bae, Y.-J.; Lee, S.-B.; Lai, V.D.; Wang, S.-J.; Mo, I.-P. Amino acid substitutions in low pathogenic avian influenza virus strains isolated from wild birds in Korea. *Virus Genes* **2018**, *54*, 397–405. [CrossRef] [PubMed]
73. Yeo, S.-J.; Than, D.-D.; Park, H.-S.; Sung, H.W.; Park, H. Molecular Characterization of a Novel Avian Influenza A (H2N9) Strain Isolated from Wild Duck in Korea in 2018. *Viruses* **2019**, *11*, 1046. [CrossRef] [PubMed]

74. Nguyen, A.T.V.; Hoang, V.T.; Sung, H.W.; Yeo, S.-J.; Park, H. Genetic Characterization and Pathogenesis of Three Novel Reassortant H5N2 Viruses in South Korea, 2018. *Viruses* **2021**, *13*, 2192. [CrossRef] [PubMed]
75. Lee, D.-H.; Lee, H.-J.; Lee, Y.-N.; Jeong, O.-M.; Kang, H.-M.; Kim, M.-C.; Kwon, J.-S.; Kwon, J.-H.; Lee, J.-B.; Park, S.-Y.; et al. Application of DNA Barcoding Technique in Avian Influenza Virus Surveillance of Wild Bird Habitats in Korea and Mongolia. *Avian Dis.* **2010**, *54*, 677–681. [CrossRef]
76. Na, E.-J. *Genotyping and Genetic Characteristics of Avian Influenza Virus isolated from Wild Birds in South Korea, 2019–2020*; Jeonbuk National University: Iksan, Republic of Korea, 2020.
77. Animal and Plant Quarantine Agency. *Survey of HPAI Infection Status in Wild Birds*; Animal and Plant Quarantine Agency: Gimcheon, Republic of Korea, 2018.
78. Kim, H.K.; Kim, H.J.; Noh, J.Y.; Van Phan, L.; Kim, J.H.; Song, D.; Na, W.; Kang, A.; Nguyen, T.L.; Shin, J.H. Serological evidence of H5-subtype influenza A virus infection in indigenous avian and mammalian species in Korea. *Arch. Virol.* **2018**, *163*, 649–657. [CrossRef]
79. Takekawa, J.Y.; Heath, S.R.; Douglas, D.C.; Perry, W.M.; Javed, S.; Newman, S.; Suwal, R.N.; Rahmani, A.R.; Choudhury, B.C.; Prosser, D.J. Geographic variation in Bar-headed geese Anser indicus: Connectivity of wintering and breeding grounds across a broad front. *Wildfowl* **2009**, *59*, 100–123.
80. NIBR (National Institute of Biological Resources). *Winter Waterbird Census of Korea*; National Institute of Biological Resources: Incheon, Republic of Korea, 2020.
81. Chen, X.; Li, C.; Sun, H.-T.; Ma, J.; Qi, Y.; Qin, S.-Y. Prevalence of avian influenza viruses and their associated antibodies in wild birds in China: A systematic review and meta-analysis. *Microb. Pathog.* **2019**, *135*, 103613. [CrossRef]
82. Chen, X.; Qi, Y.; Wang, H.; Wang, Y.; Wang, H.; Ni, H. Prevalence of Multiple Subtypes of Avian Influenza Virus Antibodies in Egg Yolks of Mallards (Anas platyrhynchos) and White-winged Terns (*Chlidonias leucopterus*) in the Northeastern Republic of China. *J. Wildl. Dis.* **2018**, *54*, 834–837. [CrossRef]
83. Muzaffar, S.B.; Ydenberg, R.C.; Jones, I.L. Avian influenza: An ecological and evolutionary perspective for waterbird scientists. *Waterbirds* **2006**, *29*, 243–257. [CrossRef]
84. Lee, Y.-J.; Kang, H.-M.; Lee, E.K.; Song, B.M.; Jeong, J.; Kwon, Y.K.; Kim, H.R.; Lee, K.J.; Hong, M.S.; Jang, I. Novel reassortant influenza A (H5N8) viruses, South Korea, 2014. *Emerg. Infect. Dis.* **2014**, *20*, 1087. [CrossRef] [PubMed]
85. Tavakoli, A.; Rezaei, F.; Nasab, G.S.F.; Adjaminezhad-Fard, F.; Noroozbabaei, Z.; Mokhtari-Azad, T. The Comparison of Sensitivity and Specificity of ELISA-based Microneutralization Test with Hemagglutination Inhibition Test to Evaluate Neutralizing Antibody against Influenza Virus (H1N1). *Iran J. Public Health* **2017**, *46*, 1690–1696. [PubMed]
86. Rowe, T.; Abernathy, R.A.; Hu-Primmer, J.; Thompson, W.W.; Lu, X.; Lim, W.; Fukuda, K.; Cox, N.J.; Katz, J.M. Detection of antibody to avian influenza A (H5N1) virus in human serum by using a combination of serologic assays. *J. Clin. Microbiol.* **1999**, *37*, 937–943. [CrossRef] [PubMed]

viruses

Article

Drivers of Spatial Expansions of Vampire Bat Rabies in Colombia

Zulma E. Rojas-Sereno [1,*], Daniel G. Streicker [2,3], Andrea Tatiana Medina-Rodríguez [4] and Julio A. Benavides [1,5,*]

1   Centro Para la Investigación de la Sustentabilidad y Doctorado En Medicina de la Conservación, Facultad Ciencias de la Vida, Universidad Andres Bello, República 440, Santiago 8320000, Chile
2   School of Biodiversity, One Health and Veterinary Medicine, University of Glasgow, Glasgow G12 8QQ, UK
3   Centre for Virus Research, MRC-University of Glasgow, Glasgow G61 1QH, UK
4   Dirección Técnica de Vigilancia Epidemiológica, Subgerencia de Protección Animal, Instituto Colombiano Agropecuario ICA, Bogotá 110931, Colombia
5   MIVEGEC, IRD, CNRS, Université de Montpellier, 34394 Montpellier, France
*   Correspondence: z.rojassereno@uandresbello.edu (Z.E.R.-S.); julio.benavides@ird.fr (J.A.B.)

Abstract: Spatial expansions of vampire bat-transmitted rabies (VBR) are increasing the risk of lethal infections in livestock and humans in Latin America. Identifying the drivers of these expansions could improve current approaches to surveillance and prevention. We aimed to identify if VBR spatial expansions are occurring in Colombia and test factors associated with these expansions. We analyzed 2336 VBR outbreaks in livestock reported to the National Animal Health Agency (Instituto Colombiano Agropecuario—ICA) affecting 297 municipalities from 2000–2019. The area affected by VBR changed through time and was correlated to the reported number of outbreaks each year. Consistent with spatial expansions, some municipalities reported VBR outbreaks for the first time each year and nearly half of the estimated infected area in 2010–2019 did not report outbreaks in the previous decade. However, the number of newly infected municipalities decreased between 2000–2019, suggesting decelerating spatial expansions. Municipalities infected later had lower cattle populations and were located further from the local reporting offices of the ICA. Reducing the VBR burden in Colombia requires improving vaccination coverage in both endemic and newly infected areas while improving surveillance capacity in increasingly remote areas with lower cattle populations where rabies is emerging.

Keywords: Latin America; *Desmodus rotundus*; livestock; passive surveillance; spatial epidemiology; zoonosis

Citation: Rojas-Sereno, Z.E.; Streicker, D.G.; Medina-Rodríguez, A.T.; Benavides, J.A. Drivers of Spatial Expansions of Vampire Bat Rabies in Colombia. *Viruses* **2022**, *14*, 2318. https://doi.org/10.3390/v14112318

Academic Editor: Myriam Ermonval and Serge Morand

Received: 21 September 2022
Accepted: 17 October 2022
Published: 22 October 2022

1. Introduction

Vampire bat-transmitted rabies (VBR) is a lethal zoonosis that represents a public health problem and an economic burden for the livestock sector in Latin America [1,2]. Cattle are particularly affected by VBR and the costs of vaccination and rabies mortality amount to substantial economic losses to small-scale farmers [3]. Since VBR is transmitted by spillover from bats to livestock, outbreaks of VBR usually occur in farming areas with low vaccination coverage of cattle where the virus circulates among vampire bat populations [4–7]. Improving our understanding of the spatio-temporal dynamics of VBR in bats could support more effective measures to reduce the burden of VBR in livestock; for example, optimizing the spatial or temporal distribution of vaccines to livestock [1,8–10].

Spatial expansions of VBR were reported at the range limits of common vampire bats and within their core distribution, generating novel public health and agricultural risks [1,2,11]. For example, in Mexico, VBR has lagged behind the gradual northward expansion of bats towards the United States of America and wave like expansions into historically rabies-free vampire bat populations were reported in multiple regions of Peru,

Argentina, and at a small local scale in Brazil [2,5,12–14]. However, the drivers of the occurrence, velocity, or direction of spatial expansions remain unknown but could be affected by environmentally driven changes in the dynamics of rabies within bat populations, influencing bat distribution and mobility across the landscape [1,2,4,5,15–17]. For example, differences in cattle density across the landscape could influence the number and size of bat colonies and thus influence viral spread across colonies [15]. An increase in VBR outbreaks was suspected to result from an increase in juvenile dispersal during the wet season in Argentina [18], while bat dispersal is suspected to increase VBR outbreaks during the dry seasons in Colombia [4]. A major challenge is that countries that differ in these environmental conditions also have surveillance systems with differing sensitivity, limiting the value of international comparisons to understand the drivers of the observed variation in VBR spread.

Colombia has reported hundreds of cases of VBR in livestock annually to the Epidemiological Surveillance and Information System of the Instituto Colombiano Agropecuario (ICA) since 1982 [4,19]. The country is composed of five biogeographic regions (Andean, Caribbean, Pacific, Orinoquía, and Amazon), which differ in temperature, precipitation, topography, and livestock density [20–22]. For example, the Andean region includes mountains up to 5800mts [21], which may form natural barriers to bat dispersal and therefore VBR propagation. Variable precipitation regimes across regions (e.g., unimodal, bimodal, mixed, and aseasonal) [22], could also influence temporal patterns of bat dispersal and thus VBR spatio-temporal dynamics such as seasonality [4,18]. Since the common vampire bat *Desmodus rotundus* is the main reservoir of rabies in all these biogeographic regions [23], Colombia represents an ideal system to study how different environmental factors could influence VBR spatio-temporal dynamics.

Previous studies of VBR in Colombia found that most outbreaks occurred in areas of high cattle density in the northern region of the Caribbean and the eastern region of Orinoquía [4,6,20,24]. The number of outbreaks in livestock doubled from 2010 to 2019 with three outbreak peaks in 1985, 2010, and 2014 [4,19]. The reasons for these changes in the annual number of outbreaks remain unknown; both spatial expansions and changes in the incidence within endemic areas were hypothesized [2,14,25]. VBR outbreaks were also speculated to be seasonal, with a higher number of outbreaks during the drier months [4]. However, previous studies have not formally tested whether viral expansions are occurring and could explain the observed inter-annual or seasonal changes in VBR incidence. This study aimed to identify whether spatial expansions of VBR are occurring in Colombia and their relationship to the reported burden of rabies, and to test whether biotic or abiotic conditions are associated with viral expansions.

2. Materials and Methods

2.1. National Surveillance VBR Data

Data on VBR outbreaks were provided by the Epidemiological Surveillance Technical Direction from the ICA, which performs passive surveillance of VBR in livestock. Suspected cases on the basis of neurological signs and mortality are reported by farmers, organizations, veterinary professionals, and other actors related to primary production to specific personnel involved in rabies surveillance at each of the 172 local reporting offices across the country (Figure 1A). Yearly campaigns are carried out to sustain an 'Early Warning System' of rabies surveillance in each region by training local stakeholders involved in livestock production to maintain constant surveillance of vampire bat bites and clinical signs compatible with rabies, and to engage in timely reporting of suspected rabies cases to local authorities [19]. In this study, a VBR outbreak was operationally defined when at least one animal from a cluster of animals presenting clinical signs compatible with rabies was confirmed to have died from rabies using the direct fluorescent antibody test [26]. The farm of origin was considered the epidemiological unit for each outbreak. At least 1 veterinarian in charge of rabies surveillance was present in each local reporting office for field assistance during the study period, supervised by thirteen regional epidemiologists, and two national coordinators of rabies surveillance and disease control. No major changes in the number of personnel or reporting offices were observed during our study period. After suspected outbreaks reports were entered into the Official Control Disease National Information System (SINECO), laboratory confirmation and information validation were monitored at the national level [19]. A total of 2336 laboratory-confirmed VBR outbreaks across Colombia occurring between 2000 and 2019 were analyzed. Information on the department, municipality, village, and livestock species was obtained for all outbreaks. GPS coordinates were available for outbreaks occurring from 2010 to 2019. Outbreaks occurring from 2000 to 2009 (856 outbreaks, 37%) had no GPS coordinates recorded and were assigned to the centroid of the nearest village where the farm was located using QGIS 3.3.4 [27].

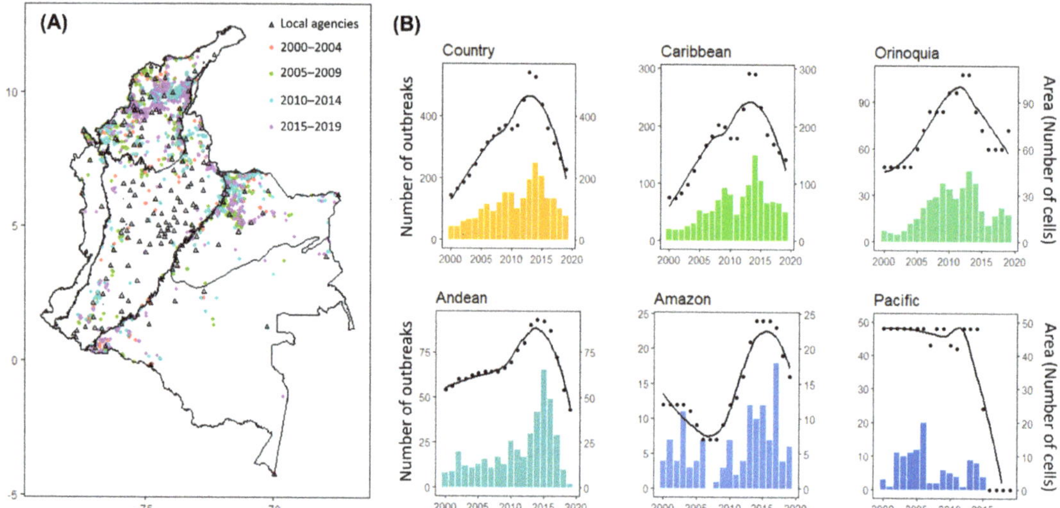

Figure 1. Spatial and temporal distribution of VBR outbreaks in Colombia from 2000 to 2019: (**A**) Locations of VBR outbreaks across the five biogeographic regions of Colombia. Point colors illustrate the five-year periods of outbreak occurrence. Triangles represent the local reporting offices of the ICA's national rabies surveillance system; (**B**) Bars represent the total number of VBR outbreaks per year. Points correspond to the estimated annual infected area, expressed in number of cells. The number of cells was estimated as the sum of cell grids with a density level higher than the 95th percentile value (i.e., density = 0.01) from a Kernel estimation choosing a bandwidth = 0.01 and a grid = 200 × 200 cells. Lines (black) represent the model prediction tendency estimated using the method 'loess' in the geom_smooth function of the ggplot2 in R. Each plot represents estimates at the country or biogeographic region level.

2.2. Colombian Biogeographical Regions and Municipality Data

The five biogeographical regions in Colombia are separated by Andean mountains and include the (I) Caribbean region (dry forest and tropical desert), (II) Pacific region (tropical rainforest), (III) Andean Region (low and high elevation tropical forest), (IV) Amazon region (tropical rainforest), and (V) Orinoquía region (dry tropical grass plains) [21,28]. Each municipality was assigned to a biogeographical region using polygons delimiting each region obtained from the administrative and biogeographic open access shapefiles at the Environmental Information System from Colombia—SIAC (www.siac.gov.co/, accessed on 29 June 2020). Altitude, mean annual precipitation and mean annual temperature raster files of Colombia with a resolution of 0.8 km^2 were obtained from Bioclim (https://www.worldclim.org/data/bioclim.html, accessed on 29 June 2020) and imported to QGIS 3.3.4. The Point Sampling Tool plugin was used to assign regional and environmental variables values to each outbreak. Addresses of local reporting offices of ICA were obtained from the ICA website (https://www.ica.gov.co/, accessed on 29 June 2020). Annual livestock (horses, pigs, sheep, goats, and buffalos) populations for each municipality were only available from ICA from 2006 to 2019. To avoid excluding outbreaks from 2000 to 2005 in the analysis, the livestock population estimated in 2006 was assigned to those years (2000 to 2005) for each municipality. However, the livestock population remained relatively constant from 2006 to 2017 and only increased after 2017 (Figure S1), so we expect that this extrapolation would not considerably affect the robustness of our results.

2.3. Identifying Changes in the Area Affected by VBR

Changes in the infected area, i.e., differences in the geographic extent of locations reporting VBR outbreaks over time, were evaluated using two approaches. First, we calculated the annual number of municipalities that reported VBR for the first time, referred to as 'new municipalities' [2]. The area of each municipality was estimated using the areaPolygon function from the Geosphere package in R. Second, we used GPS locations of each VBR outbreak to approximate the annual and monthly infected area. Specifically, we estimated VBR-infected areas at the national and regional levels using kernel density estimation using the bkde2D function from the KernSmooth package in R [29]. We chose a bandwidth = 0.01 (10 km^2 radius) and a grid = 200 × 200 cells that generated 10 km^2 cells, an area compatible with the distance of bat movements previously reported [30]. The VBR infected area was estimated as the sum of cell grids with a density level higher than the 95th percentile value (i.e., density = 0.01) and was expressed as a number of infected cells. To verify that the conclusions of our analyses regarding newly infected areas remained similar when the estimated area affected around a location was larger, we also estimated the infected area using a less restricted bandwidth = 0.1 that generates a 100 km^2 radius and a density level higher than the 88th percentile value (i.e., density = 0.005). We tested the correlation between the annual number of outbreaks and the estimated infected area using a Spearman's correlation test with the cor.test function in R. Annual changes in the infected area were estimated by comparing the infected area in a specific year to the infected area of the previous year. Large-scale changes in the infected area over time were assessed by comparing the infected area in the second decade of the study (2010–2019) to the infected area in the first decade (2000–2009). 'New infected areas' were considered as cells infected in the second period that were not infected in the first period, whereas cells infected in both periods were considered 'endemic areas'.

2.4. Seasonality Analysis

Generalized Additive Models (GAMs) were developed using the gam function in R to test the non-linear annual and monthly variation in the number of outbreaks, the kernel density estimates of the infected area at the national and regional levels, and the annual number of new municipalities reporting outbreaks. GAMs included month (ks = 12 and bs = "cc") and year (k = 20 and bs = "ps") as smoothed variables, where k (knots) = temporal dimension used for the spline function and bs = the spline basis, using a cyclic cubic regression (cc) for month and a *p*-spline (ps) for year.

2.5. Identifying Drivers of Potential Spatial Expansions

For each new municipality infected, we recorded the 'time to first outbreak', referring here to the number of years between the start of our dataset (2000) and the year of the first outbreak reported in that municipality. We then tested whether environmental and anthropogenic drivers could explain the time to the first outbreak using a multivariate regression model. The model was composed of local environmental factors including the estimated cattle population at the year of the first outbreak (or the total livestock population in a separate model) within each municipality, the first outbreak's altitude, and mean annual precipitation. Since the temperature was highly correlated to altitude (Spearman's correlation test: Rho = −0.89, *p*-value ≤ 0.01), the regression model only included altitude and precipitation as environmental variables. Since VBR outbreaks are often underreported in Latin America and under-reporting can be correlated with distance to the reporting office [3], we also included distances from the first outbreak in each municipality to the closest reporting office, calculated using the geodist function in R. Given the count nature of the response variable (i.e., years ranging from 0 to 19) and since data were overdispersed (dispersion value = 4.06, *p*-value $\leq$ 0.01, DHARMa nonparametric dispersion test in R), we built a quasi-Poisson Generalized Linear Mixed Model using a Penalized Quasi-Likelihood with the function glmmPQL in R, including the biogeographical region as a random effect.

To identify whether spatial expansions occurred in a 'wave-like' spread where the time of arrival to a municipality increases with its distance to the first case, we intended to include in the model the distance between the first outbreak in each municipality and the first outbreak reported in each biogeographic region [2]. However, the presence of multiple, geographically distant outbreaks in the first month of our dataset suggested that no reliable origin for a 'wave-like' spread could be identified, particularly in regions where outbreaks likely occurred prior to 2000 (e.g., the Caribbean). Thus, this variable was ultimately excluded from the analysis.

3. Results

3.1. Spatio-Temporal Distribution of VBR

From a total of 2336 VBR outbreaks, most were reported in cattle (2037 outbreaks, 87%), followed by horses (285 outbreaks, 12%), pigs (5 outbreaks, 0.2%), sheep (4 outbreaks, 0.2%), goats (2 outbreaks, 0.1%), and buffalos (2 outbreaks, 0.1%). VBR outbreaks were reported in all five biogeographic regions of the country (Figure 1A and Figure S2A) including 27% (297 out of 1123) of all municipalities (Figure S2B). Half of the VBR outbreaks were reported in the Caribbean region (1243 outbreaks, 53%, 132 out of the 297 municipalities reporting outbreaks). The Orinoquía (437 outbreaks, 19%, 34 municipalities) and Andean (421, 18%, 113 municipalities) regions reported a similar number of outbreaks. Outbreaks were only occasionally reported in the Amazon (126, 5%, 31 municipalities) and Pacific (109, 5%, nine municipalities) regions.

The number of VBR outbreaks varied significantly across years and months at the national level (GAM: year effect estimate: edf = 10.25, F = 21.45, *p*-value $\leq$ 0.01; month effect estimate: edf = 2.66, F = 0.73, *p*-value = 0.03, Table S1), with a peak in 2014 (Figure 1B) and the month of August (Figure S3). However, outbreaks peaked at different years in each biogeographic region and no seasonality was detected when separate models were built for each region (Table S2). There was no correlation between the annual number of VBR outbreaks and the annual total number of livestock (Spearman's correlation: rho = 0.02, *p*-value = 0.25) between 2006 and 2019. The Pacific was the only region not reporting outbreaks after 2015. The national annual number of outbreaks was strongly correlated with the VBR infected area estimated by Kernel densities (Spearman's correlation using 10 km^2 radius: rho = 0.93, *p*-value $\leq$ 0.01, Spearman's correlation using 100 km^2 radius: rho = 0.90, *p*-value $\leq$ 0.01, Figure 1B), suggesting that the annual burden of VBR may be more tightly linked to the spatial extent of the virus than to local variation in incidence.

3.2. Changes in the VBR-Infected Area

From a total of 297 municipalities reporting VBR outbreaks between 2000 and 2019, the annual number of infected municipalities ranged from 30 to 72 (mean $\pm$ SD = 47 $\pm$ 12). The highest number of municipalities reporting outbreaks coincided with a peak in the number of outbreaks in 2014 (72 municipalities and 247 outbreaks), when all regions reported outbreaks (Figure 2A). An average of 15 municipalities [SD = 7, range: 6–31] reported VBR outbreaks for the first time each year. Strikingly, these 'new municipalities' accounted for 24% (554 outbreaks) of all outbreaks and an average of 27% [SD = 18, range: 10–69] of outbreaks per year. Additionally, an average of 38% [SD = 15, range: 10–58] of the area infected per year was reported among 'new municipalities', implying that a considerable portion of the rabies burden occurs in areas that would have been considered rabies-free the prior year. Although new municipalities were reported throughout the study, we noted a marked decrease in the rate that new municipalities were reporting VBR over time (GAM: edf = 1.00, F = 9.49, *p*-value ≤ 0.01, Table S3; Figure 2B). Within biogeographic regions, the number of new municipalities significantly decreased in the Caribbean (Caribbean GAM: edf = 1.71, F = 5.24, *p*-value = 0.01, Table S4, Figure S4) but not significantly in other regions when considered alone.

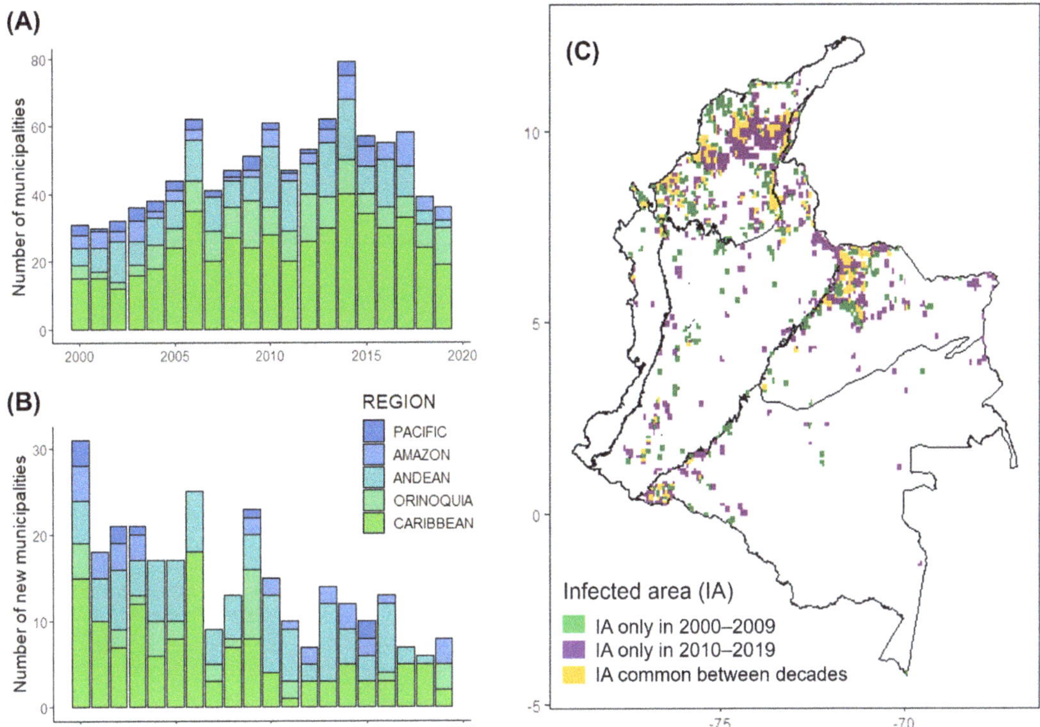

Figure 2. Annual number of municipalities and infected areas in Colombia from 2000 to 2019: (**A**) Bars represent the annual number of municipalities reporting VBR outbreaks; (**B**) Bars represent the annual number of 'new municipalities' reporting VBR outbreaks for the first time. Each bar division illustrates a biogeographic region; (**C**) Comparison between the infected area (IA) in the second (2010–2019) and the first (2000–2009) decade of the study. 'New infected areas' were considered as cells reporting VBR outbreaks in the second decade that did not report outbreaks in the first decade (cells in purple), whereas cells reporting outbreaks in both decades were considered as 'endemic areas' (cells in yellow). The infected area was estimated using a bandwidth = 0.01 that generated a 10 km^2 radius.

As measured by kernel densities, the VBR-infected area varied significantly across years and months (GAM: year effect estimate: edf = 8.75, F = 18.23, *p*-value ≤ 0.01; month effect estimate: edf = 1.95, F = 0.63, *p*-value = 0.02, Table S5), with different peaks in each region (e.g., Caribbean, Orinoquía, Andean and Pacific) (Figures 1B and S5). However, the Caribbean was the only region with a significant monthly variation (Caribbean GAM: edf = 1.82, F = 0.54, *p*-value = 0.03, Table S6), with a peak in the infected area during June (Figure S5). An average of 57% (SD = 16, ranging between 30–88 using a 10 km^2 radius) or 31% (SD = 12, ranging between 14–59 using a 100 km^2 radius) of the annual area infected across the country originated from new cells reporting VBR outbreaks. Comparing the infected area between 2000–2009 and 2010–2019, 49% (1025 out of 2083 infected cells using a 10 km^2 radius) or 29% (1374 out of 4716 infected cells using a 100 km^2 radius) of the total area infected only reported outbreaks during the second decade (Figures 2C, S6 and S7). Cells infected for the first time in the second decade were also identified in municipalities previously infected during the first decade (Figure S7).

3.3. Potential Drivers of Spatial Expansions

The time to the first outbreak in each municipality was negatively correlated with the municipality's number of cattle (glmmPQL, Estimate = -2.0×10^{-6}, *p*-value = 0.04, Figure 3A) and was positively correlated with the distance from the outbreak to the reporting office (glmmPQL: Estimate = 2.34×10^{-3}, *p*-value = 0.01, Table 1, Figures 3B and S8). In contrast, the time to the first outbreak was not significantly correlated with altitude, annual precipitation, or the total livestock population (tested in a different model instead of 'number of cattle', glmmPQL: Estimate = -1.2×10^{-6}, *p*-value = 0.12).

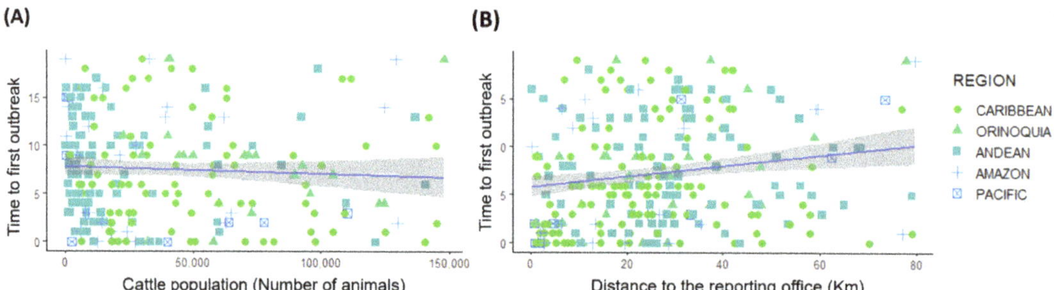

Figure 3. Influence of significant biotic and abiotic conditions on the time of arrival of the first VBR outbreak at each municipality: (**A**) Correlation between the time of the first outbreak (e.g., number of years since 2000) in a municipality and its cattle population. Total livestock population was tested in a different model instead of 'number of cattle' without statistical significance (glmmPQL: Estimate = -1.2×10^{-6}, *p*-value = 0.12). Each point shape represents a biogeographical region; (**B**) Correlation between the time of the first outbreak in a municipality and the geographical distance of that outbreak to the closest reporting office. Lines (blue) represent the tendency estimated for the model and the shadow represents the confidence interval (CI = 95%, grey) estimated using the method 'lm' in the geom_smooth function ggplot2 in R.

Table 1. Drivers correlated with the time to the first VBR outbreak appearance in each outbreak location: Results from a quasi-Poisson Generalized Linear Mixed Model using a Penalized Quasi-Likelihood with the glmmPQL function in R, including region as a random effect. Time to the first outbreak was not significantly correlated to the total livestock population tested in a different model instead of 'number of cattle' (glmmPQL: Estimate = -1.2×10^{-6}, *p*-value = 0.12). Asterisks identified values considered as statistically significant at *p*-value < 0.05.

Variable	Value	Std. Error	DF	t-Value	*p*-Value
(Intercept)	1.88	0.15	239	12.39	0.00
Altitude	1.59×10^{-5}	8.02×10^{-5}	239	0.20	0.84
Precipitation	5.16×10^{-5}	5.66×10^{-5}	239	0.91	0.36
Cattle population	-2.00×10^{-6}	9.50×10^{-7}	239	-2.12	0.04 *
Distance to the reporting office	2.34×10^{-3}	9.43×10^{-4}	239	2.48	0.01 *

4. Discussion

Although VBR is considered an endemic disease affecting livestock in Colombia, a poor understanding of its spatio-temporal dynamics limits the effectiveness of measures to anticipate and prevent VBR outbreaks. Our analyses of VBR outbreaks in livestock passively reported to ICA between 2000 and 2019 revealed that temporal changes in VBR-infected areas were consistent with VBR spatial expansions, with more than 30% of the annual infected area originating from newly infected cells and municipalities, and 49% of the infected area in 2010−2019 not reported as infected in the previous decade. Spatial expansions of VBR in Colombia appeared to be decelerating and the virus arrived later in areas with lower cattle populations and located far from local reporting offices. Given our finding that a considerable portion of the national burden of VBR mortality in livestock was attributable to spatial expansions of the virus, our results highlight the need to fortify rabies surveillance and spillover prevention at viral range limits.

Over the 20 years of surveillance data that we analyzed, changes in the VBR-infected area were observed at the country and regional levels (Figure 1B). However, no monotonic trend was observed, with peaks in the infected area during different years across different regions. Changes in the area reporting VBR in livestock could be explained by annual changes in the spread of VBR among bat populations, in part due to spatial expansions as previously observed in Brazil, Mexico, and Peru [2,5,14,31]. For example, changes in bat dispersal or behavior due to environmental changes or culling could affect viral transmission between bat colonies [4,32,33]. New lineages of VBR-colonizing bat populations could also generate changes in the spread of VBR and thus in the area reporting outbreaks in livestock [7,25]. Vampire bat distribution could also be expanding in some areas as a consequence of climate change, an increasing number of human-made structures that could be used as roosts (e.g., mines, tunnels, and abandoned houses), and expansion of livestock populations [16,34]. However, previous studies have shown that VBR spatial expansions such as traveling waves occur in areas with already established vampire bat populations rather than by the expansion of vampire bat populations to new areas [2]. Alternatively to changes in rabies circulation among bats or expansion in the bat distribution, annual variation in the area covered by livestock vaccination could also result in changes in the VBR mortality reported [3], with a higher infected area reported during years of lower vaccination coverage. Thus, future research including estimates of vaccination coverage at the municipality level could contribute to a better understanding of what drives the observed changes in the area reporting VBR outbreaks.

Regardless of the mechanism underlying temporal variation in the affected area of Colombia, our analysis suggests that the burden of VBR is tightly linked to the spatial extent of the virus at any point in time. Specifically, we found a close correspondence between estimates of the infected area and the number of outbreaks reported, suggesting that the burden of rabies in livestock is predominately driven by the presence or absence of the virus rather than local variation in incidence within endemically infected vampire bat populations. Biologically, this pattern might arise through the combination of the extremely low incidence of rabies in vampire bat populations (such that variation in levels of viral circulation is trivial) and a high rate of spillover when the virus is circulating locally due to the high contact rates between bats and livestock [35]. If verified, this finding would have important implications for rabies management since interventions that aim to reduce rabies incidence in vampire bat populations, such as bat population control, might be less effective than direct measures at the bat–livestock interface such as livestock vaccination. We also found that a substantial fraction of all outbreaks occurred in previously rabies-free areas, which is surprising given the relative size of these areas compared to where the virus has been historically endemic. This relatively high burden in historically rabies-free areas is likely to be exacerbated by low or absent livestock vaccination in areas that have not historically been affected, highlighting the potential gains from fortifying rabies surveillance and vaccination efforts at viral range limits [3].

Our analyses suggest that spatial expansions of VBR in Colombia are occurring but have decelerated through time. The slowing rate of expansion could reflect the exhaustion of available rabies-free municipalities suited for both vampire bats and cattle. For example, the Caribbean region had already reported VBR outbreaks in half of its municipalities by the end of our study period, although a few new municipalities are still reported annually [1,4,9]. Our inability to infer the geographic origins or routes of viral expansions, as was possible elsewhere in South America [2,13], also supports the conclusion that at the start of our study, VBR had already established within some or all biogeographical regions, with the observed expansions perhaps emanating in non-linear routes or from multiple origins within a region. Despite this complexity, our study was, for the first time, able to identify correlates of the timeline of VBR arrival to locations. We found that spatial expansions arrived later in municipalities with low cattle populations, which we hypothesize may reflect the epidemiological isolation of these areas arising from smaller or more sparsely connected vampire bat populations [15]. It is alternatively possible that areas with low livestock density instead have fewer real or reported spillovers due to the smaller number of susceptible animals; however, a study in Peru found that reporting was negatively (not positively) related to herd size and was strongly driven by the historical presence of VBR, both arguing against effects of surveillance bias alone [3,36]. We also found evidence for delayed arrival (or reporting) of VBR in municipalities far from reporting offices. This result is likely to reflect both the true geographic isolation of these areas concerning VBR spread and lower reporting rates due to logistical constraints, such that VBR may take more time to be detected [2]. Unfortunately, data on reporting efforts at the municipality level were not available for our study, requiring future research to discriminate between these two alternative scenarios. Regardless, increased recognition of VBR in the most isolated areas adds logistical challenges to the national rabies program relying on livestock vaccination campaigns to reduce the VBR burden [3,36].

Instead of resulting from spatial expansions, temporal changes in the VBR-infected area could result from VBR endemic circulation across the country combined with heterogeneous surveillance efforts that increased over time, generating new VBR reports in areas previously considered as VBR-free when surveillance became sufficient to detect an outbreak. Although data on surveillance efforts during our study period are not available for Colombia to test this hypothesis, no changes in surveillance strategy or capacity were reported at such a large scale that would be compatible with an increase in almost half of the area reporting VBR in the second decade of our study period [19,37]. Further, earlier studies argued that data collected through similar passive surveillance systems provide an accurate reflection of viral arrival to new areas suggesting, as described above, that we would have been unlikely to find a negative relationship between the time until VBR arrival and livestock density if observations of outbreaks were based on reporting alone [2,13]. Finally, given that outbreaks in newly infected areas are likely to be larger due to low vaccination coverage, we speculate that such a sudden increase in mortality would be readily reported [38].

Despite the occurrence of spatial expansions of VBR outbreaks in Colombia, around half of annual outbreaks still occur in 'endemic areas'. Thus, VBR burden reduction will require improving vaccination coverage in both endemic areas and newly infected municipalities/areas. Our findings suggest that vaccination programs in municipalities previously reporting outbreaks were insufficient to eliminate livestock deaths in the following years. Mandatory vaccination in high-risk zones was established by the ICA's animal health program in 2003 but has been firmly implemented since 2015 [39,40].However, vaccination coverage remains limited for restricted laboratory production and other reasons that remain unclear. Thus, identifying factors that limit vaccination would help reduce the VBR burden in endemic areas. For example, farmers' low knowledge of veterinary public authorities and farms located at a higher elevation, rather than a low socio-economic status, were associated with low rabies vaccination intake in Peru [3]. In contrast, reducing the burden of rabies in newly infected areas, which comprised on average at least 27% of outbreaks per year, requires improved epidemiological capacity to forecast the routes and velocity of viral invasions and how these are affected by environmental conditions. As such, our study is a starting point toward the eventual aim of preventive vaccination in areas with emerging risks.

5. Conclusions

Overall, our study suggests that part of the changes in the area reporting VBR in the livestock of Colombia could result from spatial expansions of VBR. Our findings suggest that reducing the VBR burden will require improving vaccination coverage in both endemic areas and newly infected areas. This study supports previous work identifying spatial expansions in Brazil, Mexico, and Peru, and suggests that VBR is slowing in speed across Colombia. Delayed and still ongoing viral detections in municipalities with lower cattle populations and those farthest from the local reporting offices indicate surveillance gaps in remote areas that are likely to decrease outbreak predictability, which may intensify losses in newly infected areas.

Supplementary Materials: The following supporting information can be downloaded at: https://www.mdpi.com/article/10.3390/v14112318/s1, Figure S1. Annual number of livestock in Colombia from 2006 to 2019; Table S1: GAM model for seasonality of the number of VBR outbreaks at the national level from 2000 to 2019; Figure S2: Spatial distribution of VBR outbreaks in Colombia from 2000 to 2019; Table S2: GAM model for seasonality of the number of VBR outbreaks at the regional level from 2000 to 2019; Figure S3: Annual and monthly distribution of the number of VBR outbreaks from 2000 to 2019; Table S3: GAM model for seasonality of the number of new municipalities reporting VBR outbreaks for the first time at the national level from 2000 to 2019; Figure S4: Annual distribution of the number of new municipalities reporting VBR outbreaks for the first time at the national level from 2000 to 2019; Table S4: GAM model for seasonality of the new municipalities reporting VBR outbreaks for the first time at the regional level from 2000 to 2019; Figure S5: Annual and monthly distribution of the infected area covered from 2000 to 2019; Table S5: GAM model for seasonality of the VBR infected area covered at the national level from 2000 to 2019; Figure S6: Comparison between the infected area in the second decade of the study (2010–2019) and in the first decade (2000–2009); Table S6: GAM model for seasonality of the VBR infected area covered at the regional level from 2000 to 2019; Figure S7: Spatial distribution of the first VBR outbreak occurrence in Colombia from 2000 to 2019; Figure S8: Drivers correlated to the time of arrival of the first VBR outbreak at a municipality according to biogeographic region.

Author Contributions: Conceptualization, J.A.B., D.G.S. and Z.E.R.-S.; methodology, J.A.B., D.G.S. and Z.E.R.-S.; software, Z.E.R.-S. and J.A.B.; validation, Z.E.R.-S. and J.A.B.; formal analysis, Z.E.R.-S., J.A.B., and D.G.S.; investigation, Z.E.R.-S.; resources, A.T.M.-R., Z.E.R.-S., J.A.B. and D.G.S.; data curation, Z.E.R.-S.; writing—original draft preparation, Z.E.R.-S., J.A.B., and D.G.S.; writing—review and editing, Z.E.R.-S., J.A.B., D.G.S. and A.T.M.-R.; visualization, Z.E.R.-S.; supervision, J.A.B. and D.G.S.; project administration, Z.E.R.-S.; funding acquisition, Z.E.R.-S., J.A.B., D.G.S. All authors have read and agreed to the published version of the manuscript.

Funding: Analysis and writing work were funded by the National Agency for Research and Development (ANID) Beca Doctorado Nacional 21201076, awarded to Z.E.R.-S., D.G.S was supported by a Wellcome Senior Research Fellowship (217221/Z/19/Z)

Institutional Review Board Statement: Not applicable.

Informed Consent Statement: Not applicable.

Data Availability Statement: VBR outbreaks and annual livestock population data were obtained from the Epidemiological Surveillance Technical Direction of the ICA. Data can be publicly requested directly from ICA.

Acknowledgments: The authors thank the Epidemiological Surveillance and Animal Health Technical Direction of ICA for sharing their data and for providing useful comments regarding animal health surveillance in Colombia and feedback on the results of these analyses.

Conflicts of Interest: The authors declare no conflict of interest. The funders had no role in the design of the study; in the collection, analyses, or interpretation of data; in the writing of the manuscript; or in the decision to publish the results.

References

1. Johnson, N.; Aréchiga-Ceballos, N.; Aguilar-Setien, A. Vampire Bat Rabies: Ecology, Epidemiology and Control. *Viruses* **2014**, *6*, 1911–1928. [CrossRef] [PubMed]
2. Benavides, J.A.; Valderrama, W.; Streicker, D.G. Spatial Expansions and Travelling Waves of Rabies in Vampire Bats. *Proc. R. Soc. B Biol. Sci.* **2016**, *283*, 20160328. [CrossRef]
3. Benavides, J.A.; Rojas Paniagua, E.; Hampson, K.; Valderrama, W.; Streicker, D.G. Quantifying the Burden of Vampire Bat Rabies in Peruvian Livestock. *PLoS Negl. Trop. Dis.* **2017**, *11*, e0006105. [CrossRef] [PubMed]
4. Brito-Hoyos, D.M.; Sierra, E.B.; Álvarez, R.V. Distribución Geográfica Del Riesgo de Rabia de Origen Silvestre y Evaluación de Los Factores Asociados Con Su Incidencia En Colombia, 1982–2010. *Rev. Panam. Salud Publica/Pan Am. J. Public Health* **2013**, *33*, 8–14. [CrossRef]
5. Bárcenas-Reyes, I.; Loza-Rubio, E.; Zendejas-Martínez, H.; Luna-Soria, H.; Cantó-Alarcón, G.J.; Milián-Suazo, F. Comportamiento Epidemiológico de La Rabia Paralítica Bovina En La Región Central de México, 2001-2013. *Rev. Panam. Salud Pública* **2015**, *38*, 396–402. [PubMed]
6. Cárdenas-Contreras, Z. Análisis Espacio Temporal de La Rabia Bovina de Origen Silvestre En Colombia (2005–2014). Master's Thesis, Universidad Autónoma de Barcelona, Barcelona, Spain, 2017.
7. Botto Nuñez, G.; Becker, D.J.; Plowright, R.K. The Emergence of Vampire Bat Rabies in Uruguay within a Historical Context. *Epidemiol. Infect.* **2019**, *147*, e180. [CrossRef]
8. Blackwood, J.C.; Streicker, D.G.; Altizer, S.; Rohani, P. Resolving the Roles of Immunity, Pathogenesis, and Immigration for Rabies Persistence in Vampire Bats. *Proc. Natl. Acad. Sci. USA* **2013**, *110*, 20837–20842. [CrossRef] [PubMed]
9. Escobar, L.E.; Peterson, T.; Favi, M.; Yung, V.; Medina-Vogel, G. Bat-Borne Rabies in Latin America. *Rev. Inst. Med. Trop. Sao Paulo* **2015**, *57*, 63–72. [CrossRef]
10. McClure, K.M.; Gilbert, A.T.; Chipman, R.B.; Rees, E.E.; Pepin, K.M. Variation in Host Home Range Size Decreases Rabies Vaccination Effectiveness by Increasing the Spatial Spread of Rabies Virus. *J. Anim. Ecol.* **2020**, *89*, 1375–1386. [CrossRef]
11. Mungrue, K.; Mahabir, R. The Rabies Epidemic in Trinidad of 1923 to 1937: An Evaluation with a Geographic Information System. *Wilderness Environ. Med.* **2011**, *22*, 28–36. [CrossRef]
12. Torres, C.; Lema, C.; Gury Dohmen, F.; Beltran, F.; Novaro, L.; Russo, S.; Freire, M.C.; Velasco-Villa, A.; Mbayed, V.A.; Cisterna, D.M. Phylodynamics of Vampire Bat-Transmitted Rabies in Argentina. *Mol. Ecol.* **2014**, *23*, 2340. [CrossRef] [PubMed]
13. Streicker, D.G.; Winternitzc, J.C.; Satterfield, D.A.; Condori-Condori, R.E.; Broos, A.; Tello, C.; Recuenco, S.; Velasco-Villa, A.; Altizer, S.; Valderrama, W. Host-Pathogen Evolutionary Signatures Reveal Dynamics and Future Invasions of Vampire Bat Rabies. *Proc. Natl. Acad. Sci. USA* **2016**, *113*, 10926–10931. [CrossRef] [PubMed]
14. Santos, B.L.; Bruhn, F.R.P.; Coelho, A.C.B.; Estima-Silva, P.; Echenique, J.V.; Sallis, E.S.V.; Schild, A.L. Epidemiological Study of Rabies in Cattle in Southern Brazil: Spatial and Temporal Distribution from 2008 to 20171. *Pesqui. Vet. Bras.* **2019**, *39*, 460–468. [CrossRef]
15. Delpietro, H.A.; Marchevsky, N.; Simonetti, E. Relative Population Densities and Predation of the Common Vampire Bat (Desmodus Rotundus) in Natural and Cattle-Raising Areas in North-East Argentina. *Prev. Vet. Med.* **1992**, *14*, 13–20. [CrossRef]
16. Lee, D.N.; Papeş, M.; Van Den Bussche, R.A. Present and Potential Future Distribution of Common Vampire Bats in the Americas and the Associated Risk to Cattle. *PLoS ONE* **2012**, *7*, e42466. [CrossRef] [PubMed]
17. Streicker, D.G.; Lemey, P.; Velasco-Villa, A.; Rupprecht, C.E. Rates of Viral Evolution Are Linked to Host Geography in Bat Rabies. *PLoS Pathog.* **2012**, *8*, e1002720. [CrossRef] [PubMed]
18. Lord, R.D. Seasonal Reproduction of Vampire Bats and Its Relation to Seasonality of Bovine Rabies. *J. Wildl. Dis.* **1992**, *28*, 292–294. [CrossRef] [PubMed]
19. ICA. *Sanidad Animal 2016*; Dirección Tecnica de Vigilancia Epidemiológica. Subgerencia de Protección Animal. Instituto Colombiano Agropecuario; Produmedios: Bogotá, Colombia, 2019.
20. Etter, A.; McAlpine, C.; Wilson, K.; Phinn, S.; Possingham, H. Regional Patterns of Agricultural Land Use and Deforestation in Colombia. *Agric. Ecosyst. Environ.* **2006**, *114*, 369–386. [CrossRef]
21. Gonzáles, M.; García, H.; Corzo, G.; Madriñan, S. Ecosistemas Terrestres de Colombia y El Mundo. In *Biodiversidad, Conservación y Desarrollo*; Sánchez, J.A., Mandriñán, S., Eds.; Universidad de los Andes, Ediciones Uniandes: Bogotá, Colombia, 2012; Volume I, pp. 69–113.
22. Urrea, V.; Ochoa, A.; Mesa, O. Seasonality of Rainfall in Colombia. *Water Resour. Res.* **2019**, *55*, 4149–4162. [CrossRef]
23. Mantilla-Meluk, H.; Mantilla-Meluk, H.; Jimenez-Ortega, A.M.; Baker, R.J. *Phyllostomid Bats of Colombia: Annotated Checklist, Distribution, and Biogeography*; Museum of Texas Tech University: Lubbock TX, USA, 2009; Volume 56.
24. Escobar, L.E.; Peterson, A.T. Spatial Epidemiology of Bat-Borne Rabies in Colombia. *Rev. Panam. Salud Pública* **2013**, *34*, 135–136.
25. Streicker, D.G.; Fallas González, S.L.; Luconi, G.; Barrientos, R.G.; Leon, B. Phylodynamics Reveals Extinction–Recolonization Dynamics Underpin Apparently Endemic Vampire Bat Rabies in Costa Rica. *Proc. R. Soc. B Biol. Sci.* **2019**, *286*. [CrossRef] [PubMed]
26. OIE Rabies (Infection with Rabies Virus and Other Lyssaviruses). Available online: https://www.oie.int/fileadmin/Home/eng/Health_standards/tahm/3.01.17_RABIES.pdf (accessed on 5 August 2020).
27. QGIS.org. QGIS Geographic Information System. Available online: http://www.qgis.org (accessed on 29 June 2020).

28. IGAC Grandes Biomas y Biomas Continentales de Colombia. Insituto Geográfico Agustin Codazzi. Available online: https://geoportal.igac.gov.co/sites/geoportal.igac.gov.co/files/geoportal/grandes_biomas.pdf (accessed on 29 June 2020).

29. Wand, M. KernSmooth: Functions for Kernel Smoothing Supporting Wand & Jones, 1995. R Package Version 2.23-20. Available online: https://cran.r-project.org/package=KernSmooth (accessed on 1 September 2021).

30. Ávila-Flores, R.; Bolaina-Badal, A.L.; Gallegos-Ruiz, A.; Sánchez-Gómez, W.S. Use of Linear Features by the Common Vampire Bat (*Desmodus Rotundus*) in a Tropical Cattle-Ranching Landscape. *Therya* **2019**, *10*, 229–234. [CrossRef]

31. Ulloa-Stanojlovic, F.M.; Dias, R.A. Spatio-temporal Description of Bovine Rabies Cases in Peru, 2003–2017. *Transbound. Emerg. Dis.* **2020**, *67*, 1688–1696. [CrossRef] [PubMed]

32. George, D.B.; Webb, C.T.; Farnsworth, M.L.; O'Shea, T.J.; Bowen, R.A.; Smith, D.L.; Stanley, T.R.; Ellison, L.E.; Rupprecht, C.E. Host and Viral Ecology Determine Bat Rabies Seasonality and Maintenance. *Proc. Natl. Acad. Sci. USA* **2011**, *108*, 10208–10213. [CrossRef]

33. Streicker, D.G.; Recuenco, S.; Valderrama, W.; Benavides, J.G.; Vargas, I.; Pacheco, V.; Condori Condori, R.E.; Montgomery, J.; Rupprecht, C.E.; Rohani, P.; et al. Ecological and Anthropogenic Drivers of Rabies Exposure in Vampire Bats: Implications for Transmission and Control. *Proc. R. Soc. B Biol. Sci.* **2012**, *279*, 3384–3392. [CrossRef]

34. Benavides, J.A.; Valderrama, W.; Recuenco, S.; Uieda, W.; Suzán, G.; Avila-Flores, R.; Velasco-Villa, A.; Almeida, M.; de Andrade, F.A.G.; Molina-Flores, B.; et al. Defining New Pathways to Manage the Ongoing Emergence of Bat Rabies in Latin America. *Viruses* **2020**, *12*, 1002. [CrossRef]

35. Griffiths, M.E.; Bergner, L.M.; Broos, A.; Meza, D.K.; Filipe, A. da S.; Davison, A.; Tello, C.; Becker, D.J.; Streicker, D.G. Epidemiology and Biology of a Herpesvirus in Rabies Endemic Vampire Bat Populations. *Nat. Commun.* **2020**, *11*, 5951. [CrossRef]

36. Anderson, A.; Shwiff, S.; Gebhardt, K.; Ramírez, A.J.; Shwiff, S.; Kohler, D.; Lecuona, L. Economic Evaluation of Vampire Bat (Desmodus Rotundus) Rabies Prevention in Mexico. *Transbound. Emerg. Dis.* **2014**, *61*, 140–146. [CrossRef]

37. ICA. *Sanidad Animal 2005*; Boletines Epidemiológicos; Dirección Tecnica de Vigilancia Epidemiológica. Subgerencia de Protección Animal. Instituto Colombiano Agropecuario; Imprenta Nacional de Colombia: Bogotá, Colombia, 2007.

38. Seetahal, J.F.R.; Vokaty, A.; Carrington, C.V.F.; Adesiyun, A.A.; Mahabir, R.; Hinds, A.Q.J.; Rupprecht, C.E. The History of Rabies in Trinidad: Epidemiology and Control Measures. *Trop. Med. Infect. Dis.* **2017**, *2*, 27. [CrossRef]

39. ICA. *No Resolución 00383 de 2015*; Ministerio de Agricultura y Desarrollo Rural: Bogotá, Colombia, 2015; p. 20.

40. ICA. *Resolución 23132 de 2018*; Ministerio de Agricultura y Desarrollo Rural: Bogotá, Colombia, 2018; p. 37.

Article

Swine-to-Ferret Transmission of Antigenically Drifted Contemporary Swine H3N2 Influenza A Virus Is an Indicator of Zoonotic Risk to Humans

Carine K. Souza [1], J. Brian Kimble [1,†], Tavis K. Anderson [1], Zebulun W. Arendsee [1], David E. Hufnagel [1], Katharine M. Young [1], Phillip C. Gauger [2], Nicola S. Lewis [3], C. Todd Davis [4], Sharmi Thor [4] and Amy L. Vincent Baker [1,*]

[1] Virus and Prion Research Unit, National Animal Disease Center, United States Department of Agriculture-Agricultural Research Service, Ames, IA 50010, USA
[2] Department of Veterinary Diagnostic and Production Animal Medicine, College of Veterinary Medicine, Iowa State University, Ames, IA 50011, USA
[3] Department of Pathology and Population Sciences, Royal Veterinary College, University of London, Hertfordshire, London NW1 0TU, UK
[4] Influenza Division, National Center for Immunization and Respiratory Diseases, Centers for Disease Control and Prevention, Atlanta, GA 30333, USA
* Correspondence: amy.vincent@usda.gov
† Current address: Foreign Arthropod-Borne Animal Disease Research Unit, National Bio and Agro-Defense Facility, USDA-ARS, Manhattan, KS 66502, USA.

Citation: Souza, C.K.; Kimble, J.B.; Anderson, T.K.; Arendsee, Z.W.; Hufnagel, D.E.; Young, K.M.; Gauger, P.C.; Lewis, N.S.; Davis, C.T.; Thor, S.; et al. Swine-to-Ferret Transmission of Antigenically Drifted Contemporary Swine H3N2 Influenza A Virus Is an Indicator of Zoonotic Risk to Humans. *Viruses* 2023, 15, 331. https://doi.org/ 10.3390/v15020331

Academic Editors: Myriam Ermonval and Serge Morand

Received: 8 December 2022
Revised: 21 January 2023
Accepted: 22 January 2023
Published: 24 January 2023

Abstract: Human-to-swine transmission of influenza A (H3N2) virus occurs repeatedly and plays a critical role in swine influenza A virus (IAV) evolution and diversity. Human seasonal H3 IAVs were introduced from human-to-swine in the 1990s in the United States and classified as 1990.1 and 1990.4 lineages; the 1990.4 lineage diversified into 1990.4.A–F clades. Additional introductions occurred in the 2010s, establishing the 2010.1 and 2010.2 lineages. Human zoonotic cases with swine IAV, known as variant viruses, have occurred from the 1990.4 and 2010.1 lineages, highlighting a public health concern. If a variant virus is antigenically drifted from current human seasonal vaccine (HuVac) strains, it may be chosen as a candidate virus vaccine (CVV) for pandemic preparedness purposes. We assessed the zoonotic risk of US swine H3N2 strains by performing phylogenetic analyses of recent swine H3 strains to identify the major contemporary circulating genetic clades. Representatives were tested in hemagglutination inhibition assays with ferret post-infection antisera raised against existing CVVs or HuVac viruses. The 1990.1, 1990.4.A, and 1990.4.B.2 clade viruses displayed significant loss in cross-reactivity to CVV and HuVac antisera, and interspecies transmission potential was subsequently investigated in a pig-to-ferret transmission study. Strains from the three lineages were transmitted from pigs to ferrets via respiratory droplets, but there were differential shedding profiles. These data suggest that existing CVVs may offer limited protection against swine H3N2 infection, and that contemporary 1990.4.A viruses represent a specific concern given their widespread circulation among swine in the United States and association with multiple zoonotic cases.

Keywords: influenza A virus; H3N2; swine; pandemic preparedness; zoonosis; variant; antigenic drift

1. Introduction

The H3N2 subtype of influenza A virus (IAV) has caused influenza morbidity and mortality in humans globally, with more severe annual epidemics than H1N1 or influenza B virus since the 1968 pandemic [1]. From 1968 to the present, human seasonal H3N2 strains have been introduced into pig populations, establishing endemic lineages that cause an important respiratory disease in pigs, impacting the swine industry, and posing a zoonotic health concern for humans. Pigs may serve as an intermediate host of IAV due to the expression of both α2,6 and α2,3 linked sialic acids on receptors, which also serve as IAV receptors in humans and birds, respectively [2]. Pigs may be infected by avian or human-derived IAV, although host barriers appear to limit the frequency of these interspecies events, particularly from birds to pigs [2]. When there is interspecies transmission, there may be rapid evolution within the swine host due to reassortment events and the error-prone replication of the virus. The evolution of human and avian IAV within the swine host has led to the emergence of novel strains with pandemic potential [3,4]. A notable example of the two-way transmission of IAV between humans and swine populations worldwide was the emergence of the 2009 pandemic H1N1 (H1N1pdm09) of swine-origin and subsequent transmission from humans to swine, resulting in reassortment with endemic swine IAV, including North American swine H3N2 viruses [5,6].

In the 1990s, the introduction of a human seasonal H3N2 virus into swine and the subsequent reassortment into viruses containing a triple reassortant internal gene cassette (TRIG) [4,7] dramatically increased endemic swine IAV diversity in North America [6,8,9]. This virus lineage had gene segments derived from classical swine lineage H1N1, human seasonal H3N2 viruses, and North American avian IAV [6,8,9]. The H3 genes from human seasonal introductions maintained in swine are phylogenetically distinct and referred to as the 1990.1 and 1990.4 clades using global nomenclature based on decade of introduction [10]. The 1990.4 lineage diversified into six phylogenetic clades (1990.4.A–F) [11,12]. More recent human-to-swine H3N2 transmission events in the 2010 decade led to the 2010.1 and 2010.2 lineages in U.S. swine from incursions occurring in 2010–11 and in 2016–17, respectively [13–15].

Human infections with IAV of swine origin, termed variant viruses to differentiate them from human seasonal IAV, are sporadically detected in the U.S. To date, a total of 434 H3N2 variant cases have been detected, with most cases involving direct or indirect contact with infected swine at interfaces such as agricultural fairs and livestock shows [13]. Most of these cases occurred in 2011–2012 ($n = 306$) and were caused by the 1990.4.A swine lineage, but there have been variant cases detected from other 1990.4 genetic clades along with the 1990.1 and 2010.1 lineages [10]. Although human-to-human transmission of variant IAV strains is rare, the zoonotic potential and/or pandemic risk of swine-origin IAV should not be underestimated, as exemplified by the H1N1pdm09 pandemic [3]. Variant IAV cases in the U.S. are monitored and reported by the Centers for Disease Control and Prevention (CDC), which then reports novel influenza viruses to the World Health Organization (WHO). Animal influenza activity, including variant detections, is reviewed twice per year during the WHO consultation meetings on the composition of IAV vaccines. If these variant viruses are genetically and antigenically distinct from current human seasonal vaccines and existing pre-pandemic candidate vaccine viruses (CVVs), a representative strain from the clade may be considered for the development of a new CVV [16]. Over the past decade, several avian and swine IAV strains from human zoonotic infections have been selected as CVVs and are available within the WHO Global Influenza Surveillance and Response System (GISRS). This initiative helps the international community prepare for the public health risks of animal influenza viruses with a potential global impact [17].

We previously identified swine H3N2 strains against which humans are likely to lack population immunity or are not protected by a current human seasonal vaccine, or CVV [18]. Adult human sera revealed limited immunity against the 1990.1, 1990.4.A, and 1990.4.B swine lineages, especially in individuals born after 1970 [18]. To further understand the zoonotic potential of swine IAV, we analyzed contemporary swine H3 HA genes collected in 2020 in the U.S. and selected representatives from each detected clade. Then, we used serological methods to test ferret antisera raised against human vaccines or related variant CVV strains to identify swine strains with limited cross-reactivity. Lastly, three representative H3N2 strains that showed a substantial reduction in cross-reactivity to a relevant CVV or a human seasonal vaccine were tested for zoonotic potential in a pig-to-ferret interspecies transmission model.

2. Material and Methods

2.1. Genetic Analysis and Strain Selection

At the biannual WHO information meeting on the composition of influenza virus vaccines, animal influenza activity data for 6-month periods are presented and compared against human IAV vaccine components and pre-pandemic CVVs. All swine H3 HA sequences collected between 1 January 2020, and 30 June 2020 were downloaded from GISAID [19]. Sequences were compiled with human seasonal vaccine and CVV strains and aligned using MAFFT v7.453 [20]. We inferred a maximum-likelihood phylogeny for the HA nucleotide alignment using IQ-TREE v2 implementing automatic model selection [21,22] with subsequent tree visualization and annotation in Smot v.1.0.0 [23]. The HA genes and associated available whole genome data were classified to a genetic clade or evolutionary lineage using the octoFLU pipeline [24], and a consensus HA1 for each identified clade was generated from the translated amino acid sequence data using flutile (https://github.com/flu-crew/flutile, accessed on 5 December 2022). We identified lineages circulating in the U.S. and selected a representative strain for each genetic clade by generating a pairwise distance matrix and choosing the best match between the HA1 clade consensus sequence and a virus isolate available in the USDA IAV swine virus repository. We excluded the 1970.1, 2000.3, and Other-Human-1990 lineages as they were not detected in the U.S. We did not identify field isolates for additional assessment from the 2010.2 lineage because it was less frequently detected and retained cross-reactivity with human seasonal vaccine anti-sera [14] or the 2010.1 lineage because it includes a within-clade CVV that was previously characterized in our swine-to-ferret model [25], and human sera contained cross-reactive antibodies to this clade [18]. Selected viruses were tested by hemagglutinin inhibition (HI) assay (using guinea pig red blood cells) against ferret antisera raised against human seasonal vaccine viruses and CVVs. For swine-to-ferret transmission studies, we expanded the selection criteria to ensure the identified strain included a representative neuraminidase (NA) and internal gene constellation through identifying the predominant evolutionary lineages and genetic clades paired to the HA genes using octoFLUshow [24]. We identified the amino acid differences between the characterized strains, clade consensus HA1 sequences, and within-clade CVVs, or human seasonal vaccine strains, using flutile (https://github.com/flu-crew/flutile, accessed on 5 December 2022).

2.2. Viruses and Ferret Antisera

Swine H3N2 isolates were selected from clades 1990.1 (A/swine/Missouri/A02257614/2018), 1990.4.A (A/swine/North Carolina/A02245294/2019) and 1990.4.B.2 (A/swine/Illinois/A02479007/2020) and provided by the National Veterinary Services Laboratories (NVSL) through the U.S. Department of Agriculture (USDA) IAV swine surveillance system in conjunction with the USDA-National Animal Health Laboratory Network (NAHLN). The virus isolates A/Minnesota/11/2010 × 203 CVV, IDCDC-RG55C (A/Ohio/28/2016-like) CVV, A/Indiana/27/2018 variant, and A/Iowa/60/2018, a human H3N2 seasonal vaccine strain (the A (H3N2) component of non-egg-based vaccines used in the 2020 Southern Hemisphere vaccine), along with ferret antisera produced against these strains, were provided from the CDC, Atlanta, Georgia, U.S.A. The viruses was propagated on Madin-Darby canine kidney (MDCK) cells grown in Opti-MEM (Life Technologies, Waltham, MA, USA). Virus growth media contained antibiotics/antimycotics and 1 µg/mL of tosylsulfonyl phenylalanyl chloromethyl ketone (TPCK)-trypsin (Worthington Biochemical Corp., Lakewood, NJ, USA).

2.3. Hemagglutination Inhibition

Prior to hemagglutination inhibition (HI) assays, ferret antisera were heat inactivated at 56 °C for 30 min then treated with a 20% Kaolin suspension (Sigma-Aldrich, St. Louis, MO, USA), followed by an adsorption with 0.75% guinea pig red blood cells to remove nonspecific hemagglutination inhibitors, as previously described [26]. In order to address coverage of contemporary U.S. swine H3N2 viruses by the nearest CVV or HuVac, HI assays were performed using ferret antisera in the presence of 20 nM of oseltamivir carboxylate [10].

Pig antisera were treated with receptor-destroying enzyme (RDE) (Denka Seiken Co., LTD., Tokyo, Japan), heat inactivated at 56 °C for 30 min, and adsorbed with 50% turkey red blood cells for nonspecific hemagglutination inhibitors.

2.4. Swine-to-Ferret Transmission Study Design

A total of 20 three-week-old pigs obtained from a herd free of IAV and porcine reproductive and respiratory syndrome virus were housed in a biosafety level 2 containment facility in compliance with an approved USDA-ARS NADC animal care and use protocol. Upon arrival, pigs were treated prophylactically with ceftiofur (Zoetis, Florham Park, NJ, USA), according to the label directions, to reduce potential respiratory bacterial pathogens. Sixteen 4–6-month-old male and female ferrets were obtained from an influenza-free, high health source for use as transmission contacts. Upon arrival, all pigs and ferrets were screened for antibody against influenza A nucleoprotein (NP) by a commercial enzyme-linked immunosorbent assay (ELISA) (MultiS ELISA; Idexx, Westbrook, ME, USA) to ensure the absence of preexisting immunity from prior exposure or passively acquired maternal antibody. Pigs were randomly assigned into four groups of five (three experimental groups, one naïve control group) and placed into separate containment rooms.

Pigs in experimental groups were challenged with 2 mL intranasally of 1×10^6 50% tissue culture infectious dose $(TCID_{50})/mL$ [27] of each strain diluted in phosphate-buffered saline (PBS). At 2 days post infection (dpi), four ferrets per group were housed individually in open-front isolators were placed approximately 4 feet from the pig deck, as previously described [25]. All animals received a subcutaneous radio frequency microchip (pigs: Deston Fearing, Dallas, TX; Ferrets: Biomedic Data Systems Inc., Seaford, DE, USA) for identification and body temperature monitoring purposes. The body temperatures of contact ferrets were monitored from −2 to 12 dpc. A febrile response was considered when a ferret displayed a temperature greater than 2 standard deviations above the mean of ferret temperatures before exposure (>39.5 °C). Ferrets were provided routine care and handled before pigs, with a change in outer gloves and surface decontamination of gowns and equipment with 70% ethanol between individual ferrets.

Nasal swab samples (FLOQSwabs; Copan Diagnostics, Murrieta, CA, USA) were collected from pigs at 0, 1, 3, and 5 dpi, and then three pigs from each group were humanely euthanized and necropsied at 5 dpi to evaluate lung lesions, as previously described [28]. The remaining two pigs in each group were swabbed at 7 and 9 dpi, then humanely euthanized and necropsied at 14 dpi. Broncho-alveolar fluid (BALF) and serum samples were collected at each of the necropsy timepoints described above. To assess virus replication in contact ferrets, nasal wash samples were collected at 0, 1, 3, 5, 7, 9, 11, and 12 days post-contact (dpc), following methods previously described [25]. BALF and plasma samples from ferrets were collected at necropsy on 12 dpc [25]. To confirm the pig-to-ferret transmission by the seroconversion of contact ferrets, the sera from contact ferrets collected at 12 dpc were analyzed by HI assay and a commercial blocking enzyme-linked immunosorbent assay (ELISA) against influenza A nucleoprotein (NP) (MultiS ELISA; Idexx, Westbrook, ME, USA) with a positive optical density (O.D.) cut-off of <0.6.

2.5. Virus Replication and Shedding

Nasal swab, nasal wash, and BALF samples were titrated on MDCK cells to evaluate virus replication in the nose and lungs, as previously described for pigs [28] and ferrets [25]. MDCK-inoculated monolayers were evaluated for cytopathic effect (CPE) between 48 and 72 h post-infection, fixed with 4% phosphate-buffered formalin, and stained using immunocytochemistry (ICC) with an anti-influenza A NP monoclonal antibody [29]. A $TCID_{50}$ titer per mL was calculated for each positive sample. Nasal wash samples that were positive in the virus isolation but negative in the virus titration were confirmed by qPCR VetMAX™-Gold SIV Detection Kit (ThermoFisher Scientific, MA, USA) and run with a qPCR standard curve included with the kit, ranging from 10–1,000,000 copies per μL.

2.6. Pathologic Examination

At 5 dpi, the percentage of the lung affected by purple-red consolidation typical of influenza virus in swine was visually estimated [28]. The percentage of the lung affected with purple-red consolidation typical of influenza virus in ferrets was visually estimated at 12 dpc to assess disease resolution if infected. Tissue samples from the trachea and right middle or affected lung lobe were fixed in 10% buffered formalin for histopathologic examination. Tissues were processed by routine histopathologic procedures, and slides were stained with hematoxylin and eosin (H&E) or immunohistochemistry [30].

2.7. Microbiological Assays

BALF samples were cultured for aerobic bacteria on blood agar and Casmin (NAD-enriched) plates to indicate the presence of concurrent bacterial pneumonia. PCR assays for porcine circovirus 2 (PCV2) were conducted for swine BALF samples [31]. To exclude other causes of pneumonia in pigs, commercial assays for *Mycoplasma hyopneumoniae* and PRRSV were conducted according to the manufacturer's recommendations (VetMax, Life Technologies, Carlsbad, CA, USA).

2.8. Data Analysis

Results were analyzed with Prism 8 (GraphPad, San Diego, CA, USA) with analysis of variance (ANOVA), with $p < 0.05$ considered significant. Variables with significant effects by treatment group were subjected to pairwise mean comparisons using the Tukey-Kramer test. Clinical data associated with this study are available for download from the USDA Ag Data Commons at https://doi.org/10.15482/USDA.ADC/1528327, accessed on 5 December 2022 and the phylogenetic analyses are available from https://github.com/flu-crew/datasets, accessed on 5 December 2022.

3. Results

3.1. Genetic Characterization of Dominant U.S. Swine H3N2 Strains

There were 180 H3 HA genes collected in swine between 1 January 2020, and 30 June 2020. These HA genes represented six evolutionary lineages, eight genetic clades, and a single human-to-swine spillover: 1970.1 ($n = 6$, 3.3%), 1990.1 ($n = 4$, 2.2%), 1990.4.A ($n = 93$, 51.7.%), 1990.4.B.2 ($n = 3$, 1.7%), 1990.4.I ($n = 1$, 0.6%), 2000.3 ($n = 3$, 1.7%), 2010.1 ($n = 63$, 35%), 2010.2 ($n = 6$, 3.3%), and Other-Human-1990 ($n = 1$, 0.6%). Figure 1 shows the evolutionary relationships of the selected contemporary representative H3N2 and the most similar CVV or a human seasonal H3N2 vaccine and if a variant case occurred. We excluded the globally detected 1970.1, 2000.3, and Other-Human-1990 lineages from further analyses as they were not detected in U.S. swine. The genome constellations of internal genes have different lineage designations, "T" for the TRIG lineage, "P" for the H1N1pdm09 lineage, and "V" for the lineage derived from a swine live attenuated influenza virus vaccine (LAIV). The selected H3N2 viruses have different genome constellations: The A/swine/Missouri/A02257614/2018 strain has the HA gene from the 1990.1 swine LAIV origin lineage, the NA gene from the 1998-N2 swine LAIV origin, and the internal gene constellation is TVVVPT in the order of PB2, PB1, PA, NP, M, and NS gene segments. The HA segment of this virus has 52 amino acid residues different from HuVac A/Iowa/60/2018. The A/swine/North Carolina/A02245294/2019 strain has the HA from 1990.4.A lineage, NA from the 2002-N2 lineage, and an internal gene constellation of TTTPPT. The HA segment of this virus has nine residues different from the within-clade CVV A/Minnesota/11/2010. The A/swine/Illinois/A02479007/2020 strain has the HA from the 1990.4.B.2 lineage, NA from the 2002-N2 lineage, and internal genes of TTPTPT. The HA segment of this virus has 27 residues different from the CVV A/Minnesota/11/2010. The amino acid differences between the representative HA gene and the within-clade CVV, or human seasonal vaccine, are displayed in Tables S1–S4.

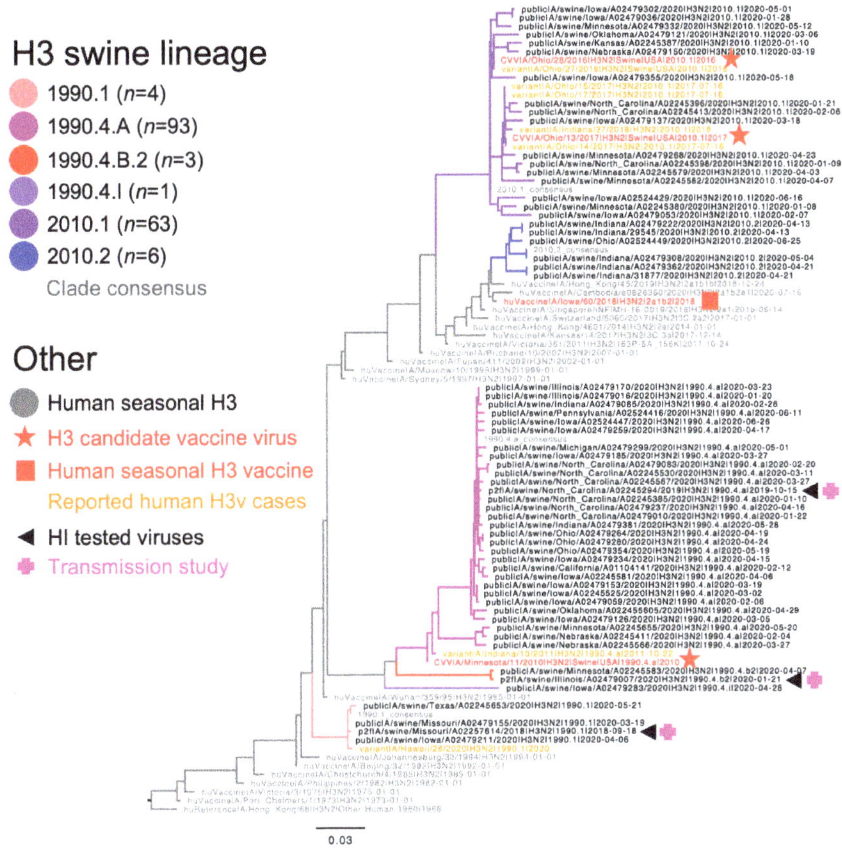

Figure 1. Evolutionary relationships of contemporary H3 swine influenza A viruses. A representative random sample of H3 swine hemagglutinin (HA) genes collected between January 2020 and June 2020. Reference human HA genes, candidate vaccine viruses (CVV), and variant cases are indicated by branch color or shape. Swine IAV strains tested in hemagglutination inhibition assays are marked by a black triangle, and those used in transmission studies by a pink plus sign (+). The numbers in parentheses indicate the number of each genetic clade detected during the sampling period.

3.2. Loss in Cross-Reactivity between Dominant Swine H3N2 Strains and CVV or HuVac

The HA1 domain of selected representative swine H3N2 strains was compared to the HuVac and CVV vaccine strains. A range in percentages of amino acid identity was observed (Table 1). To determine if CVV or HuVac antisera had cross-reactivity against contemporary swine H3N2 strains, we tested the representative U.S. dominant swine H3N2 strains against reference ferret antisera generated against CVVs from the same genetic clade, or the most genetically similar CVV if there was no within-clade CVV. Ferret antiserum tested with the clade 1990.1 representative strain, A/swine/Missouri/A02257614/2018, demonstrated a 32-fold decrease in HI titer from the most similar clade 1990.4.A CVV, A/Minnesota/11/2010, and a >32-fold decrease from the HuVac A/Iowa/60/2018. Ferret antiserum tested with this strain had a 32-fold decrease from H3N2v A/Indiana/27/2018, a representative of contemporary swine strains. Antiserum was tested with the representative strain of clade 1990.4.A A/swine/North Carolina/A02245294/2019 demonstrated a 16-fold-decrease in HI titer compared to the within-clade CVV A/Minnesota/11/2010 and a greater than 32-fold decrease from the HuVac A/Iowa/60/2018. Antiserum tested with this strain also had a 32-fold decrease from the H3N2v virus, A/Indiana/27/2018. The 1990.4.B.2

clade represented by A/swine/Illinois/A02479007/2020 lacks a CVV, and ferret antiserum from the most similar CVV, A/Minnesota/11/2010, demonstrated a 32-fold-decrease and 16-fold decrease with the HuVac A/Iowa/60/2018 antiserum, respectively. The H3N2v A/Indiana/27/2018 antiserum had a 64-fold decrease with this strain. None of the selected representative swine strains cross-reacted with the IDCDC-RG55C A/Ohio/28/2016-like CVV. These data demonstrated that contemporary representative H3N2 viruses from the 1990.1, 1990.4.A, and 1990.4.B.2 clades frequently detected in U.S. pig populations had limited cross-reactivity against HuVac and CVV sera (Table 2).

Table 1. Pairwise amino acid sequence similarity of the HA1 domain from swine H3 clade consensus sequences from the U.S. to candidate vaccine viruses or human seasonal vaccine viruses and clade representative viruses used in this study. Within-clade comparisons are highlighted in grey.

	1990.1 Consensus	A/swine/Missouri/A02257614/2018	1990.4.A Consensus	A/Minnesota/11/2010 CVV	A/swine/North Carolina/A02245294/2019	1990.4.B.2. Consensus	A/swine/Illinois/A02479007/2020	2010.1 Consensus	A/Ohio/28/2016	A/Indiana/27/2018 *	A/Iowa/60/2018 HuVac
1990.1 consensus		99.70	85.71	87.23	85.71	87.23	86.02	84.19	84.19	83.89	83.89
A/swine/Missouri/A02257614/2018	99.70		86.02	87.54	86.02	87.54	86.32	84.19	84.19	83.89	84.19
1990.4.A consensus	85.71	86.02		97.26	100	90.88	89.67	83.59	82.98	82.37	81.46
A/Minnesota/11/2010 CVV	87.23	87.54	97.26		97.26	93.01	91.79	83.89	83.89	82.98	83.28
A/swine/North Carolina/A02245294/2019	85.71	86.02	100	97.26		90.88	89.67	83.59	82.98	82.37	81.46
1990.4.B.2 consensus	87.23	87.54	90.88	93.01	90.88		95.14	84.5	85.11	82.98	83.89
A/swine/Illinois/A02479007/2020	86.02	86.32	89.67	91.79	89.67	95.14		82.98	83.28	82.37	82.67
2010.1 consensus	84.19	84.19	83.59	83.89	83.59	84.5	82.98		99.39	98.18	89.36
A/Ohio/28/2016	84.19	84.19	82.98	83.89	82.98	85.11	83.28	99.39		97.57	89.67
A/Indiana/27/2018 *	83.89	83.89	82.37	82.98	82.37	82.98	82.37	98.18	97.57		88.15
A/Iowa/60/2018 HuVac	83.89	84.19	81.46	83.28	81.46	83.89	82.67	89.36	89.67	88.15	

* A/Indiana/27/2018 H3N2v is a variant strain representing contemporary 2010.1 swine strains.

Table 2. Hemagglutination inhibition assays with ferret antisera raised against CVV and vaccine viruses tested for the ability to inhibit hemagglutination of contemporary swine viruses.

Strain	Lineage	Antigenic Motif	A/Minnesota/11/2010 × 203	IDCDC-RG55C A/Ohio/28/2016-like	A/Indiana/27/2018 *	A/Iowa/60/2018
A/swine/Missouri/A02257614/2018	1990.1	KHKEYS	40	<10	20	<10
A/Minnesota/11/2010 × 203	1990.4.A	NYNNYK	**1280**	<10	10	<10
A/swine/North Carolina/A02245294/2019	1990.4.A	NYHNYK	80	<10	20	<10
A/swine/Illinois/A02479007/2020	1990.4.B.2	SYHNYK	40	<10	10	20
IDCDC-RG55C A/Ohio/28/2016-like	2010.1	KTHNFK	<10	**1280**	80	20
A/Indiana/27/2018 *	2010.1	NTRDFT	<10	10	**640**	10
A/Iowa/60/2018	HuVac	STHNYK	<10	<10	10	**320**

* A/Indiana/27/2018 H3N2v is representative of contemporary 2010.1 swine strains. Vaccine strains and homologous titers are bolded; gray highlighted cells indicate the within-clade titer of a contemporary swine strain.

3.3. Swine-to-Ferret Transmission of Antigenically Drifted Swine H3N2 Lineages

Pigs infected with the 1990.1 clade virus demonstrated a profile of modest shedding with group mean titers of 1.78×10^2 TCID$_{50}$/mL at 1 dpi and reached a peak group mean titer of 3.16×10^3 TCID$_{50}$/mL at 5 dpi (Figure 2A). In contrast, pigs infected with the 1990.4.A clade virus demonstrated high virus shedding with group mean titers of 1.78×10^5 TCID$_{50}$/mL at 1 dpi and 2.23×10^4 TCID$_{50}$/mL at 5 dpi (Figure 2B). Similarly, pigs infected with the 1990.4.B.2 clade virus shed high virus titers at 1 dpi with group mean titers of 1.78×10^5 TCID$_{50}$/mL, 1.78×10^3 TCID$_{50}$/mL at 3 dpi, and shed virus until 5 dpi with a group mean titer of 1.78×10^4 TCID$_{50}$/mL. Only one of the remaining pigs was infected with the 1990.4.B.2 virus was positive in nasal swabs at 7 dpi (Figure 2C). The other remaining pigs from all groups had no detectable virus in nasal swabs after 5 dpi. Pigs in all groups had virus detected in the lungs at 5 dpi, with group mean titers of 1.78×10^4 TCID$_{50}$/mL for the clade 1990.1 virus, 5.62×10^5 TCID$_{50}$/mL for the clade 1990.4.A virus and 3.16×10^5 TCID$_{50}$/mL for the clade 1990.4.B.2 virus (Figure 2D), but macroscopic and microscopic lesions in the lungs were minimal in all groups (Table S5). The two remaining pigs from each group seroconverted at 14 dpi with HI titers ranging from 160–320, confirming infection (Table S6).

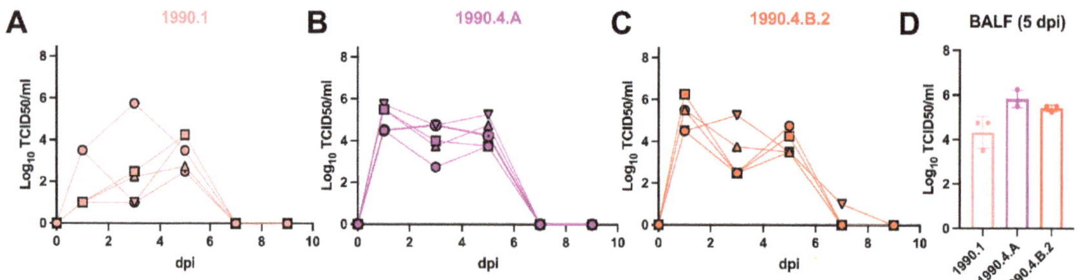

Figure 2. Infected pig nasal virus titers and BALF titers. Nasal shedding of H3N2 clades: 1990.1 (**A**), 1990.4.A (**B**), 1990.4.B.2 (**C**) was measured on 1, 3, 5, 7 and 9 dpi, and BALF (**D**) were collected at 5 dpi. Nasal swab samples and BALF were measured by $TCID_{50}$ in MDCK cells and recorded as log_{10} $TCID_{50}$/mL.

Two out of four contact ferrets from the 1990.1 group started shedding virus at low levels, with an average of 10^1 $TCID_{50}$/mL in nasal washes at 1 dpc and at 3 dpc and an average titer of 5.62×10^2 $TCID_{50}$/mL (Table 3, Figure 3). Two out of four contact ferrets had peak virus shedding at 5 dpc with an average titer of 3.16×10^3 $TCID_{50}$/mL and continued to shed virus until 7 dpc with 5.62×10^2 $TCID_{50}$/mL. Detection of virus by cell culture in the nasal washes of F#55 at 3 dpc and F#56 at 1 dpc was confirmed by qPCR, with C_t values of 36, approximately 100 copies/μL. This may represent surface contamination of their noses rather than a productive infection since the virus was not detected on other days and these two ferrets remained seronegative. Only two out of four contact ferrets of the 1990.1 clade seroconverted by 1 dpc. F#55 and F#56 had HI titers below 40 and were also negative by NP-ELISA (Table 3, Figure 3A).

Table 3. Cumulative clinical, viral, and serological measures of ferrets exposed to H3N2-infected pigs.

Viral Clade	Ferret #	Change in Bodyweight (%) from 0–12 dpc	dpc with Nasal Detection	Peak Titer log_{10} $TCID_{50}$/mL	12 dpc HI Titer	NP-ELISA S/N
1990.1	53	7.9	3,5,7	3.5	320	0.171
	55	12.1	3	0.5	10	1.003
	56	9.4	1	0.5	20	1.116
	64	3.5	1,3,5,7	4.4	640	0.188
1990.4.A	57	4.9	3,5,7	5.5	160	0.335
	58	5.5	3,5,7,9	5.8	160	0.314
	59	10.5	3,5,7,9,11	5.2	160	0.410
	60	1.5	3,5,7	4.8	160	0.272
1990.4.B.2	61	10.3	3,5,7,9	4.8	640	0.213
	62	2.0	3,5,7	6.5	640	0.265
	63	3.9	3,5,7,9	5.5	1280	0.511
	54	11.8	1,3,5,7,9	6.5	320	0.555
No virus	13	5.2	none	0	<10	1.082
	14	3.5	none	0	<10	0.829

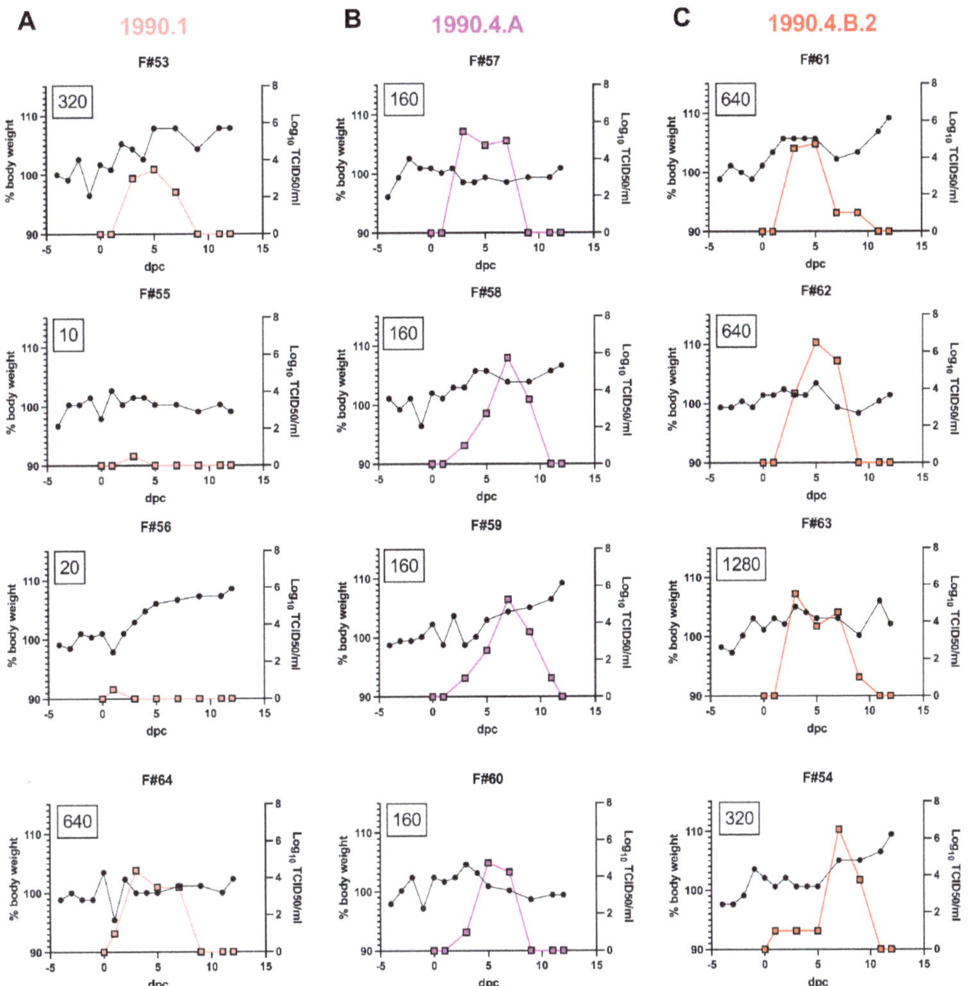

Figure 3. Swine H3N2 transmitted to ferrets. Percent body weight change and nasal wash virus titers were recorded for individual ferrets exposed to 1990.1 (**A**), 1990.4.A (**B**), and 1990.4.B.2 (**C**) swine H3N2 strains. Ferret body weight was taken daily and recorded as a percentage of the −3 day average body weight prior to exposure (left axis, black circles). Nasal washes were measured for viral shedding by $TCID_{50}$ in MDCK cells and recorded as $\log_{10} TCID_{50}/mL$ (right axis, color square). The black boxed numbers indicate the HI titers of sera collected at 12 days post contact (dpc).

All four contact ferrets of the 1990.4.A group started shedding virus at 3 dpc with an average titer of $1.78 \times 10^2\ TCID_{50}/mL$. Two out of four ferrets peaked at 7 dpc (F#58 and F#59), one peaked at 3 dpc (F#57), and another peaked at 5 dpc (F#60). The virus titer group average was $5.62 \times 10^3\ TCID_{50}/mL$ at 5 dpc and $1.78 \times 10^5\ TCID_{50}/mL$ at 7 dpc. Two out of four contact ferrets continued shedding virus with an average titer of $3.16 \times 10^3\ TCID_{50}/mL$ at 9 dpc, and only one ferret shed virus until 11 dpc at $10^1\ TCID_{50}/mL$. All four contact ferrets from the 1990.4.A group seroconverted by 12 dpc (Table 3, Figure 3B).

One out of four contact ferrets of 1990.4.B.2 group started shedding virus at 1 dpc with a titer of 10^1 TCID$_{50}$/mL. All four contact ferrets shed virus at 3 dpc with an average titer of 5.62×10^3 TCID$_{50}$/mL, increased virus titers at 5 dpc with an average titer of 1.78×10^4 TCID$_{50}$/mL and maintained virus shedding at 7 dpc with an average titer of 3.16×10^4 TCID$_{50}$/mL. Then, after the virus peak, three out of four contact ferrets had lower virus shedding, with an average titer of 1.78×10^2 TCID$_{50}$/mL until 9 dpc. At 12 dpc, no virus was detected in nasal washes or BALF samples from ferrets. All four contact ferrets from the 1990.4.B.2 group seroconverted by 12 dpc (Table 3, Figure 3C).

There were minimal clinical signs in ferrets, with the exception of one ferret (F#54) from the 1990.4.B.2 group, in which coughing was observed and it had visible lung lesions at 12 dpc. None of the contact ferrets from the 1990.1 group demonstrated a febrile response at any time. In the 1990.4.A group, only F#59 displayed a febrile response of 39.6 °C at 9 and 12 dpc. In the 1990.4.B.2 clade group, F#54 displayed a febrile response at 4 dpc of 39.7 °C. Body weight change was not significantly different between groups; however, ferrets positive for virus in nasal washes showed a trend toward static weight during the virus shedding period (Figure 3).

4. Discussion

Swine IAVs are considered a threat to public health due to swine populations worldwide harboring a vast array of antigenically diverse IAVs that occasionally spill over into humans. Consequently, swine-origin variant virus cases in humans have raised public health concerns, and there is a critical need to assess the zoonotic potential of these viruses before swine-to-human epidemics or pandemics occur [32]. Zoonotic interspecies IAV transmission requires important factors, such as virus adaptation, exposure of a susceptible human, virus load, and close contact with infected swine that increases the risk of infection [33–36]. The US CDC have reported variant cases since 2005, with most cases identified in individuals (mainly children) at swine exhibitions who had close contact with swine at agricultural fairs. Within this period, H3N2v virus infections have been the most frequently detected in the U.S. [32]. Although person-to-person transmission of variant viruses is rare, variants have the potential to evolve and may acquire the ability to transmit from human-to-human, as occurred with the H1N1pdm09 [37]. In response to the recognition of the public health risk of swine IAV, human pandemic preparedness efforts have expanded to include the characterization of swine-origin variant strains and, when appropriate, the development of CVV. However, the genetic and antigenic diversity of swine H3 clades [18,38] requires regular characterization to identify swine IAV that represent a pandemic threat. Here, we identified U.S. swine H3N2 lineages with zoonotic potential based on the loss of cross-reactivity of post-infection ferret antisera raised to relevant CVV or human seasonal vaccines with an integrated assessment using an interspecies swine-to-ferret transmission model.

The 1990.1, 1990.4.A, and 1990.4.B.2 clades characterized in this study currently circulate in U.S. pig populations. The 1990.4.A clade had a significant increase in detection frequency from 7% in 2017 to 32% in 2019 [38]. Since 2019, the 1990.4.A clade has been one of the most frequent IAV lineages detected in the U.S. swine population, comprising ~52% of detections between 2020 and 2022 [39]. Although there is a within-clade CVV (A/Minnesota/11/2010 × 203), a contemporary representative strain of the 1990.4.A clade displayed a 16-fold reduction in HI cross-reactivity, suggesting a loss in protection and potential zoonotic risk. The other two swine H3N2 strains (1990.1 and 1990.4.B.2) represented 2.9% and 3% of detections in the U.S. in 2020, respectively. Although these two clades were infrequently detected, there have been sustained detections over the past decade, and there are no available CVVs within these clades. Additionally, our serologic assessment indicated very limited cross-reactivity to an available CVV and/or human seasonal vaccines. Given limited cross-reactivity observed in our study, the 1990.1 and 1990.4.B.2 are not covered by existing pandemic preparedness CVVs or human seasonal vaccines. Additionally, in a previous study, 1990.1, 1990.4.A, and 1990.4.B strains demonstrated limited cross-reactivity to post-infection and post-vaccination adult human cohort sera, with results suggesting that older subjects have reduced immunity to these swine IAV clades [18]. Consequently, we selected contemporary strains from these clades to assess zoonotic potential in an interspecies transmission model.

To determine the potential for interspecies transmission of the selected H3N2 swine viruses, we used a swine-to-ferret transmission model [25,40]. Ferrets are widely used as an animal model for pathogenesis and transmission studies of influenza viruses and vaccine efficacy studies and are a useful animal model for humans because of their anatomic and physiological similarities [41]. Both 1990.4.A and 1990.4.B.2 clades replicated efficiently in the respiratory tract of pigs and transmitted high virus titers in nasal washes to contact ferrets, as confirmed by seroconversion of contact ferrets at 12 dpc. A previous ferret-to-ferret transmission study with an H3N2v virus demonstrated that high virus shedding titers in ferret nasal washes correlated with high growth in human airway epithelial cells, suggesting that these viruses can infect cells of the human airway [42]. Although the body weight and temperature were not significantly different in ferrets infected with all the viruses we tested, the 1990.4.A and 1990.4.B.2 contact groups displayed a trend of weight loss or no weight gain when virus shedding was detected. In contrast, the 1990.1 strain demonstrated a lower profile of virus shedding and transmission, as demonstrated by the absence of seroconversion in two of the four contact ferrets and low virus titers in nasal shedding. Although two ferrets did not seroconvert by 12 dpc, transient virus in their nasal cavities was detected at 1 dpc and 3 dpc, confirmed by qPCR. Because this study was focused on transmission, we did not characterize lung lesions or viruses in the lungs in ferrets until necropsy by 12 dpc, reflecting illness recovery by that timepoint. Only one of the ferrets infected with 1990.4.B.2 presented clinical respiratory signs, such as coughing and dyspnea, and displayed lung lesions at the necropsy on 12 dpc. A previous swine-to-ferret study of a 2010.1 lineage of H3N2 demonstrated efficient interspecies transmission [25]. Viruses from this 2010.1 swine lineage have resulted in 73 variant cases with genetic sequence data available between 2010 and 2022 [25,32]. Our data suggest that H3 viruses introduced to US swine in the 1990s (1990.4.A and 1990.4.B.2) and those introduced in the 2010s (2010.1) have retained the ability to transmit and replicate in humans and have zoonotic potential.

Although HA plays a key role in the restriction of interspecies transmission, efficient virus infection and transmission require balanced actions of HA receptor-binding and NA sialidase activity [43]. The HA protein of IAV is the primary target of protective immune responses and is a major component of vaccines. Thus, HA cross-neutralizing antibodies are important for a protective immune response. Substitutions in seven amino acid positions near the receptor binding site of the HA (145, 155, 156, 158, 159, and 189; H3 numbering) are key for antigenic drift in human [44,45] and swine IAV [12], and determine the antigenic phenotype of the virus. Our data support the proposition that genetic changes (Tables 1 and S1–S4) in the HA1 for the 1990.1, 1990.4.A, and 1990.4.B.2 swine clades resulted in significant antigenic drift from the human-seasonal vaccines and the tested CVVs.

Reassortant swine IAV with different gene constellations have demonstrated different profiles of transmission efficiency in pigs [46,47]. The 1990.1 isolate selected for this study contained PB1, PA, and NP genes derived from a commercial live attenuated influenza vaccine (LAIV) for swine (Boehringer Ingelheim, St. Joseph, MO, USA) [48,49]. Reassortment between the LAIV vaccine and endemic swine IAV field strains circulating in the U.S. was detected in 2018 [50] and vaccine use was discontinued. The 1990.1 virus displayed a lower virus shedding profile and a lower number of contact ferret seroconversions by 12 dpc compared to the 1990.4.A and 1990.4.B.2 groups. Despite differences in genome constellation, all three clades from the 1990s (1990.1, 1990.4.A, and 1990.4.B.2) contained the M segment from the H1N1pdm09 lineage, which has previously been associated with high transmission efficiency in a guinea pig model [51]. Notably, variant 1990.4.A H3N2 viruses detected in 2011–2012 contained the M from the pandemic and the rest of the internal genes from the TRIG lineage [52]. PB2 and PA segments are important components of the viral polymerase complex and essential for viral replication. The viruses from genetic clades represented in this study, 1990.1, 1990.4.A, and 1990.4.B.2, contained residues encoded in the PB2 gene previously described to enhance replication in humans: 271A, 590S, 591R, and 661A, as well as the residue 669V encoded in the PA gene [53–56]. These genetic markers support the proposition that these swine viruses represent an increased zoonotic risk.

Taken together, our study indicated that the swine H3 1990.1, 1990.4.A, and 1990.4.B.2 clades are a zoonotic risk as they displayed reduced cross-reactivity with ferret antisera raised to human seasonal vaccines and/or CVVs and transmitted from pig-to-ferrets. Our findings suggest that existing CVVs should be updated to reflect contemporary swine IAV diversity; specifically, we suggest that the 1990.4.A clade that is widespread in U.S. swine populations and that has previously caused a significant number of H3 variant human cases requires revision. In addition, our study suggests that minor swine IAV clades that are regularly detected in US swine populations have zoonotic potential and should be considered in risk assessments of swine H3N2 IAV for pandemic preparedness strategies.

Supplementary Materials: The following supporting information can be downloaded at: https://www.mdpi.com/article/10.3390/v15020331/s1, Table S1: Amino acid differences between human seasonal vaccine A/Iowa/60/2018 and the 1990.1 clade represented by A/swine/Missouri/A02257614/2018 strain. Annotations include mutations within putative epitopes and receptor binding sites; Table S2: Amino acid differences between A/Minnesota/11/2010 CVV and the 1990.4.A is the clade represented by A/swine/North Carolina/A02245294/2019 and the 1990.4.B.2 is the clade represented by A/swine/Illinois/A02479007/2020. Annotations include mutations within putative epitopes and receptor binding sites; Table S3: Amino acid differences between A/Ohio/28/2016 CVV and the 2010.1 clade represented by the A/Indiana/27/2018 variant strain. Annotations include mutations within putative epitopes and receptor binding sites; Table S4: Amino acid differences between A/Ohio/28/2016 CVV and A/Iowa/60/2018 HuVac. Annotations include mutations within the putative epitope and receptor binding site; Table S5: Percentage of macroscopic lung lesions and microscopic lung and trachea scores in pigs; Table S6: Pig HI data at 0, 5, and 14 dpi; Figure S1: Detection proportions of swine H3N2 collected in 2020 in the USDA influenza A virus in swine surveillance system.

Author Contributions: Conceptualization, C.K.S., J.B.K., T.K.A. and A.L.V.B.; methodology, C.K.S., J.B.K., T.K.A., N.S.L. and A.L.V.B.; formal analysis, C.K.S., T.K.A., D.E.H., Z.W.A. and P.C.G.; investigation, J.B.K., C.K.S., K.M.Y. and A.L.V.B.; resources, S.T. and C.T.D.; writing—original draft preparation, C.K.S.; writing—review and editing, J.B.K., T.K.A., P.C.G. and A.L.V.B.; supervision, T.K.A. and A.L.V.B.; funding acquisition, T.K.A. and A.L.V.B. All authors have read and agreed to the published version of the manuscript.

Funding: This work was supported in part by the U.S. Department of Agriculture (USDA) Agricultural Research Service (ARS project number 5030-32000-231-000-D); the National Institute of Allergy and Infectious Diseases, National Institutes of Health, Department of Health and Human Services (contract numbers HHSN272201400008C and 75N93021C00015); the Centers for Disease Control and Prevention (contract number 21FED2100395IPD); the USDA Agricultural Research Service Research Participation Program of the Oak Ridge Institute for Science and Education (ORISE) through an interagency agreement between the U.S. Department of Energy (DOE) and the USDA Agricultural Research Service (contract number DE-AC05-06OR23100); and the SCINet project of the USDA Agricultural Research Service (ARS project number 0500-00093-001-00-D). The funders had no role in study design, data collection and interpretation, or the decision to submit the work for publication. Mention of trade names or commercial products in this article is solely for the purpose of providing specific information and does not imply recommendation or endorsement by the USDA, CDC, DOE, or ORISE. The USDA and CDC are equal opportunity providers and employers. The findings and conclusions in this report are those of the authors and do not necessarily represent the views of the Centers for Disease Control and Prevention or the Agency for Toxic Substances and Disease Registry.

Institutional Review Board Statement: The study was approved by the USDA-ARS National Animal Disease Center Institutional Animal Care and Use Committee (protocol codes ARS-2017-660 and ARS-2019-779, and dates of approval were 13 November 2017 and 9 March 2019).

Informed Consent Statement: Not applicable.

Data Availability Statement: **The** clinical data associated with this study are available for download from the USDA Ag Data Commons at https://doi.org/10.15482/USDA.ADC/1528327 (Accessed on 7 December 2022) and the phylogenetic analyses are available at https://github.com/flu-crew/datasets (Accessed on 7 December 2022).

Acknowledgments: We thank Nick Otis for laboratory assistance and NADC Animal Resources Unit caretaker staff for assistance with animal studies. We gratefully acknowledge pork producers, swine veterinarians, and laboratories for participating in the USDA Influenza A Virus in Swine Surveillance System and publicly sharing sequences. We also gratefully acknowledge all data contributors, i.e., the authors and their originating laboratories responsible for obtaining the specimens, and their submitting laboratories responsible for generating the genetic sequence and metadata and sharing via the GISAID Initiative, on which components of this research are based.

Conflicts of Interest: The authors declare that they have no competing interests.

References

1. Jester, B.J.; Uyeki, T.M.; Jernigan, D.B. Fifty Years of Influenza A(H3N2) Following the Pandemic of 1968. *Am. J. Public Health* **2020**, *110*, 669–676. [CrossRef] [PubMed]
2. Long, J.S.; Mistry, B.; Haslam, S.M.; Barclay, W.S. Publisher Correction: Host and viral determinants of influenza A virus species specificity. *Nat. Rev. Microbiol.* **2019**, *17*, 124, Correction in *Nat. Rev. Microbiol.* **2019**, *17*, 67–81. [CrossRef]
3. Garten, R.J.; Davis, C.T.; Russell, C.A.; Shu, B.; Lindstrom, S.; Balish, A.; Sessions, W.M.; Xu, X.; Skepner, E.; Deyde, V.; et al. Antigenic and Genetic Characteristics of Swine-Origin 2009 A(H1N1) Influenza Viruses Circulating in Humans. *Science* **2009**, *325*, 197–201. [CrossRef]
4. Olsen, C.W.; Karasin, A.I.; Carman, S.; Li, Y.; Bastien, N.; Ojkic, D.; Alves, D.; Charbonneau, G.; Henning, B.M.; Low, D.E.; et al. Triple Reassortant H3N2 Influenza A Viruses, Canada, 2005. *Emerg. Infect. Dis.* **2006**, *12*, 1132–1135. [CrossRef]
5. Rajao, D.S.; Vincent, A.L.; Perez, D. Adaptation of Human Influenza Viruses to Swine. *Front. Veter- Sci.* **2018**, *5*, 347. [CrossRef]
6. Zhou, N.N.; Senne, D.A.; Landgraf, J.S.; Swenson, S.L.; Erickson, G.; Rossow, K.; Liu, L.; Yoon, K.; Krauss, S.; Webster, R.G. Genetic reassortment of avian, swine, and human influenza A viruses in American pigs. *J. Virol.* **1999**, *73*, 8851–8856. [CrossRef] [PubMed]
7. Webby, R.J.; Swenson, S.L.; Krauss, S.L.; Gerrish, P.J.; Goyal, S.M.; Webster, R.G. Evolution of Swine H3N2 Influenza Viruses in the United States. *J. Virol.* **2000**, *74*, 8243–8251. [CrossRef] [PubMed]
8. Nelson, M.I.; Vincent, A.L.; Kitikoon, P.; Holmes, E.C.; Gramer, M.R. Evolution of Novel Reassortant A/H3N2 Influenza Viruses in North American Swine and Humans, 2009–2011. *J. Virol.* **2012**, *86*, 8872–8878. [CrossRef]
9. Vincent, A.L.; Ma, W.; Lager, K.M.; Janke, B.H.; Richt, J.A. Swine influenza viruses a North American perspective. *Adv. Virus Res.* **2008**, *72*, 127–154.
10. Anderson, T.K.; Chang, J.; Arendsee, Z.W.; Venkatesh, D.; Souza, C.K.; Kimble, J.B.; Lewis, N.S.; Davis, C.T.; Vincent, A.L. Swine Influenza A Viruses and the Tangled Relationship with Humans. *Cold Spring Harb. Perspect. Med.* **2020**, *11*, a038737. [CrossRef] [PubMed]
11. Bolton, M.J.; Abente, E.J.; Venkatesh, D.; Stratton, J.A.; Zeller, M.; Anderson, T.K.; Lewis, N.S.; Vincent, A.L. Antigenic evolution of H3N2 influenza A viruses in swine in the United States from 2012 to 2016. *Influ. Other Respir. Viruses* **2018**, *13*, 83–90. [CrossRef]
12. Lewis, N.S.; Anderson, T.K.; Kitikoon, P.; Skepner, E.; Burke, D.F.; Vincent, A.L. Substitutions near the Hemagglutinin Receptor-Binding Site Determine the Antigenic Evolution of Influenza A H3N2 Viruses in U.S. Swine. *J. Virol.* **2014**, *88*, 4752–4763. [CrossRef]
13. Rajão, D.S.; Gauger, P.C.; Anderson, T.K.; Lewis, N.S.; Abente, E.J.; Killian, M.L.; Perez, D.R.; Sutton, T.C.; Zhang, J.; Vincent, A.L. Novel Reassortant Human-Like H3N2 and H3N1 Influenza A Viruses Detected in Pigs Are Virulent and Antigenically Distinct from Swine Viruses Endemic to the United States. *J. Virol.* **2015**, *89*, 11213–11222. [CrossRef] [PubMed]
14. Sharma, A.; Zeller, M.A.; Souza, C.K.; Anderson, T.K.; Vincent, A.L.; Harmon, K.; Li, G.; Zhang, J.; Gauger, P.C. Characterization of a 2016–2017 Human Seasonal H3 Influenza A Virus Spillover Now Endemic to U.S. Swine. *Msphere* **2022**, *7*, e00809-21. [CrossRef] [PubMed]
15. Zeller, M.; Li, G.; Harmon, K.M.; Zhang, J.; Vincent, A.L.; Anderson, T.K.; Gauger, P.C. Complete Genome Sequences of Two Novel Human-Like H3N2 Influenza A Viruses, A/swine/Oklahoma/65980/2017 (H3N2) and A/Swine/Oklahoma/65260/2017 (H3N2), Detected in Swine in the United States. *Genome Announc.* **2018**, *7*, e01203-18. [CrossRef]
16. Robertson, J.S.; Nicolson, C.; Harvey, R.; Johnson, R.; Major, D.; Guilfoyle, K.; Roseby, S.; Newman, R.; Collin, R.; Wallis, C.; et al. The development of vaccine viruses against pandemic A(H1N1) influenza. *Vaccine* **2011**, *29*, 1836–1843. [CrossRef]
17. Burke, S.A.; Trock, S.C. Use of Influenza Risk Assessment Tool for Prepandemic Preparedness. *Emerg. Infect. Dis.* **2018**, *24*, 471–477. [CrossRef]
18. Souza, C.K.; Anderson, T.K.; Chang, J.; Venkatesh, D.; Lewis, N.S.; Pekosz, A.; Shaw-Saliba, K.; Rothman, R.E.; Chen, K.-F.; Vincent, A.L. Antigenic Distance between North American Swine and Human Seasonal H3N2 Influenza A Viruses as an Indication of Zoonotic Risk to Humans. *J. Virol.* **2022**, *96*, e01374-21. [CrossRef]
19. Shu, Y.; McCauley, J. GISAID: Global initiative on sharing all influenza data—From vision to reality. *Euro Surveill* **2017**, *22*, 30494.

20. Katoh, K.; Standley, D.M. MAFFT Multiple Sequence Alignment Software Version 7: Improvements in Performance and Usability. *Mol. Biol. Evol.* **2013**, *30*, 772–780. [CrossRef]
21. Hoang, D.T.; Chernomor, O.; Von Haeseler, A.; Minh, B.Q.; Vinh, L.S. UFBoot2: Improving the Ultrafast Bootstrap Approximation. *Mol. Biol. Evol.* **2018**, *35*, 518–522. [CrossRef] [PubMed]
22. Minh, B.Q.; Schmidt, H.A.; Chernomor, O.; Schrempf, D.; Woodhams, M.D.; von Haeseler, A.; Lanfear, R. IQ-TREE 2: New Models and Efficient Methods for Phylogenetic Inference in the Genomic Era. *Mol. Biol. Evol.* **2020**, *37*, 1530–1534. [CrossRef]
23. Arendsee, Z.W.; Baker, A.L.V.; Anderson, T.K. smot: A python package and CLI tool for contextual phylogenetic subsampling. *J. Open Source Softw.* **2022**, *7*, 4193. [CrossRef]
24. Chang, J.; Anderson, T.K.; Zeller, M.; Gauger, P.C.; Vincent, A.L. octoFLU: Automated Classification for the Evolutionary Origin of Influenza A Virus Gene Sequences Detected in U.S. Swine. *Microbiol. Resour. Announc.* **2019**, *8*, e00673-19. [CrossRef] [PubMed]
25. Kaplan, B.S.; Kimble, J.B.; Chang, J.; Anderson, T.K.; Gauger, P.C.; Janas-Martindale, A.; Killian, M.L.; Bowman, A.S.; Vincent, A.L. Aerosol Transmission from Infected Swine to Ferrets of an H3N2 Virus Collected from an Agricultural Fair and Associated with Human Variant Infections. *J. Virol.* **2020**, *94*, e01009-20. [CrossRef] [PubMed]
26. Kitikoon, P.; Gauger, P.C.; Vincent, A.L. Hemagglutinin Inhibition Assay with Swine Sera. *Methods Mol. Biol.* **2014**, *1161*, 295–301. [CrossRef] [PubMed]
27. Reed, L.J.; Muench, H. A simple method of estimating fifty per cent endpoints. *Am. J. Epidemiol.* **1938**, *27*, 493–497. [CrossRef]
28. Gauger, P.C.; Vincent, A.L.; Loving, C.L.; Henningson, J.N.; Lager, K.M.; Janke, B.H.; Kehrli, M.E., Jr.; Roth, J.A. Kinetics of Lung Lesion Development and Pro-Inflammatory Cytokine Response in Pigs with Vaccine-Associated Enhanced Respiratory Disease Induced by Challenge with Pandemic (2009) A/H1N1 Influenza Virus. *Vet. Pathol.* **2012**, *49*, 900–912. [CrossRef]
29. Kitikoon, P.; Nilubol, D.; Erickson, B.J.; Janke, B.H.; Hoover, T.C.; Sornsen, S.A.; Thacker, E.L. The immune response and maternal antibody interference to a heterologous H1N1 swine influenza virus infection following vaccination. *Vet. Immunol. Immunopathol.* **2006**, *112*, 117–128. [CrossRef]
30. Vincent, L.L.; Janke, B.H.; Paul, P.S.; Halbur, P.G. A Monoclonal-Antibody-Based Immunohistochemical Method for the Detection of Swine Influenza Virus in Formalin-Fixed, Paraffin-Embedded Tissues. *J. Vet. Diagn. Investig.* **1997**, *9*, 191–195. [CrossRef]
31. Opriessnig, T.; Yu, S.; Gallup, J.M.; Evans, R.B.; Fenaux, M.; Pallares, F.; Thacker, E.L.; Brockus, C.W.; Ackermann, M.R.; Thomas, P.; et al. Effect of Vaccination with Selective Bacterins on Conventional Pigs Infected with Type 2 Porcine Circovirus. *Vet. Pathol.* **2003**, *40*, 521–529. [CrossRef] [PubMed]
32. Centers for Disease Control and Prevention (CDC) Reported Infections with Variant Influenza Viruses in the United States. Available online: https://www.cdc.gov/flu/swineflu/variant-cases-us.htm (accessed on 3 January 2020).
33. Henritzi, D.; Petric, P.P.; Lewis, N.S.; Graaf, A.; Pessia, A.; Starick, E.; Breithaupt, A.; Strebelow, G.; Luttermann, C.; Parker, L.M.K.; et al. Surveillance of European Domestic Pig Populations Identifies an Emerging Reservoir of Potentially Zoonotic Swine Influenza A Viruses. *Cell Host Microbe* **2020**, *28*, 614–627.e6. [CrossRef] [PubMed]
34. Kessler, S.; Harder, T.C.; SchwemmLe, M.; Ciminski, K. Influenza A Viruses and Zoonotic Events-Are We Creating Our Own Reservoirs? *Viruses* **2021**, *13*, 2250. [PubMed]
35. Nelson, M.I.; Worobey, M. Origins of the 1918 Pandemic: Revisiting the Swine "Mixing Vessel" Hypothesis. *Am. J. Epidemiol.* **2018**, *187*, 2498–2502. [CrossRef]
36. Short, K.R.; Richard, M.; Verhagen, J.H.; van Riel, D.; Schrauwen, E.J.; Brand, J.M.V.D.; Mänz, B.; Bodewes, R.; Herfst, S. One health, multiple challenges: The inter-species transmission of influenza A virus. *One Health* **2015**, *1*, 1–13. [CrossRef]
37. Novel Swine-Origin Influenza A (H1N1) Virus Investigation Team. Emergence of a Novel Swine-Origin Influenza A (H1N1) Virus in Humans. *N. Engl. J. Med.* **2009**, *360*, 2605–2615. [CrossRef]
38. Neveau, M.N.; Zeller, M.A.; Kaplan, B.S.; Souza, C.K.; Gauger, P.C.; Vincent, A.L.; Anderson, T.K. Genetic and Antigenic Characterization of an Expanding H3 Influenza A Virus Clade in U.S. Swine Visualized by Nextstrain. *Msphere* **2022**, *7*, e00994-21. [CrossRef]
39. Zeller, M.; Anderson, T.K.; Walia, R.W.; Vincent, A.L.; Gauger, P.C. ISU FLUture: A veterinary diagnostic laboratory web-based platform to monitor the temporal genetic patterns of Influenza A virus in swine. *BMC Bioinform.* **2018**, *19*, 1–10. [CrossRef]
40. Kimble, J.B.; Souza, C.K.; Anderson, T.K.; Arendsee, Z.W.; Hufnagel, D.E.; Young, K.M.; Lewis, N.S.; Davis, C.T.; Thor, S.; Baker, A.L.V. Interspecies Transmission from Pigs to Ferrets of Antigenically Distinct Swine H1 Influenza A Viruses with Reduced Reactivity to Candidate Vaccine Virus Antisera as Measures of Relative Zoonotic Risk. *Viruses* **2022**, *14*, 2398. [CrossRef]
41. DiPiazza, A.T.; Richards, K.A.; Liu, W.-C.; Albrecht, R.A.; Sant, A.J. Analyses of Cellular Immune Responses in Ferrets following Influenza Virus Infection. *Methods Mol. Biol.* **2018**, *1836*, 513–530. [CrossRef] [PubMed]
42. Pearce, M.B.; Jayaraman, A.; Pappas, C.; Belser, J.A.; Zeng, H.; Gustin, K.M.; Maines, T.R.; Sun, X.; Raman, R.; Cox, N.J.; et al. Pathogenesis and transmission of swine origin A(H3N2)v influenza viruses in ferrets. *Proc. Natl. Acad. Sci. USA* **2012**, *109*, 3944–3949. [CrossRef]
43. Neumann, G.; Kawaoka, Y. Host range restriction and pathogenicity in the context of influenza pandemic. *Emerg. Infect. Dis.* **2006**, *12*, 881–886. [CrossRef] [PubMed]
44. Koel, B.F.; Burke, D.F.; Bestebroer, T.M.; van der Vliet, S.; Zondag, G.C.M.; Vervaet, G.; Skepner, E.; Lewis, N.S.; Spronken, M.I.J.; Russell, C.A.; et al. Substitutions Near the Receptor Binding Site Determine Major Antigenic Change during Influenza Virus Evolution. *Science* **2013**, *342*, 976–979. [CrossRef] [PubMed]

45. Burke, D.F.; Smith, D.J. A Recommended Numbering Scheme for Influenza A HA Subtypes. *PLoS ONE* **2014**, *9*, e112302. [CrossRef] [PubMed]
46. Ma, J.; Shen, H.; Liu, Q.; Bawa, B.; Qi, W.; Duff, M.; Lang, Y.; Lee, J.; Yu, H.; Bai, J.; et al. Pathogenicity and Transmissibility of Novel Reassortant H3N2 Influenza Viruses with 2009 Pandemic H1N1 Genes in Pigs. *J. Virol.* **2015**, *89*, 2831–2841. [CrossRef]
47. Rajão, D.S.; Walia, R.R.; Campbell, B.; Gauger, P.C.; Janas-Martindale, A.; Killian, M.L.; Vincent, A.L. Reassortment between Swine H3N2 and 2009 Pandemic H1N1 in the United States Resulted in Influenza A Viruses with Diverse Genetic Constellations with Variable Virulence in Pigs. *J. Virol.* **2017**, *91*, e01763-16. [CrossRef]
48. Genzow, M.; Goodell, C.; Kaiser, T.J.; Johnson, W.; Eichmeyer, M. Live attenuated influenza virus vaccine reduces virus shedding of newborn piglets in the presence of maternal antibody. *Influ. Other Respir. Viruses* **2017**, *12*, 353–359. [CrossRef]
49. Kaiser, T.J.; Smiley, R.A.; Fergen, B.; Eichmeyer, M.; Genzow, M. Influenza A virus shedding reduction observed at 12 weeks post-vaccination when newborn pigs are administered live-attenuated influenza virus vaccine. *Influ. Other Respir. Viruses* **2019**, *13*, 274–278. [CrossRef]
50. Sharma, A.; Zeller, M.; Li, G.; Harmon, K.M.; Zhang, J.; Hoang, H.; Anderson, T.K.; Vincent, A.L.; Gauger, P.C. Detection of live attenuated influenza vaccine virus and evidence of reassortment in the U.S. swine population. *J. Vet. Diagn. Investig.* **2020**, *32*, 301–311. [CrossRef]
51. Chou, Y.-Y.; Albrecht, R.A.; Pica, N.; Lowen, A.C.; Richt, J.A.; García-Sastre, A.; Palese, P.; Hai, R. The M Segment of the 2009 New Pandemic H1N1 Influenza Virus Is Critical for Its High Transmission Efficiency in the Guinea Pig Model. *J. Virol.* **2011**, *85*, 11235–11241. [CrossRef]
52. Epperson, S.; Jhung, M.; Richards, S.; Quinlisk, P.; Ball, L.; Moll, M.; Boulton, R.; Haddy, L.; Biggerstaff, M.; Brammer, L.; et al. Human Infections with Influenza A(H3N2) Variant Virus in the United States, 2011–2012. *Clin. Infect. Dis.* **2013**, *57*, S4–S11. [CrossRef] [PubMed]
53. Arai, Y.; Kawashita, N.; Daidoji, T.; Ibrahim, M.S.; Elgendy, E.; Takagi, T.; Takahashi, K.; Suzuki, Y.; Ikuta, K.; Nakaya, T.; et al. Novel Polymerase Gene Mutations for Human Adaptation in Clinical Isolates of Avian H5N1 Influenza Viruses. *PLoS Pathog.* **2016**, *12*, e1005583. [CrossRef] [PubMed]
54. Hayashi, T.; Wills, S.; Bussey, K.A.; Takimoto, T. Identification of Influenza A Virus PB2 Residues Involved in Enhanced Polymerase Activity and Virus Growth in Mammalian Cells at Low Temperatures. *J. Virol.* **2015**, *89*, 8042–8049. [CrossRef] [PubMed]
55. Liu, Q.; Qiao, C.; Marjuki, H.; Bawa, B.; Ma, J.; Guillossou, S.; Webby, R.J.; Richt, J.A.; Ma, W. Combination of PB2 271A and SR polymorphism at positions 590/591 is critical for viral replication and virulence of swine influenza virus in cultured cells and in vivo. *J. Virol.* **2012**, *86*, 1233–1237. [CrossRef] [PubMed]
56. Sun, X.; Pulit-Penaloza, J.A.; Belser, J.A.; Pappas, C.; Pearce, M.B.; Brock, N.; Zeng, H.; Creager, H.M.; Zanders, N.; Jang, Y.; et al. Pathogenesis and Transmission of Genetically Diverse Swine-Origin H3N2 Variant Influenza A Viruses from Multiple Lineages Isolated in the United States, 2011–2016. *J. Virol.* **2018**, *92*, e00665-18. [CrossRef]

Article

SARS-CoV-2 Omicron (B.1.1.529) Infection of Wild White-Tailed Deer in New York City

Kurt J. Vandegrift [1,2,*], Michele Yon [3], Meera Surendran Nair [3,4], Abhinay Gontu [3,4], Santhamani Ramasamy [3,4], Saranya Amirthalingam [3,4], Sabarinath Neerukonda [5], Ruth H. Nissly [3,4], Shubhada K. Chothe [3,4], Padmaja Jakka [3,4], Lindsey LaBella [3,4], Nicole Levine [6], Sophie Rodriguez [6], Chen Chen [7], Veda Sheersh Boorla [7], Tod Stuber [8], Jason R. Boulanger [9], Nathan Kotschwar [9], Sarah Grimké Aucoin [10], Richard Simon [10], Katrina L. Toal [10], Randall J. Olsen [11,12], James J. Davis [13,14], Dashzeveg Bold [15], Natasha N. Gaudreault [15], Krishani Dinali Perera [15], Yunjeong Kim [15], Kyeong-Ok Chang [15], Costas D. Maranas [7], Juergen A. Richt [15], James M. Musser [11,12], Peter J. Hudson [1,2], Vivek Kapur [2,6,*] and Suresh V. Kuchipudi [2,3,4,*]

[1] Department of Biology, The Pennsylvania State University, University Park, PA 16802, USA
[2] The Center for Infectious Disease Dynamics, Huck Institutes of the Life Sciences, The Pennsylvania State University, University Park, PA 16802, USA
[3] Animal Diagnostic Laboratory, Department of Veterinary and Biomedical Sciences, The Pennsylvania State University, University Park, PA 16802, USA
[4] Department of Veterinary and Biomedical Sciences, The Pennsylvania State University, University Park, PA 16802, USA
[5] United States Department of Health and Human Services, Silver Spring, MD 20993, USA
[6] Department of Animal Science, The Pennsylvania State University, University Park, PA 16802, USA
[7] Department of Chemical Engineering, The Pennsylvania State University, University Park, PA 16802, USA
[8] National Veterinary Services Laboratories, Veterinary Services, U.S. Department of Agriculture, Ames, IA 50010, USA
[9] White Buffalo, Inc., Chester, CT 06412, USA
[10] City of New York Parks & Recreation, New York, NY 10029, USA
[11] Laboratory of Molecular and Translational Human Infectious Disease Research, Center for Infectious Diseases, Department of Pathology and Genomic Medicine, Houston Methodist Research Institute and Houston Methodist Hospital, Houston, TX 77030, USA
[12] Departments of Pathology and Laboratory Medicine and Microbiology and Immunology, Weill Cornell Medical College, New York, NY 10021, USA
[13] Consortium for Advanced Science and Engineering, University of Chicago, Chicago, IL 60637, USA
[14] Division of Data Science and Learning, Argonne National Laboratory, Argonne, IL 60439, USA
[15] Department of Diagnostic Medicine/Pathobiology, Kansas State University, Manhattan, KS 66506, USA
[*] Correspondence: kjv1@psu.edu (K.J.V.); vkapur@psu.edu (V.K.); skuchipudi@psu.edu (S.V.K.); Tel.: +1-814-574-9852 (K.J.V.); +1-814-865-9788 (V.K.); +1-814-863-4436 (S.V.K.)

Citation: Vandegrift, K.J.; Yon, M.; Surendran Nair, M.; Gontu, A.; Ramasamy, S.; Amirthalingam, S.; Neerukonda, S.; Nissly, R.H.; Chothe, S.K.; Jakka, P.; et al. SARS-CoV-2 Omicron (B.1.1.529) Infection of Wild White-Tailed Deer in New York City. *Viruses* **2022**, *14*, 2770. https://doi.org/10.3390/v14122770

Academic Editors: Myriam Ermonval, Serge Morand and Qiang Ding

Received: 7 October 2022
Accepted: 1 December 2022
Published: 12 December 2022

Publisher's Note: MDPI stays neutral with regard to jurisdictional claims in published maps and institutional affiliations.

Abstract: There is mounting evidence of SARS-CoV-2 spillover from humans into many domestic, companion, and wild animal species. Research indicates that humans have infected white-tailed deer, and that deer-to-deer transmission has occurred, indicating that deer could be a wildlife reservoir and a source of novel SARS-CoV-2 variants. We examined the hypothesis that the Omicron variant is actively and asymptomatically infecting the free-ranging deer of New York City. Between December 2021 and February 2022, 155 deer on Staten Island, New York, were anesthetized and examined for gross abnormalities and illnesses. Paired nasopharyngeal swabs and blood samples were collected and analyzed for the presence of SARS-CoV-2 RNA and antibodies. Of 135 serum samples, 19 (14.1%) indicated SARS-CoV-2 exposure, and 11 reacted most strongly to the wild-type B.1 lineage. Of the 71 swabs, 8 were positive for SARS-CoV-2 RNA (4 Omicron and 4 Delta). Two of the animals had active infections and robust neutralizing antibodies, revealing evidence of reinfection or early seroconversion in deer. Variants of concern continue to circulate among and may reinfect US deer populations, and establish enzootic transmission cycles in the wild: this warrants a coordinated One Health response, to proactively surveil, identify, and curtail variants of concern before they can spill back into humans.

Keywords: SARS-CoV-2; omicron; white-tailed deer; reinfection; reservoir competence; variant of concern; *Odocoileus virginianus*; spillover; zoonotic; disease ecology; enzootic transmission

1. Introduction

Recent investigations have established that severe acute respiratory syndrome coronavirus-2 (SARS-CoV-2) infects a wide range of non-human animal hosts, including farmed mink, companion animals (e.g., cats, dogs, ferrets), and zoo animals (e.g., tigers, lions, cougars, snow leopards, gorillas, otters, and hippopotami) [1–5]. White-tailed deer (*Odocoileus virginianus*) are highly susceptible to SARS-CoV-2, as evidenced by experimental infection studies [6,7] and documentation of widespread natural infections in Iowa [8], Ohio [9], Pennsylvania [10], and several other States in the US, including New York [11]. Most recently, SARS-CoV-2 infections in deer from Canada have been confirmed, with evidence for long-term evolution within deer, and spillback to humans, further heightening concerns about the potential of deer to serve as a reservoir of SARS-CoV-2 [12].

Recent SARS-CoV-2 variants, such as Delta and Omicron, are more highly transmissible between humans than those previously described [13,14]; indeed, the effective reproduction number of Omicron in humans is estimated to be nearly threefold greater than the Delta variant [15]. While there are reports of Delta variant spillover into multiple animal hosts—including cats, dogs, pumas, and lions in a zoo in South Africa [16], as well as white-tailed deer across many US states, and Syrian hamsters in pet shops in Hong Kong [17]—spillover of the Omicron variant to non-human animal species has not yet been documented. There is a lack of clarity on the origins of Omicron, with competing hypotheses including emergence from a chronically infected human host, silent spread in a cryptic human population, or emergence from a yet-unknown non-human animal population [18].

While the widespread SARS-CoV-2 spillover infection of white-tailed deer across North America [8–12,19] has raised the possibility that deer could serve as a SARS-CoV-2 reservoir, there are several unanswered questions: firstly, whether or not Omicron has spilled over to free-living deer; secondly, while experimental studies show that deer infected with SARS-CoV-2 develop neutralizing antibodies [6,7], an open question remains as to whether deer, like humans, can be reinfected with SARS-CoV-2, even in the presence of neutralizing antibodies; thirdly, while white-tailed deer that were experimentally infected with the SARS-CoV-2 wild-type or the Alpha variant remained largely asymptomatic [6,7], it remains unclear whether free-ranging deer infected with the more recent SARS-CoV-2 variants (e.g, Delta or Omicron) exhibit discernible clinical signs. To address these three questions, we collected and tested nasal swabs and serum samples, and we provided a clinical examination of free-ranging white-tailed deer in Staten Island, New York, between December 2021 and February 2022.

Here, we report SARS-CoV-2 Delta and Omicron variant spillover infection in the white-tailed deer population inhabiting Staten Island, which is a borough of New York City. To our knowledge, this is the first report of Omicron infection in a wildlife species. The results show that a majority of the SARS-CoV-2 seropositive deer in Staten Island exhibited strong serum reactivity to the wild-type SARS-CoV-2 B.1 lineage or the Alpha variant. Furthermore, robust levels of neutralizing antibodies were found in two deer that were positive for Delta variant RNA in their nasal passages, suggesting that deer, like humans [20], may be reinfected with SARS-CoV-2 or exhibit early seroconversion from ongoing Delta infections. Taken together, our studies suggest that white-tailed deer are emerging as a wildlife reservoir of SARS-CoV-2, and this highlights an urgent need to better assess the spillback risks and the evolutionary trajectories of SARS-CoV-2 variants in non-human animal reservoirs.

2. Materials and Methods

2.1. Samples

White-tailed deer were opportunistically darted and anesthetized by a veterinarian during an ongoing deer sterilization program implemented by the City of New York Parks & Recreation (NYC Parks). Once darted, the animals were tracked and processed at a nearby sampling site. The anesthetized animals were given ear tags, and were sampled and released. The GPS location of the sampling site for each individual was recorded, and the animal's sex and age were determined. All captured deer were examined by a qualified veterinarian for gross abnormalities or illnesses; the examination included a standard temperature, pulse, and respiration reading and pulse oximeter readings using the lingual technique. Patient monitoring continued during the surgical process, and discontinued when the patient was again ambulatory. Blood samples were collected from the jugular vein, into serum separator tubes, and the serum was frozen at $-20\,^{\circ}$C until testing. Nasal swabs were collected by inserting Copan floQ swabs (Copan Diagnostics Inc.) into the nostril and touching the sides of the nasal wall. The swab was rotated for 30 s in each nostril, and was placed directly in Universal Transport Media (Copan Diagnostics Inc). The samples were submitted to the Penn State Animal Diagnostic Laboratory (ADL) for diagnostic testing, and were subsequently used in this study.

2.2. Surrogate Virus Neutralization Test (sVNT)

The serum samples were screened, using a surrogate virus neutralization test (sVNT) assay that had previously been validated for detecting SARS-CoV-2 antibodies in deer [21]. The sVNT assay uses cPass™ technology (Genscript), and detects total neutralizing antibodies measured as percent inhibition [22]. Animals with inhibition above 30% are considered positive.

2.3. Generation of SARS-CoV-2 S Pseudotyped Viruses

Pseudoviruses were produced, using a third-generation human immunodeficiency virus packaging system, as previously described [23]. Three plasmids—the transfer plasmid encoding luciferase and ZsGreen (BEI Resources Cat no: NR-52516), the helper plasmid encoding Gag/pol (BEI Resources Cat no: NR-52517), and the spike encoding plasmid of variants described in the study—were co-transfected in HEK 293T cells propagated in DMEM with 10% FBS, and maintained at 37 °C. Pseudovirus-containing supernatants were collected after 48 h, filtered through 0.45 μM low-protein binding filters, and were aliquoted and stored at $-80\,^{\circ}$C until further use.

2.4. Neutralization Assay of Deer Sera against Pseudotyped Virus Expressing the Spike Protein of SARS-CoV-2 Wild-Type or Variants

The deer serum samples that tested positive in the sVNT assay were examined for neutralizing activity with a SARS-CoV-2 pseudovirus neutralization assay (pVNT), using pseudotyped viruses that carried the S protein—which represented the wild-type SARS-CoV-2 B.1 lineage, Alpha, Beta, Gamma, Delta, or Omicron variants—to test the relative neutralizing titers against the SARS-CoV-2 variants. Pseudovirus neutralization assays were performed, using HEK 293T cells expressing ACE2, and TMPRSS2 cells (293T ACE2/TMPRSS2; BEI Resources Cat no: NR-55293), as described previously. Briefly, the pseudoviruses were incubated with threefold serial dilutions of sera for an hour at 37 °C. The pseudovirus/sera mixtures were subsequently inoculated into 96-well plates seeded with 3.0×10^4 293T ACE2/TMPRSS2 cells/well, a day before the assay. The residual pseudovirus infectivity was determined 48 h later, by quantifying the luciferase activity. The percentage neutralization was calculated, upon normalization to a virus-only control. Each serum was run in duplicate, in two independent experiments against each pseudovirus, to determine the 50% neutralization titer (NT_{50}). The curves were fitted using a nonlinear regression curve, for which GraphPad Prism Software version 6 (San Diego, CA, USA) was employed; connected scatterplots were made using R software (R version 4.1.3).

2.5. Antigen Cartography

The antigen cartography was created using the NT_{50} measurements of the serum samples against the SARS-CoV-2 B.1 lineage, Alpha, Beta, Gamma, Delta, and Omicron variants in the pseudovirus neutralization assays. The distance between the SARS-CoV-2 (B.1 lineage and variants) and serum samples was calculated, and antigenic maps were generated, as described previously, using antigen cartography software accessed on 30 May 2022 (https://acmacs-web.antigenic-cartography.org/) [24,25]. The SARS-CoV-2 B.1 lineage, Alpha, Beta, Gamma, Delta, and Omicron variants, and the serum samples were positioned on a two-dimensional (2D) antigen map, based on the distances calculated by the antigen cartography algorithm, as described by Smith et al. [24]. The confidence area of the positions of the SARS-CoV-2 B.1 lineage, Alpha, Beta, Gamma, Delta, and Omicron variants, and the serum samples, were indicated as blobs, as estimated with stress parameter 0.1 [25]. The distance in antigenic units between the SARS-CoV-2 and serum samples was plotted, using GraphPad software version 9.0.0 (San Diego, CA, USA). A statistical analysis was performed, using a two-tailed unpaired Student's t test with Welch's correction. A *p* value of <0.05 indicated that the mean distance between the SARS-CoV-2 (B.1 lineage and variants) and the serum samples was significant.

2.6. RNA Extraction and RT-PCR for SARS-CoV-2 Detection

The swab samples were processed, and real-time RT-PCR was undertaken, following the standardized protocols for SARSR-CoV-2 detection in animal samples at Penn State's ADL. RNA was extracted from 400 µL of swab samples, using a KingFisher Flex machine (ThermoFisher Scientific, Waltham, MA, USA) and a MagMAX Viral/Pathogen extraction kit (ThermoFisher Scientific, Waltham, MA, USA), following the manufacturer's instructions. The presence of SARS-CoV-2 viral RNA was further tested, using the OPTI Medical SARS-CoV-2 RT-PCR kit, which is a highly sensitive assay that targets the N gene [26,27]. The RT-PCR assays were carried out on an ABI 7500 Fast instrument (ThermoFisher Scientific, Waltham, MA, USA). The internal control RNase P was utilized, to confirm that the samples were not contaminated with human tissue or fluids during harvesting or processing. The samples were also tested using a TaqPath kit (ThermoFisher Scientific, Waltham, MA, USA), which targets the SARS-CoV-2 ORF1ab, N gene, and S gene [27,28] as the first screen for Omicron, which typically presents as an S gene drop out in these assays [28].

2.7. SARS-CoV-2 Genome Sequencing

The total RNA extracted from the swab samples was used for the whole genome sequencing of the SARS-CoV-2, as previously described [8,29–32], and the sequencing libraries were prepared according to version 4.1 of the ARTIC nCoV-2019 protocol (https://artic.network/ncov-2019, accessed on 15 February 2022). We used a semi-automated workflow, which employed BioMek i7 liquid-handling workstations (Beckman Coulter Life Sciences) and MANTIS automated liquid handlers (FORMULATRIX). Using a NovaSeq 6000 instrument (Illumina), we generated short sequence reads, to ensure a very high depth of coverage. The sequencing libraries were prepared in duplicate, and were sequenced with an SP 300 cycle reagent kit.

2.8. SARS-CoV-2 Genome Sequence Analysis and Identification of Variants

The viral genomes were assembled, using the BV-BRC SARS-CoV-2 assembly service [32,33], which uses a pipeline that is similar to the One Codex SARS-CoV-2 variant-calling pipeline [34]. Briefly, the pipeline uses seqtk version 1.3-r116 for sequence trimming [35], minimap version 2.1 [36] for aligning the reads against the reference genome Wuhan-Hu-1 NC_045512.2 [36,37], samtools version 1.11 [35] for sequence and file manipulation [38], and iVar version 1.2.2 [39] for primer trimming and variant calling [40]. To increase stringency, the minimum read depth for the assemblies (based on samtools mpileup) was set at three, to determine consensus. Genetic lineages, variants being monitored, and variants were identified and designated by Pangolin version 3.1.11, with the pangoLEARN module 2021-08-024, using the previously described genome sequence analysis pipelines [29,32,41]. Single Nucleotide Polymorphisms (SNPs) were identified, using the vSNP (https://github.com/USDA-VS/vSNP, accessed on 6 March 2022) SNP analysis program.

2.9. Data Analysis and Visualization

QGIS mapping software version 3.16.10 was used, to visually portray the geographic location of the white-tailed deer that were sampled [38].

3. Results

3.1. Molecular and Genetic Identification of SARS-CoV-2 Delta and Omicron Variants in Nasal and Tonsillar Swabs from White-Tailed Deer on Staten Island, New York

Our results show that 8 out of 71 (11.3%, 95% CI: 0.0–0.20) white-tailed deer tested positive for SARS-CoV-2 RNA (Figure 1a). To determine the identity and genetic relatedness of the circulating strains, we applied whole-genome sequencing, using a recently described pipeline [8]. Our analysis confirmed that four of the SARS-CoV-2 PCR positive samples were the Delta variant, and that the other four positive samples were the Omicron variant. Notably, the Omicron variant was the dominant circulating lineage (90%) amongst humans in New York City during the period of the deer sampling, in late 2021 and early 2022 [42]. To our knowledge, this is the first report of the Omicron variant of SARS-CoV-2 infecting white-tailed deer or any other free-living wildlife.

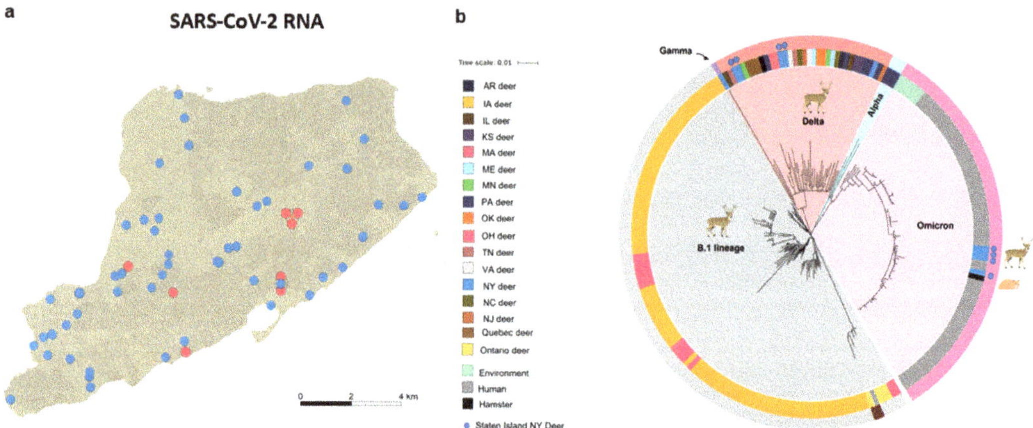

Figure 1. Distribution and whole-genome single nucleotide polymorphism (SNP)-based phylogenies of SARS-CoV-2 recovered from white-tailed deer on Staten Island, New York: (**a**) the spatial distribution of the collection sites of nasal and tonsillar swabs from white-tailed deer that were tested for the presence of SARS-CoV-2 viral RNA; red circles show sites where swabs were positive for SARS-CoV-2 viral RNA, and blue-filled circles show swabs that were negative; (**b**) whole-genome sequences of eight newly characterized white-tailed-deer-origin SARS-CoV-2 genomes were analyzed in the context of 135 publicly available white-tailed-deer-origin SARS-CoV-2 isolates, and 63 arbitrarily selected SARS-CoV-2 Omicron genomes circulating amongst humans in New York City during this same time period, as well as representative isolates from the environment or from Syrian hamsters (SI, Table S2); the genome sequences were screened for quality, for SNP positions called against the SARS-CoV-2 reference genome (NC_045512), and for SNP alignments used to generate a maximum-likelihood phylogenetic tree, using RAxML.

We performed whole-genome-sequence-based phylogenetic analyses of these newly identified Omicron and Delta sequences, with the vSNP pipeline [43] (Figure 1b). The analysis showed that the Omicron sequences recovered from the deer were clustered closely with recently reported Omicron sequences from humans in New York City, as well as with those reported from environmental sources elsewhere, but were quite distinct from the previously described isolates recovered from deer in Iowa, Ohio [8,9], 14 other US states, and Canada, from which sequences had been deposited in GISAID [11] (Figure 1b and Table S2). The Delta sequences were clustered closely with other deer Delta sequences that had been recently reported in multiple regions in North America (Figure 1b).

3.2. Delta and Omicron Infections Occur among Staten Island Deer despite Serological Evidence of Prior SARS-CoV-2 Exposure

It was not clear whether deer previously exposed to SARS-CoV-2 might be reinfected, and whether there was continued SARS-CoV-2 spillover infection of deer in urban settings after the emergence of the Omicron variant. To examine whether white-tailed deer on Staten Island had previously been exposed to SARS-CoV-2, we collected serum samples from 135 individual deer, between 12 December 2021 and 2 February 2022, and examined them for the presence of anti-SARS-CoV-2 neutralizing antibodies, using a surrogate virus neutralization assay (sVNT) [21] (Figure 2a). Due to the sampling design targeting males, most of the serum samples were from males (n = 119; 88.1%), with an age distribution skewed toward younger age classes. Eighty-six fawns constituted 63.7% of the sample, and 33 yearling deer made up 24.4%, while 16 of the 135 individuals (11.9%) were considered adults. Our sVNT results showed that 19 of the 135 (14.1% 95% CI: 0.0–0.20) serum samples were positive for SARS-CoV-2 exposure. Viral inhibition in the positive samples ranged from 33.2% to 97.0%, with a median value of 70.9% (Figure 2b and Table S1). The proportion of positive animals was comparable to the findings of Chandler et al., who identified 9 out of 29 (31%; 95% CI 17%–49%) white-tailed deer, from two other New York counties, that were seropositive to SARS-CoV-2 in 2021 [21].

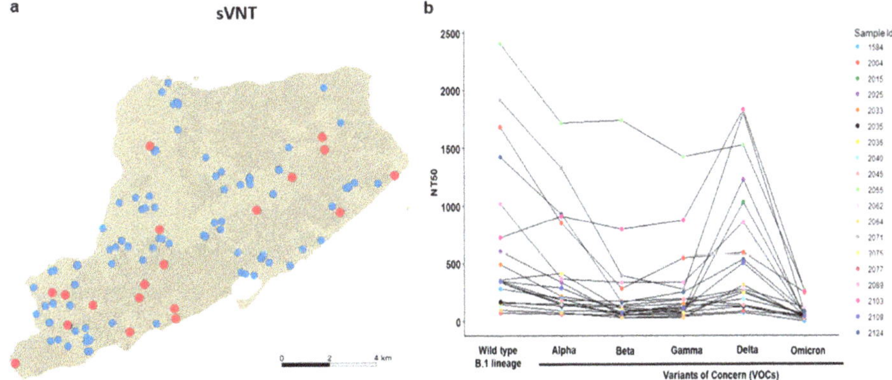

Figure 2. SARS-CoV-2 serological reactivity status of white-tailed deer on Staten Island, New York, between 12 December 2021 and 30 January 2022: (**a**) spatial distribution of sites of collection of serum samples from white-tailed deer for assessment of serological reactivity to, and neutralization of, SARS-CoV-2; red circles indicate positive detection, and blue-filled circles indicate seronegative status; (**b**) SARS-CoV-2 pseudovirus neutralization assay (pVNT) with pseudotyped viruses equipped with the spike proteins of the wild-type SARS-CoV-2 B.1 lineage, or that of the Alpha, Beta, Gamma, Delta or Omicron variants. Of the sVNT positive samples, 11 out of 19 showed stronger reactivity to the spike of the wild-type SARS-CoV-2 B.1 lineage than to any of the variants. Each serum was run in duplicate in two independent experiments against each pseudovirus, to determine the 50% neutralization titer (NT$_{50}$). Connecting lines indicate serum from the same individual.

To determine which SARS-CoV-2 variant the deer were likely exposed to previously, we utilized a SARS-CoV-2 pseudovirus neutralization assay (pVNT) with six pseudotyped viruses equipped with the spike proteins of the wild-type SARS-CoV-2 B.1 lineage or that of the Alpha, Beta, Gamma, Delta or Omicron variants. In the pVNT assay, 11 out of the 19 (58%) sVNT positive samples showed stronger reactivity to the spike of the wild-type SARS-CoV-2 B.1 lineage than to any of the variants (Figure 2b), and one of the serum samples had comparable reactivity to all the variants except Omicron. Serum samples from seven of the deer (37%) reacted most strongly to the spike of Delta, while one deer reacted most strongly to the Alpha variant (Table S1). Notably, none of the positive samples showed strong reactivity to the Omicron variant, compared with the wild-type B.1 lineage or other variants.

To investigate the antigenic relationship between the SARS-CoV-2 strains, we constructed antigen cartography, using the NT_{50} of serum samples determined by pVNT on the SARS-CoV-2 B.1 lineage, and on the Alpha, Beta, Gamma, Delta, and Omicron variants. The 2D antigenic map revealed clustering of 16 out of 19 sVNT positive samples to the B.1 lineage (antigenic units 0.26 to 0.89) and the Alpha variant (antigenic units 0.45 to 1.15), whereas 14 out of the 19 samples clustered with the Delta variant (antigenic unit < 1) (Figure 3 and Table S3). Most of the serum samples which showed strong reactivity to the B.1 lineage in pVNT also had significant reactivity to the Alpha and Delta variants, and vice versa. This indicated the cross-neutralizing potential of the serum samples between the B.1 lineage, Alpha, and Delta variants, as they were close to the B.1 lineage, Alpha, and Delta variants in the antigenic map. The mean antigenic distances between the SARS-CoV-2 Gamma, Beta, and Omicron variants and the serum samples were 1.79 ± 0.74, 2.02 ± 0.74, and 3.2 ± 0.7, respectively, and they were significantly higher ($p < 0.001$) than the antigenic distance to the B.1 lineage (Figure 3b). Both the antigen map and the antigenic units indicated that the SARS-CoV-2 Omicron variant was distantly related to other strains (Figure 3 and Table S3), as reported earlier [44]. As we used the sVNT with B.1 lineage RBD for the initial screening of the deer sera, we may have missed positive samples for Omicron antibodies.

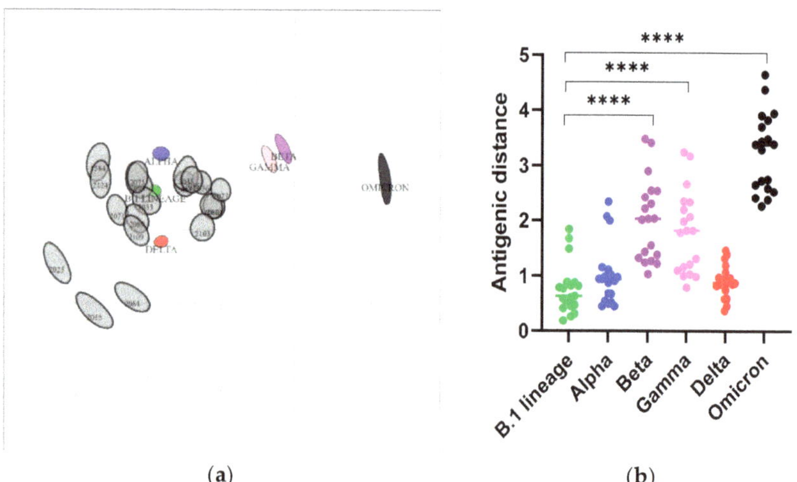

(**a**) (**b**)

Figure 3. Construction of antigen cartography based on the NT_{50} of serum samples collected from deer. (**a**) Two-dimensional antigenic map of the SARS-CoV-2 B.1 lineage, the Alpha, Beta, Gamma, Delta, and Omicron variants, and the serum samples collected from the deer. The SARS-CoV-2 strains and serum samples are shown as blobs, as estimated using stress parameter 0.1. The SARS-CoV-2 B.1 lineage, and the Alpha, Beta, Gamma, Delta, and Omicron variants are indicated in green, blue, purple, pink, red, and black, respectively, and each gray blob corresponds to serum from one deer. Both the axes of the map are antigenic distant, and each grid square represents 1 antigenic unit, which is a three-fold serum dilution (two antigenic units correspond to nine-fold serum dilution, and so on) in the pseudovirus neutralization assay. The distance between points is a measure of antigenic similarity, with closer positions indicating higher antigenic similarity. (**b**) The distances of antigenic units between SARS-CoV-2 and the serum samples. The mean distance between the serum samples and the B.1 lineage, Alpha, and Delta variants was low, compared to the Beta, Gamma, and Omicron variants. p value < 0.05 are significant, **** p < 0.0001.

3.3. Nasal Shedding of SARS-CoV-2 RNA in White-Tailed Deer with High Levels of Neutralizing Antibodies without Clinical Signs

Experimental SARS-CoV-2 infection in deer rapidly elicits a neutralizing antibody immune response [6,7], but it is not yet clear whether these antibodies prevent reinfection. While direct evidence for reinfection of experimentally or naturally SARS-CoV-2-infected white-tailed deer is lacking, our field-based studies show that two RT-PCR positive deer had relatively robust levels of neutralizing antibodies (79% and 94% inhibition, Table S1). The pVNT assay of the serum sample from deer 2089, that was found to be infected with the Delta variant, showed relatively greater reactivity to the S protein of the wild-type B1 (50% more neutralizing titer (NT_{50} 1020) than Delta (NT_{50} 858) (Table S1 and Figure 2b)). Deer 2103 showed relatively higher reactivity to the Delta S protein (NT_{50} 1829) than to the B.1 S protein (NT_{50} 726). It is possible that either one or both individuals seroconverted during previous infection by an earlier SARS-CoV-2 lineage, and were reinfected with the Delta variant, a scenario that has also been observed in humans [20,45]. Alternatively, it is possible that these young individuals with high Ct values may have rapidly seroconverted in response to SARS-CoV-2 Delta infection, even while continuing to shed viral RNA in nasal secretions, as is observed in some SARS-CoV-2-positive humans [46] and in experimental infection of deer [6,7]. However, the small sample size and the modest differences in the pVNT titers do not conclusively support either reinfection or rapid seroconversion. Therefore, these competing hypotheses need to be rigorously tested, through experimental challenge studies and intensive field investigations that should include longitudinal sampling of individual animals and deer herds. It is noteworthy, however, that most of the deer in this study that were shedding viral RNA showed no detectable neutralizing antibodies. All four Omicron RNA-positive deer, and two of the four Delta RNA-positive individuals, tested negative for serum antibodies. The ability of SARS-CoV-2 variants to overcome neutralizing antibody responses to an earlier lineage/variant in deer is important, as it may imply that deer with antibodies to one variant could be susceptible to and transmit other variants. Such a scenario has the potential for a substantial impact on the overall transmission dynamics of SARS-CoV-2 in free-ranging deer, as well as their establishment as long-term reservoir hosts.

Previous studies have suggested that white-tailed deer remain largely asymptomatic following experimental SARS-CoV-2 infection [6,7]. Clinical examination of the deer in this study showed no gross abnormalities, including the eight deer who tested positive for Delta or Omicron RNA: as the exact day of infection cannot be known, it is also possible that these deer were past the disease stage of infection. However, if SARS-CoV-2 continues to cause low-pathogenicity infections in wild deer, then it could persist, unnoticed, within the deer population, which would facilitate the occurrence of viral evolution underneath our current surveillance radar.

4. Discussion

Multiple human-to-animal spillovers, and documentation of subsequent transmission among mink and deer, highlight the "generalist" and rapidly evolving nature of SARS-CoV-2 as a pathogen of mammalian hosts, as recently described [47]. Emerging evidence suggests adaption of SARS-CoV-2 in animal hosts: for example, six specific mutations in mink (NSP9_G37E, Spike_F486L, Spike_N501T, Spike_Y453F, ORF3a_T229I, and ORF3a_L219V), and one in deer (NSP3a_L1035F), have been identified [47]. Continued replication of SARS-CoV-2 in multiple animal hosts, and variation in selection pressure, could hasten viral evolution, and increase the likelihood of a novel strain emergence [48]. Notably, a recent manuscript reported a significant divergence between two deer-derived SARS-CoV-2 Alpha variant genomes that were also divergent compared to the human Alpha variant genomes from the same region, suggesting rapid viral evolution during deer-to-deer transmission [10,48]. Additionally, highly divergent SARS-CoV-2 sequences with 76 mutations, suggestive of host adaptation under neutral selection, have been identified in white-tailed deer in Canada [12]. Based on the identification of this highly divergent SARS-CoV-2, the authors also reported the first suspected white-tailed deer-to-human transmission [12]. Taken together, there is growing evidence that SARS-CoV-2 is likely to become established within deer populations, and that deer are likely to contribute to the emergence of novel variants of SARS-CoV-2. Considering the large population size and the potential for human–deer contact in certain environments, spillback of new strains of SARS-CoV-2 to humans and/or spread to other susceptible hosts is of genuine concern: hence, longitudinal studies to monitor SARS-CoV-2 adaptation and evolution within free-living deer and other susceptible non-human animal populations are urgently needed.

Several wild and synanthropic species—including raccoons, deer mice, and skunks—are susceptible to SARS-CoV-2 infection, and share ecological space with deer [49–51]; however, widespread SARS-CoV-2 spillovers into wild animals other than white-tailed deer (and perhaps mink) are currently unknown. Understanding the role of wildlife as potential reservoirs of new variants is essential, in order to assess the risk of future outbreaks in humans—particularly when the reservoir is peridomestic and abundant in urban areas, as is the case with white-tailed deer. While SARS-CoV-2 infections in white-tailed deer have been reported in 22 states and provinces in North America [11,12,19], a recent study reported that serological screening of roe, red, and fallow deer in Germany and Austria found no evidence of SARS-CoV-2 exposure in those animals, suggesting that differences in host susceptibility and ecological factors likely contribute to the potential for establishment of cervids as reservoir hosts [52]; however, the extent of SARS-CoV-2 infections in captive or free-living cervids (and indeed most other animal hosts) across the world remains poorly understood, and warrants investigation.

Examination of the virus receptor angiotensin-converting enzyme-2 receptor indicates that many mammalian families, including cervids and mustelids [3,47,53,54], are susceptible to SARS-CoV-2 infection, and that these models should be the focus of future sampling efforts. Viral infection of a wildlife host species does not necessitate that the host species sustain the virus within its population as a reservoir, as this depends on the ecological context, i.e., the complete ecosystem within which the species lives and interacts ecologically and epidemiologically [55]; therefore, a significant first challenge is to determine whether the temperate woodland community of North America (including white-tailed deer and other species) can sustain SARS-CoV-2 in the absence of continued spillover from humans. It is also possible that a sustained wildlife reservoir for SARS-CoV-2 could be replicated in, or generalizable to, other ecological communities around the world. Our studies highlight the urgent need for stratified surveillance of at-risk wild animal species, coupled with seasonal dynamical modeling, to predict the risk of SARS-CoV-2 spillback to humans.

5. Conclusions

Natural infections of wild white-tailed deer by the Delta and Omicron variants of SARS-CoV-2, in a herd previously exposed to SARS-CoV-2, emphasizes the role of white-tailed deer as a potential reservoir species. Establishing an animal reservoir might facilitate the continued circulation of SARS-CoV-2, independent of circulation in humans; in addition, deer might transmit the infection to other susceptible wild animals—such as rodents, foxes, and raccoons—resulting in the establishment of SARS-CoV-2 enzootic transmission cycles: such a scenario might result in virus adaptation, and the emergence of novel variants that might escape the protection of current SARS-CoV-2 vaccines. Therefore, these results are highly significant, and warrant the continued monitoring of white-tailed deer and other at-risk animal hosts for SARS-CoV-2 infections.

Supplementary Materials: The following supporting information can be downloaded at: https://www.mdpi.com/article/10.3390/v14122770/s1, Table S1: Metadata associated with the deer samples collected from Staten Island, NY; Table S2: Metadata associated with the phylogenetic tree presented in Figure 1B.

Author Contributions: Conceptualization: K.J.V., S.V.K. and J.R.B.; Methodology: S.V.K., K.J.V., M.S.N., J.A.R., S.R. (Santhamani Ramasamy), S.N., V.K., C.D.M., J.M.M., R.J.O., J.J.D., R.H.N., J.R.B. and M.Y.; Investigation: K.J.V., S.V.K., J.R.B., M.Y. N.K., S.R. (Santhamani Ramasamy), S.N., M.S.N., A.G., S.A., R.H.N., L.L., S.K.C., N.L., R.S., C.C., V.S.B., C.D.M., T.S., R.J.O., J.J.D., D.B., N.N.G., K.D.P., Y.K., K.-O.C., J.A.R., J.M.M. and P.J.; Visualization: S.R. (Santhamani Ramasamy), V.K., S.R. M.S.N. and S.A.; Funding acquisition: S.V.K., K.J.V., J.M.M., R.J.O., J.J.D. and J.A.R.; Project administration: S.V.K., K.J.V., V.K., J.R.B., N.K., S.G.A., R.S., K.L.T. and R.H.N.; Supervision: S.V.K., K.J.V., V.K., J.M.M., J.A.R. and M.S.N.; Writing–original draft: K.J.V., V.K., P.J.H. and S.V.K.; Writing—review & editing: All authors. All authors have read and agreed to the published version of the manuscript.

Funding: USDA National Institute of Food and Agriculture (NIFA) Award 2020-67015-32175 (S.V.K., M.S.N., K.J.V.); NSF Ecology and Evolution of Infectious Diseases program grant 1619072 (K.J.V.); National Institute of Allergy and Infectious Diseases, National Institutes of Health grants 1R21AI156406-01 & 7R01AI134911 (K.J.V.). Houston Methodist Academic Institute Infectious Diseases Fund (J.M.M., R.J.O.). National Institute of Allergy and Infectious Diseases, National Institutes of Health, Department of Health and Human Services, Contract No. 75N93019C00076 (J.J.D.). USDA NIFA Program under award 2020-67015-33157 (J.A.R.). National Institute of General Medical Sciences (NIGMS) under award number P20GM130448 (J.A.R.). NIAID Centers of Excellence for Influenza Research and Surveillance under contract number HHSN 272201400006C (J.A.R.).

Institutional Review Board Statement: Deer blood and swab samples were submitted to the Penn State Animal Diagnostic Laboratory (ADL) for diagnostic testing, and were subsequently used in this study.

Informed Consent Statement: Not applicable.

Data Availability Statement: All data used in this study are provided in Supplemental Tables S1 and S2.

Acknowledgments: We thank Abirami Ravichandran and Nishitha Kodali for help with PCR and sample processing. We thank Matthew Ojeda Saavedra, Sindy Pena, Kristina Reppond, Madison N. Shyer, Jessica Cambric, Ryan Gadd, Rashi M. Thakur, Akanksha Batajoo, and Regan Mangham for genome sequencing. Samples used in this study came from deer handled under authorization provided through license issued by New York State Department of Environmental Conservation. We are grateful to Suelee Robbe-Austerman, Kristina Lantz, Mia Kim Torchetti, and Julianna Lenoch of the USDA APHIS for their invaluable assistance and wise counsel during the investigation. We thank Gleyder Roman-Sosa for the recombinant S and N antigen preparation. This study was made possible by the collective efforts of White Buffalo, Inc., the City of New York Parks & Recreation, and their field staff.

Conflicts of Interest: The Richt laboratory received support from Tonix Pharmaceuticals, Xing Technologies, and Zoetis, outside of the reported work. J.A.R. is an inventor of patents and patent applications on the use of antivirals and vaccines for the treatment and prevention of virus infections, owned by Kansas State University. J.R.B. and N.K. are employees of White Buffalo, Inc. a nonprofit conservation organization contracted by City of New York Parks & Recreation to implement the field studies from which the samples for the current investigation were derived. All other authors declare no financial or employment conflict of interest.

References

1. Damas, J.; Hughes, G.M.; Keough, K.C.; Painter, C.A.; Persky, N.S.; Corbo, M.; Hiller, M.; Koepfli, K.P.; Pfenning, A.R.; Zhao, H.; et al. Broad host range of SARS-CoV-2 predicted by comparative and structural analysis of ACE2 in vertebrates. *Proc. Natl. Acad. Sci. USA* **2020**, *117*, 22311–22322. [CrossRef]

2.	Hobbs, E.C.; Reid, T.J. Animals and SARS-CoV-2: Species susceptibility and viral transmission in experimental and natural conditions, and the potential implications for community transmission. *Transbound. Emerg. Dis.* **2021**, *68*, 1850–1867. [CrossRef] [PubMed]

3.	Meekins, D.A.; Gaudreault, N.N.; Richt, J.A. Natural and Experimental SARS-CoV-2 Infection in Domestic and Wild Animals. *Viruses* **2021**, *13*, 1993. [CrossRef] [PubMed]

4.	Oreshkova, N.; Molenaar, R.J.; Vreman, S.; Harders, F.; Oude Munnink, B.B.; Hakze-van der Honing, R.W.; Gerhards, N.; Tolsma, P.; Bouwstra, R.; Sikkema, R.S.; et al. SARS-CoV-2 infection in farmed minks, the Netherlands, April and May 2020. *Eurosurveillance* **2020**, *25*, 210325c. [CrossRef] [PubMed]

5.	Weiss, S.R.; Navas-Martin, S. Coronavirus pathogenesis and the emerging pathogen severe acute respiratory syndrome coronavirus. *Microbiol. Mol. Biol. Rev.* **2005**, *69*, 635–664. [CrossRef]

6.	Cool, K.; Gaudreault, N.N.; Morozov, I.; Trujillo, J.D.; Meekins, D.A.; McDowell, C.; Carossino, M.; Bold, D.; Mitzel, D.; Kwon, T.; et al. Infection and transmission of ancestral SARS-CoV-2 and its alpha variant in pregnant white-tailed deer. *Emerg. Microbes Infect.* **2022**, *11*, 95–112. [CrossRef]

7.	Palmer, M.V.; Martins, M.; Falkenberg, S.; Buckley, A.; Caserta, L.C.; Mitchell, P.K.; Cassmann, E.D.; Rollins, A.; Zylich, N.C.; Renshaw, R.W.; et al. Susceptibility of white-tailed deer (*Odocoileus virginianus*) to SARS-CoV-2. *J. Virol.* **2021**, *95*, e00083-21. [CrossRef]

8.	Kuchipudi, S.V.; Surendran-Nair, M.; Ruden, R.M.; Yon, M.; Nissly, R.H.; Vandegrift, K.J.; Nelli, R.K.; Li, L.; Jayarao, B.M.; Maranas, C.D.; et al. Multiple spillovers from humans and onward transmission of SARS-CoV-2 in white-tailed deer. *Proc. Natl. Acad. Sci. USA* **2022**, *119*, e2121644119. [CrossRef]

9.	Hale, V.L.; Dennis, P.M.; McBride, D.S.; Nolting, J.M.; Madden, C.; Huey, D.; Ehrlich, M.; Grieser, J.; Winston, J.; Lombardi, D.; et al. SARS-CoV-2 infection in free-ranging white-tailed deer. *Nature* **2022**, *602*, 481–486. [CrossRef]

10.	Marques, A.D.; Sherrill-Mix, S.; Everett, J.K.; Adhikari, H.; Reddy, S.; Ellis, J.C.; Zeliff, H.; Greening, S.S.; Cannuscio, C.C.; Strelau, K.M. Multiple Introductions of SARS-CoV-2 Alpha and Delta Variants into White-Tailed Deer in Pennsylvania. *MBio* **2022**, *13*, e0210122. [CrossRef]

11.	USDA Confirmed Cases of SARS-CoV-2 in Animals in the United States. Available online: https://www.aphis.usda.gov/aphis/dashboards/tableau/sars-dashboard (accessed on 1 February 2022).

12.	Pickering, B.; Lung, O.; Maguire, F.; Kruczkiewicz, P.; Kotwa, J.D.; Buchanan, T.; Gagnier, M.; Guthrie, J.L.; Jardine, C.M.; Marchand-Austin, A.; et al. Divergent SARS-CoV-2 variant emerges in white-tailed deer with deer-to-human transmission. *Nat. Microbiol.* **2022**, *7*, 2011–2024. [CrossRef] [PubMed]

13.	Farinholt, T.; Doddapaneni, H.; Qin, X.; Menon, V.; Meng, Q.; Metcalf, G.; Chao, H.; Gingras, M.C.; Avadhanula, V.; Farinholt, P.; et al. Transmission event of SARS-CoV-2 delta variant reveals multiple vaccine breakthrough infections. *BMC Med.* **2021**, *19*, 255. [CrossRef] [PubMed]

14.	Riediker, M.; Briceno-Ayala, L.; Ichihara, G.; Albani, D.; Poffet, D.; Tsai, D.H.; Iff, S.; Monn, C. Higher viral load and infectivity increase risk of aerosol transmission for Delta and Omicron variants of SARS-CoV-2. *Swiss Med. Wkly.* **2022**, *152*, w30133. [PubMed]

15.	Ito, K.; Piantham, C.; Nishiura, H. Relative instantaneous reproduction number of Omicron SARS-CoV-2 variant with respect to the Delta variant in Denmark. *J. Med. Virol.* **2021**, *94*, 2265–2268. [CrossRef] [PubMed]

16.	Koeppel, K.N.; Mendes, A.; Strydom, A.; Rotherham, L.; Mulumba, M.; Venter, M. SARS-CoV-2 Reverse Zoonoses to Pumas and Lions, South Africa. *Viruses* **2022**, *14*, 120. [CrossRef]

17.	Yen, H.-L.; Sit, T.H.C.; Brackman, C.J.; Chuk, S.S.Y.; Gu, H.; Tam, K.W.S.; Law, P.Y.T.; Leung, G.M.; Peiris, M.; Poon, L.L.M.; et al. Transmission of SARS-CoV-2 delta variant (AY.127) from pet hamsters to humans, leading to onward human-to-human transmission: A case study. *Lancet* **2022**, *399*, 1070–1078. [CrossRef] [PubMed]

18.	Mallapaty, S. Where did Omicron come from? Three key theories. *Nature* **2022**, *602*, 26–28. [CrossRef]

19.	Kotwa, J.D.; Massé, A.; Gagnier, M.; Aftanas, P.; Blais-Savoie, J.; Bowman, J.; Buchanan, T.; Chee, H.-Y.; Dibernardo, A.; Kruczkiewicz, P.; et al. First detection of SARS-CoV-2 infection in Canadian wildlife identified in free-ranging white-tailed deer (*Odocoileus virginianus*) from southern Québec, Canada. *bioRxiv* **2022**, 476458. [CrossRef]

20.	Zhang, J.; Ding, N.; Ren, L.; Song, R.; Chen, D.; Zhao, X.; Chen, B.; Han, J.; Li, J.; Song, Y.; et al. COVID-19 reinfection in the presence of neutralizing antibodies. *Natl. Sci. Rev.* **2021**, *8*, nwab006. [CrossRef]

21.	Chandler, J.C.; Bevins, S.N.; Ellis, J.W.; Linder, T.J.; Tell, R.M.; Jenkins-Moore, M.; Root, J.J.; Lenoch, J.B.; Robbe-Austerman, S.; DeLiberto, T.J.; et al. SARS-CoV-2 exposure in wild white-tailed deer (*Odocoileus virginianus*). *Proc. Natl. Acad. Sci. USA* **2021**, *118*, e2114828118. [CrossRef]

22.	Tan, C.W.; Chia, W.N.; Qin, X.; Liu, P.; Chen, M.I.; Tiu, C.; Hu, Z.; Chen, V.C.; Young, B.E.; Sia, W.R.; et al. A SARS-CoV-2 surrogate virus neutralization test based on antibody-mediated blockage of ACE2-spike protein-protein interaction. *Nat. Biotechnol.* **2020**, *38*, 1073–1078. [CrossRef] [PubMed]

23.	Neerukonda, S.N.; Vassell, R.; Herrup, R.; Liu, S.; Wang, T.; Takeda, K.; Yang, Y.; Lin, T.L.; Wang, W.; Weiss, C.D. Establishment of a well-characterized SARS-CoV-2 lentiviral pseudovirus neutralization assay using 293T cells with stable expression of ACE2 and TMPRSS2. *PLoS ONE* **2021**, *16*, e0248348. [CrossRef] [PubMed]

24.	Smith, D.J.; Lapedes, A.S.; de Jong, J.C.; Bestebroer, T.M.; Rimmelzwaan, G.F.; Osterhaus, A.D.; Fouchier, R.A. Mapping the antigenic and genetic evolution of influenza virus. *Science* **2004**, *305*, 371–376. [CrossRef]

25. Kendra, J.A.; Tohma, K.; Ford-Siltz, L.A.; Lepore, C.J.; Parra, G.I. Antigenic cartography reveals complexities of genetic determinants that lead to antigenic differences among pandemic GII.4 noroviruses. *Proc. Natl. Acad. Sci. USA* **2021**, *118*, e2015874118. [CrossRef] [PubMed]

26. Smithgall, M.C.; Dowlatshahi, M.; Spitalnik, S.L.; Hod, E.A.; Rai, A.J. Types of Assays for SARS-CoV-2 Testing: A Review. *Lab. Med.* **2020**, *51*, e59–e65. [CrossRef] [PubMed]

27. Verma, N.; Patel, D.; Pandya, A. Emerging diagnostic tools for detection of COVID-19 and perspective. *Biomed. Microdevices* **2020**, *22*, 83. [CrossRef] [PubMed]

28. Lee, C.K.; Tham, J.W.M.; Png, S.; Chai, C.N.; Ng, S.C.; Tan, E.J.M.; Ng, L.J.; Chua, R.P.; Sani, M.; Seow, Y.; et al. Clinical performance of Roche cobas 6800, Luminex ARIES, MiRXES Fortitude Kit 2.1, Altona RealStar, and Applied Biosystems TaqPath for SARS-CoV-2 detection in nasopharyngeal swabs. *J. Med. Virol.* **2021**, *93*, 4603–4607. [CrossRef]

29. Long, S.W.; Olsen, R.J.; Christensen, P.A.; Subedi, S.; Olson, R.; Davis, J.J.; Saavedra, M.O.; Yerramilli, P.; Pruitt, L.; Reppond, K.; et al. Sequence Analysis of 20,453 Severe Acute Respiratory Syndrome Coronavirus 2 Genomes from the Houston Metropolitan Area Identifies the Emergence and Widespread Distribution of Multiple Isolates of All Major Variants of Concern. *Am. J. Pathol.* **2021**, *191*, 983–992. [CrossRef]

30. Christensen, P.A.; Olsen, R.J.; Long, S.W.; Subedi, S.; Davis, J.J.; Hodjat, P.; Walley, D.R.; Kinskey, J.C.; Ojeda Saavedra, M.; Pruitt, L.; et al. Delta Variants of SARS-CoV-2 Cause Significantly Increased Vaccine Breakthrough COVID-19 Cases in Houston, Texas. *Am. J. Pathol.* **2022**, *192*, 320–331. [CrossRef]

31. O'Toole, A.; Scher, E.; Underwood, A.; Jackson, B.; Hill, V.; McCrone, J.T.; Colquhoun, R.; Ruis, C.; Abu-Dahab, K.; Taylor, B.; et al. Assignment of epidemiological lineages in an emerging pandemic using the pangolin tool. *Virus Evol.* **2021**, *7*, veab064. [CrossRef]

32. Olsen, R.J.; Christensen, P.A.; Long, S.W.; Subedi, S.; Hodjat, P.; Olson, R.; Nguyen, M.; Davis, J.J.; Yerramilli, P.; Saavedra, M.O.; et al. Trajectory of Growth of Severe Acute Respiratory Syndrome Coronavirus 2 (SARS-CoV-2) Variants in Houston, Texas, January through May 2021, Based on 12,476 Genome Sequences. *Am. J. Pathol.* **2021**, *191*, 1754–1773. [CrossRef]

33. Davis, J.J.; Wattam, A.R.; Aziz, R.K.; Brettin, T.; Butler, R.; Butler, R.M.; Chlenski, P.; Conrad, N.; Dickerman, A.; Dietrich, E.M.; et al. The PATRIC Bioinformatics Resource Center: Expanding data and analysis capabilities. *Nucleic Acids Res.* **2020**, *48*, D606–D612. [CrossRef] [PubMed]

34. He, C. SARS-CoV-2 Variant Calling and Consensus Assembly Pipeline. Available online: https://github.com/onecodex/sars-cov-2.git (accessed on 25 February 2022).

35. Li, H. Toolkit for Processing Sequences in FASTA/Q Formats. Available online: https://github.com/lh3/seqtk (accessed on 25 February 2022).

36. Li, H. Minimap2: Pairwise alignment for nucleotide sequences. *Bioinformatics* **2018**, *34*, 3094–3100. [CrossRef]

37. Wu, F.; Zhao, S.; Yu, B.; Chen, Y.-M.; Wang, W.; Song, Z.-G.; Hu, Y.; Tao, Z.-W.; Tian, J.-H.; Pei, Y.-Y. A new coronavirus associated with human respiratory disease in China. *Nature* **2020**, *579*, 265–269. [CrossRef] [PubMed]

38. Li, H.; Handsaker, B.; Wysoker, A.; Fennell, T.; Ruan, J.; Homer, N.; Marth, G.; Abecasis, G.; Durbin, R. The sequence alignment/map format and SAMtools. *Bioinformatics* **2009**, *25*, 2078–2079. [CrossRef] [PubMed]

39. Andersen, K.G. iVar Version 1.2.2. Available online: https://github.com/andersen-lab/ivar/releases (accessed on 25 February 2021).

40. Grubaugh, N.D.; Gangavarapu, K.; Quick, J.; Matteson, N.L.; De Jesus, J.G.; Main, B.J.; Tan, A.L.; Paul, L.M.; Brackney, D.E.; Grewal, S.; et al. An amplicon-based sequencing framework for accurately measuring intrahost virus diversity using PrimalSeq and iVar. *Genome Biol.* **2019**, *20*, 8. [CrossRef] [PubMed]

41. Long, S.W.; Olsen, R.J.; Christensen, P.A.; Bernard, D.W.; Davis, J.J.; Shukla, M.; Nguyen, M.; Saavedra, M.O.; Yerramilli, P.; Pruitt, L.; et al. Molecular Architecture of Early Dissemination and Massive Second Wave of the SARS-CoV-2 Virus in a Major Metropolitan Area. *MBio* **2020**, *11*, e0270720. [CrossRef]

42. NYCHealth Omicron Variant: NYC Report for 13 January 2022. Available online: https://www1.nyc.gov/assets/doh/downloads/pdf/covid/omicron-variant-report-jan-13--22.pdf (accessed on 1 February 2022).

43. Letunic, I.; Bork, P. Interactive Tree Of Life (iTOL): An online tool for phylogenetic tree display and annotation. *Bioinformatics* **2007**, *23*, 127–128. [CrossRef]

44. Van der Straten, K.; Guerra, D.; van Gils, M.; Bontjer, I.; Caniels, T.G.; van Willigen, H.D.; Wynberg, E.; Poniman, M.; Burger, J.A.; Bouhuijs, J.H.; et al. Mapping the antigenic diversification of SARS-CoV-2. *medRxiv* **2022**. [CrossRef]

45. Sheehan, M.M.; Reddy, A.J.; Rothberg, M.B. Reinfection Rates among Patients Who Previously Tested Positive for Coronavirus Disease 2019: A Retrospective Cohort Study. *Clin. Infect. Dis.* **2021**, *73*, 1882–1886. [CrossRef]

46. Agarwal, V.; Venkatakrishnan, A.J.; Puranik, A.; Kirkup, C.; Lopez-Marquez, A.; Challener, D.W.; Theel, E.S.; O'Horo, J.C.; Binnicker, M.J.; Kremers, W.K.; et al. Long-term SARS-CoV-2 RNA shedding and its temporal association to IgG seropositivity. *Cell Death Discov.* **2020**, *6*, 138. [CrossRef] [PubMed]

47. Tan, C.C.S.; Lam, S.D.; Richard, D.; Owen, C.; Berchtold, D.; Orengo, C.; Nair, M.S.; Kuchipudi, S.V.; Kapur, V.; van Dorp, L.; et al. Transmission of SARS-CoV-2 from humans to animals and potential host adaptation. *Nat. Commun.* **2022**, *13*, 2988. [CrossRef] [PubMed]

48. Bashor, L.; Gagne, R.B.; Bosco-Lauth, A.M.; Bowen, R.A.; Stenglein, M.; VandeWoude, S. SARS-CoV-2 evolution in animals suggests mechanisms for rapid variant selection. *Proc. Natl. Acad. Sci. USA* **2021**, *118*, e2105253118. [CrossRef] [PubMed]

49. Fagre, A.; Lewis, J.; Eckley, M.; Zhan, S.; Rocha, S.M.; Sexton, N.R.; Burke, B.; Geiss, B.; Peersen, O.; Bass, T.; et al. SARS-CoV-2 infection, neuropathogenesis and transmission among deer mice: Implications for spillback to New World rodents. *PLoS Pathog.* **2021**, *17*, e1009585. [CrossRef]

50. Francisco, R.; Hernandez, S.M.; Mead, D.G.; Adcock, K.G.; Burke, S.C.; Nemeth, N.M.; Yabsley, M.J. Experimental Susceptibility of North American Raccoons (*Procyon lotor*) and Striped Skunks (*Mephitis mephitis*) to SARS-CoV-2. *Front. Vet. Sci.* **2021**, *8*, 715307. [CrossRef]

51. Griffin, B.D.; Chan, M.; Tailor, N.; Mendoza, E.J.; Leung, A.; Warner, B.M.; Duggan, A.T.; Moffat, E.; He, S.; Garnett, L.; et al. SARS-CoV-2 infection and transmission in the North American deer mouse. *Nat. Commun.* **2021**, *12*, 3612. [CrossRef]

52. Moreira-Soto, A.; Walzer, C.; Czirják, G.Á.; Richter, M.; Marino, S.F.; Posautz, A.; de Yebra, P.R.; McEwen, G.K.; Drexler, J.F.; Greenwood, A.D. SARS-CoV-2 has not emerged in roe, red or fallow deer in Germany or Austria during the COVID 19 pandemic. *bioRxiv* **2022**. [CrossRef]

53. Chen, C.; Boorla, V.S.; Banerjee, D.; Chowdhury, R.; Cavener, V.S.; Nissly, R.H.; Gontu, A.; Boyle, N.R.; Vandegrift, K.; Nair, M.S.; et al. Computational prediction of the effect of amino acid changes on the binding affinity between SARS-CoV-2 spike RBD and human ACE2. *Proc. Natl. Acad. Sci. USA* **2021**, *118*, e2106480118. [CrossRef]

54. Fischhoff, I.R.; Castellanos, A.A.; Rodrigues, J.; Varsani, A.; Han, B.A. Predicting the zoonotic capacity of mammals to transmit SARS-CoV-2. *Proc. Royal Soc. B* **2021**, *288*, 20211651. [CrossRef]

55. Roberts, M.G.; Heesterbeek, J.A.P. Characterizing reservoirs of infection and the maintenance of pathogens in ecosystems. *J. R. Soc. Interface* **2020**, *17*, 20190540. [CrossRef]

MDPI
St. Alban-Anlage 66
4052 Basel
Switzerland
www.mdpi.com

Viruses Editorial Office
E-mail: viruses@mdpi.com
www.mdpi.com/journal/viruses

Disclaimer/Publisher's Note: The statements, opinions and data contained in all publications are solely those of the individual author(s) and contributor(s) and not of MDPI and/or the editor(s). MDPI and/or the editor(s) disclaim responsibility for any injury to people or property resulting from any ideas, methods, instructions or products referred to in the content.